The Enigma of Reason

The Enigma of Reason

A New Theory of Human Understanding

HUGO MERCIER AND
DAN SPERBER

ALLEN LANE
an imprint of
PENGUIN BOOKS

ALLEN LANE

UK | USA | Canada | Ireland | Australia
India | New Zealand | South Africa

Allen Lane Books is part of the Penguin Random House group of companies
whose addresses can be found at global.penguinrandomhouse.com.

First published in the United States of America by Harvard University Press 2017
First published in Great Britain by Allen Lane 2017
001

Printed in Great Britain by Clays Ltd, St Ives plc

A CIP catalogue record for this book is available from the British Library

ISBN: 978–1–846–14557–5

www.greenpenguin.co.uk

Contents

The Enigma of Reason

Introduction: A Double Enigma

They drink and piss, eat and shit. They sleep and snore. They sweat and shiver. They lust. They mate. Their births and deaths are messy affairs. Animals, humans are animals! Ah, but humans, and humans alone, are endowed with reason. Reason sets them apart, high above other creatures—or so Western philosophers have claimed.

The shame, the scandal of human animality, could at least be contained by invoking reason, the faculty that makes humans knowledgeable and wise. Reason rather than language—other animals seemed to have some form of language too. Reason rather than the soul—too mysterious. Endowed with reason, humans were still animals, but not beasts.

Reason: A Flawed Superpower?

With Darwin came the realization that whatever traits humans share as a species are not gifts of the gods but outcomes of biological evolution. Reason, being such a trait, must have evolved. And why not? Hasn't natural selection produced many wondrous mechanisms?

Take vision, for instance. Most animal species benefit from this amazing biological adaptation. Vision links dedicated external organs, the eyes, to specialized parts of the brain and manages to extract from patterns of retinal stimulation exquisitely precise information about the properties, location, and movement of distant objects. This is a hugely complex task—much more complex, by any account, than that of reason. Researchers in artificial intelligence have worked hard on modeling and implementing both vision and reasoning. Machine vision is still rudimentary; it comes nowhere near matching the performances of human vision. Many computer models of reasoning, on the other hand, have been claimed (somewhat optimistically) to perform even better than human reason. If vision could evolve, then why not reason?

We are told that reason, even more than vision, is a general-purpose faculty. Reason elevates cognition to new heights. Without reason, animal cognition is bound by instinct; knowledge and action are drastically limited. Enhanced with reason, cognition can secure better knowledge in all domains and adjust action to novel and ambitious goals, or so the standard story goes. But wait: If reason is such a superpower, why should it, unlike vision, have evolved in only a single species?

True, some outstanding adaptations are quite rare. Only a few species, such as bats, have well-developed echolocation systems. A bat emits ultrasounds that are echoed by surfaces in its environment. It uses these echoes to instantaneously identify and locate things such as obstacles or moving prey. Most other animals don't do anything of the sort.

Vision and echolocation have many features in common. One narrow range of radiation—light in the case of vision, ultrasounds in the case of echolocation—provides information relevant to a wide variety of cognitive and practical goals. Why, then, is vision so common and echolocation so rare? Because, in most environments, vision is much more effective. Echolocation is adaptive only in an ecological niche where vision is impossible or badly impaired—for instance, when dwelling in caves and hunting at night, as bats do.

Is reason rare—arguably unique to a single species—because it is adaptive in a very special kind of ecological niche that only humans inhabit? This intriguing possibility is well worth exploring. It is incompatible, however, with the standard approach to reason, which claims that reason enhances cognition whatever the environment it operates in and whatever the task it pursues. Understanding why only a few species have echolocation is easy. Understanding why only humans have reason is much more challenging.

Think of wheels. Animals don't have wheels. Why not?[1] After all, wheeled vehicles are much easier to construct than ones with legs or wings (just as models of reasoning seem much easier to develop than models of vision). However, artificial wheels are made separately and then added onto a vehicle, whereas biological wheels would have to grow in situ. How could a freely rotating body part either be linked to the rest of the body through nerves and blood vessels or else function without being so linked? Viable biological solutions are not easy to conceive, and that is only part of the problem.

For a complex biological adaptation to have evolved, there must have been a series of evolutionary steps, from rudimentary precursors to fully developed mechanisms, where every modification in the series has been favored (or at least not eliminated) by natural selection. The complex visual systems of insects, mollusks, or mammals, for instance, have all evolved from mere light-sensitive cells through long series of modifications, each of which was adaptive or neutral. Presumably, a similar series of adaptive steps from non-wheeled to wheeled animals was, if not impossible, at least so improbable that it never occurred.

Perhaps, then, reason is to animal cognition what wheels are to animal locomotion: an extremely improbable evolutionary outcome. Perhaps reason is so rare because it had to evolve through a series of highly improbable steps and it did so only once, only very recently in evolutionary time, and for the benefit of just one lucky species—us.

The series of steps through which reason would gradually have evolved remains a mystery. Reason seems to be hardly better integrated among the more ordinary cognitive capacities of humans than are the superpowers of Superman or Spider-Man among their otherwise ordinary human features. Of course, it could be argued that reason is a graft, an add-on, a cultural contraption—invented, some have suggested, in ancient Greece—rather than a biological adaptation. But how could a species without the superpower of reason have invented reason itself? While reason has obviously benefited from various cultural enhancements, the very ability of a species to produce, evaluate, and use reasons cries out for an evolutionary explanation. Alas, what we get by way of explanation is little more than hand waving.

The problem is even worse: the hand waving itself seems to point in a wrong direction. Imagine, by way of comparison, that, against the odds, biological wheels had evolved in one animal species. We would have no idea *how* this evolution had taken place. Still, if these wheels allowed the animals to move with remarkable efficiency in their natural environment, we would have a good idea *why* they had evolved; in other terms, we would understand their function. We might expect animal wheels, like all biological organs, to have weaknesses and to occasionally malfunction. What we would not expect, though, is to find some systematic flaw in this locomotion system that compromised the very performance of its function—for instance, a regular difference in size between wheels on opposite sides, making it hard for the animals to stay on course. A biological mechanism described as an ill-adapted adaptation is more likely to be a misdescribed mechanism. Reason as standardly described is such a case.

Psychologists claim to have shown that human reason is flawed. The idea that reason does its job quite poorly has become commonplace. Experiment after experiment has convinced psychologists and philosophers that people make egregious mistakes in reasoning. And it is not just that people reason poorly, it is that they are systematically biased. The wheels of reason are off balance.

Beyond this commonplace, polemics have flared. Reason is flawed, but how badly? How should success or failure in reasoning be assessed? What are the mechanisms responsible? In spite of their often bitter disagreements, parties to these polemics have failed to question a basic dogma. All have taken for granted that the job of reasoning is to help individuals achieve greater knowledge and make better decisions.

If you accept the dogma, then, yes, it is quite puzzling that reason should fall short of being impartial, objective, and logical. It is paradoxical that, quite commonly, reasoning should fail to bring people to agree and, even worse, that it should often exacerbate their differences. But why accept the dogma in the first place? Well, there is the weight of tradition . . . And, you might ask, what else could possibly be the function of reasoning?

Reason as standardly understood is doubly enigmatic. It is not an ordinary mental mechanism but a cognitive superpower that evolution—it used to be the gods—has bestowed only on us humans. As if this were not enigmatic enough, the superpower turns out to be flawed. It keeps leading people astray. Reason, a flawed superpower? Really?

Our goal is to resolve this double enigma. We will show how reason fits in individual minds, in social interactions, and in human evolution. To do so, we challenge the tradition, reject the dogma, and rethink both the mechanisms of reason and its function.

Where We Are Going

There have been more than two thousand years of philosophical work on reason, and more than fifty years of intense experimental work on reasoning. Some of the greatest thinkers of all time have contributed to this work. It would be beyond presumptuous to claim that most of this thinking has been on the wrong track, if it were not for the fact that both the philosophical and the psychological tradition have been vigorously contested from within.

How good is reason at guiding humans toward true knowledge and good decisions? How good are humans at using reason? We won't attempt to tell the convoluted story of these old debates that in recent times, with psychologists joining the fray, have intensified to the point of being called "rationality wars." What we will do instead in Part I of this book, "Shaking Dogma," is single out clashes that reveal how serious are the problems posed by standard approaches to reason, and how wanting the solutions. We will suggest that parties to these heated debates have managed to weaken one another to the point that the best course may well be to collect from the battlefield whatever may still be of use and to seek new adventures on more promising ground.

We are less interested anyhow in debunking shaky ideas than in developing a new scientific understanding of reason, one that solves the double enigma. Reason, we will show, far from being a strange cognitive add-on, a superpower gifted to humans by some improbable evolutionary quirk, fits quite naturally among other human cognitive capacities and, despite apparent evidence to the contrary, is well adapted to its true function.

To understand how reason could have evolved and how it works, one should pay attention not only to what makes it special but also to how it fits among other psychological capacities and how much it has in common with them. There are many mechanisms involved in drawing inferences. Reason is only one of them. In Part II, "Understanding Inference," we situate reason in relation to other inferential mechanisms, the overall picture being schematized in Figure 1.

Animals make inferences all the time: they use what they already know to draw conclusions about what they don't know—for instance, to anticipate what may happen next, and to act accordingly. Do they do this by means of some general inferential ability? Definitely not. Rather, animals use

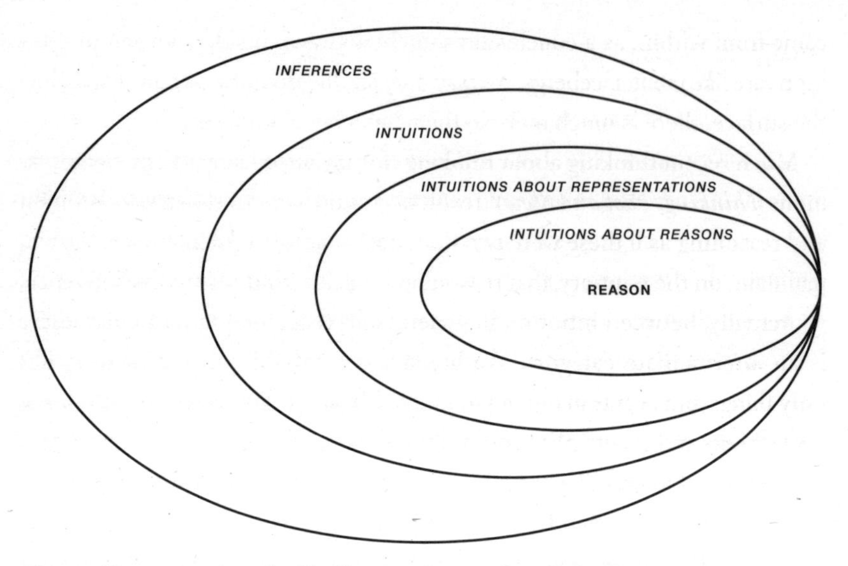

Figure 1. How reason is embedded in several categories of inference.

many different inferential mechanisms, each dealing with a distinct type of problem: What to eat? Whom to mate with? When to attack? When to flee? And so on.

Humans are like other animals: instead of one general inferential ability, they use a wide variety of specialized mechanisms. In humans, however, many of these mechanisms are not "instincts" but are acquired through interaction with other people during the child's development. Still, most of these acquired mechanisms have an instinctual basis: speaking Wolof, or English, or Tagalog, for instance, is not instinctive, but paying special attention to the sounds of speech and going through the steps necessary to acquire the language of one's community has an instinctual basis.

As far as one can tell, other animals perform all their inferences without being conscious of doing so. Humans also perform a great variety of inferences automatically and unconsciously; for instance, in acquiring their mother tongue. However, there are many inferences of which humans are partly conscious. We are talking here about intuitions. When you have an intuition—for example, the intuition that your friend Molly is upset even though she didn't say so and might even deny it—this intuition pops up fully formed in your consciousness; at the same time, however, you recognize it as something that came from within, as a conclusion somehow drawn inside your mind. Intuitions are like mental icebergs: we may only see the tip but we know that, below the surface, there is much more to them, which we don't see.

Much recent thinking about thinking (for instance Daniel Kahneman's famous *Thinking, Fast and Slow*)[2] revolves around a contrast between intuition and reasoning as if these were two quite different forms of inference. We will maintain, on the contrary, that reasoning is itself a kind of intuitive inference.

Actually, between intuition in general and reasoning in particular, there is an intermediate category. We humans are capable of representing not only things and events in our environment but also our very representations of these things and events. We have intuitions about what other people think and about abstract ideas. These intuitions about representations play a major role in our ability to understand one another, to communicate, and to share opinions and values. Reason, we will argue, is a mechanism for intuitive inferences about one kind of representations, namely, reasons.

In Part III, "Rethinking Reason," we depart in important ways from dominant approaches; we reject the standard way of contrasting reason with intuition. We treat the study of reason (in the sense of a mental faculty) and that of reasons (in the sense of justifications) as one and the same thing whereas, in both philosophy and psychology, they have been approached as two quite distinct topics.

Whereas reason is commonly viewed as a superior means to think better on one's own, we argue that it is mainly used in our interactions with others. We produce reasons in order to justify our thoughts and actions to others and to produce arguments to convince others to think and act as we suggest. We also use reason to evaluate not so much our own thought as the reasons others produce to justify themselves or to convince us.

Whereas reason is commonly viewed as the use of logic, or at least some system of rules to expand and improve our knowledge and our decisions, we argue that reason is much more opportunistic and eclectic and is not bound to formal norms. The main role of logic in reasoning, we suggest, may well be a rhetorical one: logic helps simplify and schematize intuitive arguments, highlighting and often exaggerating their force.

So, why did reason evolve? What does it provide, over and above what is provided by more ordinary forms of inference, that could have been of special value to humans and to humans alone? To answer, we adopt a much broader perspective.

Reason, we argue, has two main functions: that of producing reasons for justifying oneself, and that of producing arguments to convince others. These two functions rely on the same kinds of reasons and are closely related.

Why bother to explain and justify oneself? Humans differ from other animals not only in their hyperdeveloped cognitive capacities but also, and crucially, in how and how much they cooperate. They cooperate not only with kin but also with strangers; not only in here-and-now ventures but also in the pursuit of long-term goals; not only in a small repertoire of species-typical forms of joint action but also in jointly setting up new forms of cooperation. Such cooperation poses unique problems of coordination and trust.

A first function of reason is to provide tools for the kind of rich and versatile coordination that human cooperation requires. By giving reasons in order to explain and justify themselves, people indicate what motivates and, in their eyes, justifies their ideas and their actions. In so doing, they let others know what to expect of them and implicitly indicate what they expect of others. Evaluating the reasons of others is uniquely relevant in deciding whom to trust and how to achieve coordination.

Humans also differ from other animals in the wealth and breadth of information they share with one another and in the degree to which they rely on this communication. To become competent adults, we each had to learn a lot from others. Our skills and our general knowledge owe less to individual experience than to social transmission. In most of our daily undertakings, in family life, in work, in love, or in leisure, we rely extensively on what we have learned from others. These huge, indispensable benefits we get from communication go together with a commensurate vulnerability to misinformation. When we listen to others, what we want is honest information. When we speak to others, it is often in our interest to mislead them, not necessarily through straightforward lies but by at least distorting, omitting, or exaggerating information so as to better influence them in their opinions and in their actions.

When we listen to others, then, we should trust wisely and sometimes distrust. When we talk to others, we often have to overcome their understandable lack of trust. If we distrusted others only when they don't deserve our trust, things would be for the best. Often, however, we withhold our trust out of prudence, not because we know that others are untrustworthy but because we are not sure that we can trust them. This reticence may be wise—better safe than sorry—but still, we miss valuable information. Communication, which could be beneficial to speakers and listeners alike, often falters for lack of confidence.

The second function of reason—a function carried out through reasoning and argumentation—is, we claim, to make communication effective even when the communicators lack sufficient credibility in the eyes of their audience to be believed on trust. Reason produces reasons that communicators use as arguments to persuade a reticent audience. Reason, by the same token, helps a cautious audience evaluate these reasons, accept good arguments, and reject bad ones.

Much of our earlier joint work focused on this argumentative function of reason and developed an "argumentative theory of reasoning."[3] In this book, we broaden our perspective, consider both the argumentative and the justificatory functions of reason, and develop an interactionist approach to the mechanisms and the two functions of reason.

Part IV, "What Reason Can and Cannot Do," offers a tour of what reason does. Throughout this tour, we show how our interactionist perspective is in a good position to explain why reason behaves the way it does. We revisit some well-established but ill-explained apparent weaknesses of reason such as the confirmation bias. We also draw attention to some of its neglected strengths.

The tour starts with a pair of observations: human reason is both biased and lazy. Biased because it overwhelmingly finds justifications and arguments that support the reasoner's point of view, lazy because reason makes little effort to assess the quality of the justifications and arguments it produces. Imagine, for instance, a reasoner who happens to be partial to holidays at the beach. When reasoning about where to spend her next vacation, she will spontaneously accumulate reasons to choose a sunny place by the sea, including reasons that are manifestly poor (say, that there's a discount on the flight to the very place where she would like to go, when in fact the same discount applies to many other destinations as well).

The solitary use of reason has two typical outcomes. When the reasoner starts with a strong opinion, the reasons that come to her mind tend all to support this opinion. She is unlikely, then, to change her mind; she might even become overconfident and develop stronger opinions. But sometimes a reasoner starts with no strong opinion, or with conflicting views. In this case, reason will drive her toward whatever choice happens to be easier to justify, and this sometimes won't be the best choice. Imagine she has a choice between visiting her horrible in-laws and then vacationing at the beach, or starting with the beach and then going to see the in-laws, the latter option being somewhat cheaper. Reason will drive her toward what seems to be the rational decision: taking the cheaper option. It is likely, however, that she would come back more satisfied if she started with the in-laws instead of letting the prospect of this visit spoil her time at the beach: a better choice overall, but involving a hard-to-justify extra expense.

Psychologists generally recognize that reason is biased and lazy, that it often fails to correct mistaken intuitions, and that it sometimes makes things worse. Yet most of them also maintain that the main function of reason is to enhance individual cognition—a task it performs abysmally. The interactionist perspective, on the other hand, offers for the first time an evolutionarily plausible account of the often decried biases and shortcomings of reason.

It makes sense, we will show, for a cognitive mechanism aimed at justifying oneself and convincing others to be biased and lazy. The failures of the solitary reasoner follow from the use of reason in an "abnormal" context. Underwater, you wouldn't expect a pen—which wasn't designed to work there—or human lungs—which didn't evolve to work there either—to function properly. Similarly, take reason out of the interactive context in which it evolved, and nothing guarantees that it will yield adaptive results.

What, then, happens when reason is put back in its "normal" environment, when it gets to work in the midst of a discussion, as people exchange arguments and justifications with each other? In such a context it properly fulfills the functions for which it evolved. In particular, when people who disagree but have a common interest in finding the truth or the solution to a problem exchange arguments with each other, the best idea tends to win; whoever had it from the start or came to it in the course of the discussion is likely to convince the others. This conclusion might sound unduly optimistic, but it is supported by a wide range of evidence, from students discussing logical problems, to juries deliberating, and to forecasters trying to predict where the next war will erupt.

In the last three chapters (Part V, "Reason in the Wild") we demonstrate how robust are the features and effects of reason reviewed earlier. We find that solitary reasoning is biased and lazy, whereas argumentation is efficient not only in our overly argumentative Western societies but in all types of cultures, not only in educated adults but also in young children. Few will be surprised to hear that reason is typically biased and lazy when it is applied to moral and political issues. More surprising may be evidence that shows how, even in the moral and political realms, argumentation may work quite efficiently, allowing participants to form more accurate moral judgments and citizens to form more enlightened opinions. Such findings, however, are what one should expect in an interactionist perspective.

The last chapter (Chapter 18, "Solitary Geniuses?") is about science, generally considered the pinnacle of human reason. Science is exceptional in many ways, but is the way scientists reason itself exceptional? Scientific progress is often attributed to solitary geniuses, from Newton to Darwin or Einstein. Their superior reason, we are told, doesn't suffer from the shortcomings that plague the rest of us. Not only can these geniuses dispense with discussions with others in order to come up with new theories, such discussion might even hinder them when their revolutionary insights would be misunderstood and scorned by their not-quite-peers. Better wait for a less prejudiced new generation to see the light. Fortunately (for our theory and for scientists), science doesn't work this way. Scientists make do with the same reason that all humans use, with its biases and limitations. But they also benefit from its strengths and in particular from the fact that reason is more efficient in evaluating good arguments than in producing them: when the arguments are there, the scientific community is able to elevate the status of a new theory from fringe to textbook material in a few years.

In these five parts and eighteen chapters, what we will put to you, then, is an interactionist approach to reason that contrasts with standard intellectualist approaches: reason, we maintain, is first and foremost a social competence. We do not deny that reason can bring huge intellectual benefits, as the case of science well illustrates; on the contrary, we explain how it does this: through interaction with others.

You are unlikely to accept what we say just because we say it, so we will present you with arguments that you will be able to assess on their own merits.

We will show you how considering reason as a mechanism that draws intuitive inferences about reasons solves the first half of the enigma: reason is not a superpower implausibly grafted onto an animal mind; it is, rather, a well-integrated component of the extraordinarily developed mind that characterizes the human animal.

To resolve the second half of the enigma, we will demonstrate how apparent biases that have been described as deplorable flaws of reason are actually features well adapted to its argumentative function. A number of sometimes surprising predictions about human reason follow from our approach. The evidence we will present confirms these predictions. It is by force of argument that we hope to persuade you that the interactionist approach is right or, at least, on the right track. This, of course, makes the book itself an illustration of the perspective it defends.

I

Shaking Dogma

Reason, the faculty that gives humans superior knowledge and wisdom? This dominant view in the Western tradition has been radically undermined by fifty years of experimental research on reasoning. In Chapters 1 and 2, we show how old dogmas were shaken, but not nearly enough. The now dominant view of reasoning ("dual process" or "fast and slow thinking"), however appealing, is but a makeshift construction amid the ruins of old ideas.

1

Reason on Trial

In the cold autumn of 1619, René Descartes, then aged twenty-three and a volunteer in the armies of the Duke of Bavaria, found himself in what is now southern Germany with time to spend and nobody around he deemed worth talking to. There, in a stove-heated room, as he recounts in his *Discourse on Method,*[1] he formed the stunningly ambitious project of ridding himself of all opinions, all ideas learned from others, and of rebuilding his knowledge from scratch, step by step. Reason would be his sole guide. He would accept as true only what he could not doubt.

Descartes justified his rejection of everything he had learned from others by expressing a general disdain for collective achievements. The best work, he maintained, is made by a single master. What one may learn from books, he considered, "is not as close to the truth, composed as it is of the opinions of many different people, as the simple reasoning that any man of good sense can produce about things in his purview."[2]

Descartes would have scorned today's fashionable idea of the "wisdom of crowds." The only wisdom he recognized, at least in the sciences, was that of individual reason: "As long as one stops oneself taking anything to be true that is not true and sticks to the right order so as to deduce one thing from another, there can be nothing so remote that one cannot eventually reach it, nor so hidden that one cannot discover it."[3]

Why did Descartes decide to trust only his own mind? Did he believe himself to be endowed with unique reasoning capacities? On the contrary, he maintained that "the power of judging correctly and of distinguishing the true from the false (which is properly what is called good sense or reason) is naturally equal in all men."[4] But if we humans are all endowed with this power of distinguishing truth from falsity, how is it that we disagree so much on what is true?

"The Greatest Minds Are Capable of the Greatest Vices as Well as the Greatest Virtues"

Most of us think of ourselves as rational. Moreover, we expect others to be rational too. We are annoyed, sometimes even angry, when we see others defending opinions we think are deeply flawed. Hardly ever do we assume that those who disagree with us altogether lack reason. What aggravates us is the sense that these people do not make a proper use of the reason we assume they have. How can they fail to understand what seems so obvious (to us)?

If reason is this highly desirable power to discover the truth, why don't people endowed with it use it to the best of their capacities all the time? After all, we expect all sighted people to see what others see. Show several people a tree or a sunset, and you expect them all to see a tree or a sunset. Ask, on the other hand, several people to reason about a variety of questions, from logical problems to social issues, and what might surprise you is their coming to the same conclusions. If reason, like perception, worked to provide us with an adequate grasp of the way things really are, this should be deeply puzzling.

Descartes had an explanation: "The diversity of our opinions arises not from the fact that some of us are more reasonable than others, but solely that we have different ways of directing our thoughts, and do not take into account the same things. . . . The greatest minds are capable of the greatest vices as well as the greatest virtues."[5]

This, however, is hardly more than a restatement of the enigma, for shouldn't the way we direct our thoughts itself be guided by reason? Shouldn't reason, in the first place, protect us from intellectual vices?

Descartes was the most forceful of reason's many advocates. Reason has also had many, often passionate, detractors. Its efficacy has been questioned. Its arrogance has been denounced. The religious reformer Martin Luther was particularly scathing: "Reason is by nature a harmful whore. But she shall not harm me, if only I resist her. Ah, but she is so comely and glittering. . . . See to it that you hold reason in check and do not follow her beautiful cogitations. Throw dirt in her face and make her ugly."[6]

To be fair, Descartes's and Luther's views on reason were much richer and subtler than these isolated quotes suggest, and hence less diametrically opposed. Luther's invectives were aimed at the claims of reason in matters of faith. In a different context, the same Luther described reason, much more conventionally, as "the inventor and mentor of all the arts, medicines, laws, and of whatever wisdom, power, virtue, and glory men possess in this life" and as "the essential difference by which man is distinguished from the animals and other things."[7] Descartes for his part abstained, out of conviction or out of prudence, from critically examining faith in the light of reason.

Still, if reason were put on trial, both the prosecution and the defense could make an extraordinary case. The defense would argue, citing Descartes, Aristotle, Kant, or Popper, that humans err by not reasoning enough. The prosecution would argue, citing Luther, Hume, Kierkegaard, or Foucault, that they err by reasoning too much.

The defense and the prosecution could also produce compelling narratives to bolster their case.

Eratosthenes and the Unabomber

Do you doubt the power of reason? Just look at the sciences, the defense would exclaim. Through insightful reasoning, scientists have discovered hidden facts and deep explanations that would have been completely inaccessible otherwise. Modern science provides countless examples of the power of reason, but nothing beats, as a simple and compelling illustration, the measurement of the circumference of the earth twenty-two centuries ago, by Eratosthenes (276–195 BCE), the head librarian of the greatest library of the ancient world at Alexandria in Egypt.[8]

Already at the time, it was commonly accepted that the earth was spherical rather than flat. This best explained the curvature of the horizon at sea and the apparent movement of the sun and the stars. Still, it was, as the phrase goes, "just a theory." No one had traveled around the earth, let alone seen it from a distance as astronauts now have. How, then, could its circumference be measured?

Eratosthenes had heard that every year, on a single day, at noon, the sun shone directly to the bottom of wells in the distant town of Syene (now Aswan). This, he understood, meant that, there and then, the sun was at the zenith, vertically above the town. Syene therefore had to be on the Tropic of Cancer and that single day had to be the summer solstice (our June 21). Syene, he assumed, was due south on the same meridian as Alexandria. He knew how long it took caravans to travel from Alexandria to Syene and, on that basis, estimated the distance between the two cities to be 5,014 stades (an ancient unit of measure).

When, on the summer solstice at noon, the sun was vertically above Syene, by how many degrees was it south of the vertical in the more northern city of Alexandria? Eratosthenes measured the length of the shadow cast at that very moment by an obelisk located in front of his library (or so the story goes). He determined that the sun's rays were hitting the obelisk at an angle of 7.2 degrees south of the vertical. He understood that the sun was far enough to treat all rays that reach the earth as parallel, and that therefore the angle between the rays of the sun and the vertical at Alexandria was equal to the angle between the vertical at Alexandria and that at Syene, two lines that cross at the center of the earth (see Figure 2). In other words, that very angle of 7.2 degrees also measured the difference in degrees of latitude between Alexandria and Syene. He now had all the information he needed. Since 7.2 degrees is one-fiftieth of 360 degrees, Eratosthenes could calculate the circumference of the earth by multiplying by fifty the distance between Alexandria and Syene. The result, 252,000 stades, is 1 percent shy of the modern measurement of 24,859 miles, or 40,008 kilometers.[9]

Eratosthenes grasped the mutual relevance of apparently unrelated pieces of evidence (the pace of caravans, the sun shining to the bottom of wells, the shadow of an obelisk), of assumptions (the rotundity of the earth, its distance from the sun), and of simple geometrical ideas about angles and parallel lines. He drew on all of them to measure a circumference that he could imagine but neither see nor survey. What made his measurement not just true but convincing is—isn't it?—that it was a pure product of human reason.

How telling, the prosecution would object, that the defense of reason should choose as evidence such an exceptional achievement! It is an exception, and this is why it is still remembered after more than two thousand years. Ordi-

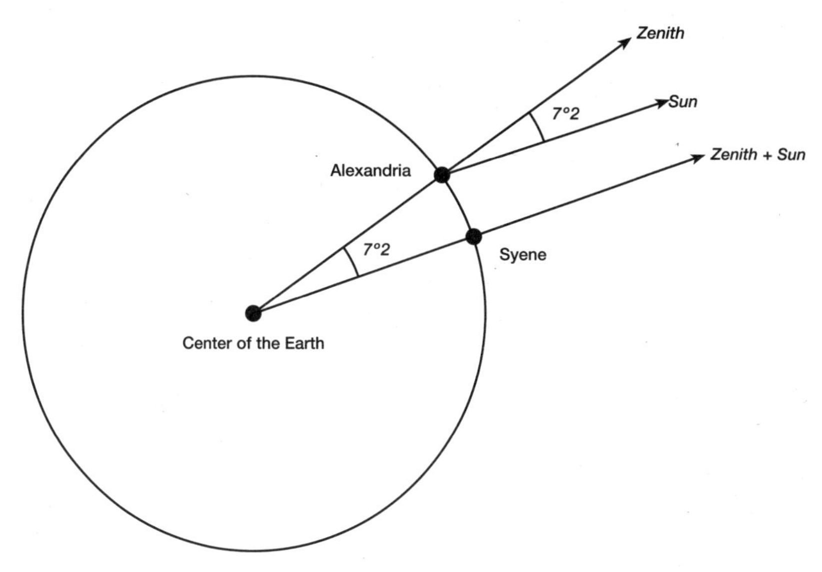

Figure 2. How Eratosthenes computed the circumference of the earth.

nary reasoning doesn't lead us far, and that is just as well, as often it leads in the wrong direction. Even extraordinary uses of reason, far from being all on the model of Eratosthenes, have led many thinkers badly astray. Publishers, newspapers, and scientific journals receive every day the thoroughly reasoned nonsense of would-be philosophers, scientists, or reformers who, failing to get their work published there, then try the World Wide Web. Some of them, however, reason not just to theoretical but also to practical absurdities, act on them, and achieve notoriety or even infamy. The prosecution might well at this juncture introduce the case of Ted Kaczynski.

As a young man, Kaczynski was unquestionably a brilliant reasoner. He had entered Harvard in 1958, at age sixteen. For his doctoral dissertation at the University of Michigan, he solved a mathematical problem that had eluded his professors for years, prompting the University of Berkeley to hire him. Two years later, however, he abandoned mathematics and academe to live in a shack in Montana, where he became an avid reader of social science and political work. Both his readings and his writings focused on what he saw as the destructive character of modern technology. Viewing technological progress as leading to disasters for the environment and for human dignity is not uncommon in Western thought, but Kaczynski went further: for him, only a violent revolution causing the collapse of modern civilization could prevent these even greater disasters.

To help trigger this revolution, Kaczynski began in 1978 to send bombs to universities, businesses, and individuals, killing three people and injuring many others. He wrote a long manifesto and managed to have it published in the *New York Times* and in the *Washington Post* in 1995 by promising that he would then "desist from terrorism." The Unabomber, as the FBI had named him, was finally arrested in 1996 and now, as we write, serves a life sentence without the possibility of parole in a Colorado jail, where he goes on reading and writing.

What had happened to the brilliant young mathematician? Had Kaczynski's reason failed him, turning him into the "raving lunatic" described by the press? Kaczynski's family arranged for his defense to try to make him plead insanity. The defense of reason would no doubt concur: *un*reason had to be the culprit. It is unlikely, however, that Kaczynski suffered at the time of his arrest from any major mental disorder. He was still a smart, highly articulate, extremely well-read man. Defective reasoning, the prosecution of reason would insist, cannot be blamed for his actions. To see this, all you need do is read the Unabomber's manifesto:

> The Industrial Revolution and its consequences have been a disaster for the human race. . . . They have destabilized society, have made life unfulfilling, have subjected human beings to indignities, . . . and have inflicted severe damage on the natural world. The continued development of technology will worsen the situation. . . . The industrial-technological system may survive or it may break down. . . . If the system survives, the consequences will be inevitable: There is no way of reforming or modifying the system so as to prevent it from depriving people of dignity and autonomy. If the system breaks down the consequences will still be very painful. But the bigger the system grows the more disastrous the results of its breakdown will be, so if it is to break down it had best break down sooner rather than later. We therefore advocate a revolution against the industrial system.[10]

This, surely, is a well-constructed argument. Most of us would disagree with the premise that technological progress is a plain disaster, but actually, many well-respected philosophers and social theorists have defended similar views. What singles out Kaczynski, the prosecution of reason would claim, is that he pushed this radically pessimistic viewpoint to its logical consequences and acted accordingly. As one of his biographers put it: "Kaczynski, in short, had become a cold-blooded killer not despite of his intellect, but *because of it.*"[11]

So, the defense of reason would counter, the prosecution wants you to believe that the problem with Ted Kaczynski, the Unabomber, is that he was reasoning too much. His manifesto is indeed more tightly reasoned than much political discourse. What made him notorious, however, were not his ideas but his crimes. Nowhere in his writings is there even the beginning of a proper argument showing that sending bombs to a few powerless academics—his former colleagues—would kick-start a "revolution against the industrial system." When you are told that excessive reliance on reasoning led someone to absurd or abhorrent conclusions, look closely at the evidence, and you will find lapses of reason: some premises were not properly examined, and some crucial steps in the argument are simply missing. Remember: a logical demonstration can never be stronger than its weakest part.

Expert Witnesses for the Prosecution

Since historical illustrations, however arresting, are not sufficient to make their cases, defense and prosecution of reason would turn to expert witnesses. Neither side would have any difficulty in recruiting psychologists to support their cause. Specialists of reasoning do not agree among themselves. Actually, the polemics in which they are engaged are hot enough to have been described as "rationality wars." This very lack of agreement among specialists who, one hopes, are all good reasoners, is particularly ironic: sophisticated reasoning on reasoning does not come near providing a consensual understanding of reasoning itself.

The prosecution of reason might feel quite smug. Experimental psychology of reasoning has been fast developing since the 1960s, exploiting a variety of ingenious experiments. The most famous of these present people with problems that, in principle, could easily be resolved with a modicum of simple reasoning. Yet most participants in these experiments confidently give mistaken answers, as if the participants were victims of some kind of "cognitive illusion." These results have been used in the rationality wars to argue that human reason is seriously defective. Reason's defenders protest that such experiments are artificial and misleading. It is as if the experiments were aimed at tricking sensible people and making them look foolish rather than aimed at understanding the ordinary workings of reason. Of course, psychologists who have devised these experiments insist that, just as visual illusions reveal important features of ordinary, accurate vision, cognitive illusions reveal important features of ordinary reasoning.[12] Philosophers, science writers, and journalists have, however, focused on the seemingly bleak implications of this research for the evaluation of human rationality and have, if anything, exaggerated their bleakness.

When you do arithmetic, it does not matter whether the numbers you add or subtract happen to be numbers of customers, trees, or stars, nor does it matter whether they are typical or surprising numbers for collections of such items. You just apply rules of arithmetic to numbers, and you ignore all the rest. Similarly, if you assume that reasoning should be just a matter of applying logic to a given set of premises in order to derive the conclusions that follow from these premises, then nothing else should interfere. Yet there is ample evidence that background knowledge and expectations do interfere in the process. This, many argue, is the main source of bad reasoning.

Here is a classic example.[13] In July 1980, Björn Borg, who was then hailed as one of the greatest tennis players of all time, won his fifth consecutive Wimbledon championship. In October of that year, Daniel Kahneman and Amos Tversky, two Israeli psychologists working in North America who would soon become world-famous, presented a group of University of Oregon students with the following problem:

> Suppose Björn Borg reaches the Wimbledon finals in 1981. Please rank order the following outcomes from most to least likely:
>
> 1. Borg will win the match.
> 2. Borg will lose the first set.

3. Borg will lose the first set but win the match.
4. Borg will win the first set but lose the match.

Seventy-two percent of the students assigned a higher probability to outcome 3 than to outcome 2. What is so remarkable about this? Well, if you have two propositions (for instance, "Borg will lose the first set" and "Borg will win the match"), then their conjunction ("Borg will lose the first set but win the match") cannot be more probable than either one of the two propositions taken separately. Borg could not both lose the first set and win the match without losing the first set, but he could lose the first set and not win the match. Failing to see this is an instance of what is known as the "conjunction fallacy." More abstractly, take two propositions that we may represent with the letters *P* and *Q*. Whenever the conjunction "*P* and *Q*" is true, so must be both *P* and *Q*, while *P* could be true or *Q* could be true and "*P* and *Q*" false. Hence, for any two propositions *P* and *Q*, claiming that their conjunction "*P* and *Q*" is more probable than either *P* or *Q* taken on its own is clearly fallacious.

Kahneman and Tversky devised many problems that caused people to commit the conjunction fallacy and other serious blunders. True, as they themselves showed, if you ask the same question not about Björn Borg at Wimbledon but rather about an unknown player at an ordinary game, then people do not commit the fallacy. They correctly rank a single event as more probable than the conjunction of that event and another event. But why on earth should people reason better about an anonymous tennis player than about a famous champion?

Here is another example from our own work illustrating how the way you frame a logical problem may dramatically affect people's performance.[14] We presented people with the following version of what, in logic, is known as a "pigeonhole problem":

> In the village of Denton, there are twenty-two farmers. All of the farmers have at least one cow. None of the farmers have more than seventeen cows. How likely is it that at least two farmers in Denton have the exact same number of cows?

Only 30 percent gave the correct answer, namely, that it is *certain*—not merely probable—that at least two farmers have the same number of cows. If you don't see this, perhaps the second version of the problem will help you.

To another group, we presented another version of the problem that, from a logical point of view, is strictly equivalent:

> In the village of Denton, there are twenty-two farmers. The farmers have all had a visit from the health inspector. The visits of the health inspector took place between the first and the seventeenth of February of this year. How likely is it that at least two farmers in Denton had the visit of the health inspector on the exact same day?

This time, 70 percent of people gave the correct answer: it is certain.

As the Borg and the farmers-cows problems illustrate, depending on how you contextualize or frame a logical problem—without touching the logic of it—most people may either fail or succeed. Isn't this, the prosecution would argue, clear evidence that human reason is seriously defective?

Expert Witnesses for the Defense

While many psychologists focused on experiments that seem to demonstrate human irrationality, other psychologists were pursuing a different agenda: to identify the mental mechanisms and procedures that allow humans to reason at all.

There is little doubt that some simple reasoning (in a wide sense of the term) occurs all the time, in particular when we talk to each other. Conjunctions such as "and," "or," and "if" and the adverb "not" elicit logical inferences of the most basic sort. Take a simple dialogue:

> *Jack (to Jill):* I lent my umbrella to you or to Susan—I don't remember whom.
> *Jill:* Well, you didn't lend it to me!
> *Jack:* Oh, then I lent to Susan.
> *Jill:* Right!

No need for Jack or Jill to have studied logic to come to the conclusion that Jack lent his umbrella to Susan.[15] But what is the psychological mechanism by means of which such inferences are being performed? According to one type of account, understanding the word "or" or the word "not" amounts to having in mind logical rules that somehow capture the meaning of such words. These rules govern deductions licensed by the presence of these "logical" words in a statement. Here is a rule for "or" (using again the letters *P* and *Q* to represent any two propositions):

> *"Or" rule:* From two premises of the form "*P* or *Q*" and "not *P*," infer *Q*.

Several psychologists (Jean Piaget, Martin Braine, and Lance Rips, in particular[16]) have argued that we perform logical deduction by means of a "mental logic" consisting in a collection of such logical rules or schemas. When Jack and Jill infer that Jack lent his umbrella to Susan, what they do is apply the "or" rule.

According to an alternative explanation, "mental model theory" (developed by Philip Johnson-Laird and Ruth Byrne),[17] no, we don't have a mental logic in our head. What we have is a procedure to represent and integrate in our mind the content of premises by means of models comparable to schematic pictures of the situation. We then read the conclusions off these models. In one model, for instance, Jack lent his umbrella to Jill. In an alternative model, he lent it to Susan. If Jack's statement is true, then the two mental models can neither be both right nor be both wrong. When we learn that the "lent to Jill" model is wrong, then we are left with just the "lent to Susan" model, and we can conclude that Jack lent his umbrella to Susan.

Much work in the psychology of reasoning has been devoted to pitting against one another the "mental logic" and the "mental models" approaches. You might wonder: What is the difference between these two accounts? Aren't they both stating the same thing in different terms? Well, true, the two theories have a lot in common. They both assume that humans have mechanisms capable of producing genuine logical inferences. Both assume that humans have the wherewithal to think in a rational manner, and in this respect, they contrast with approaches that cast doubt on human rationality.

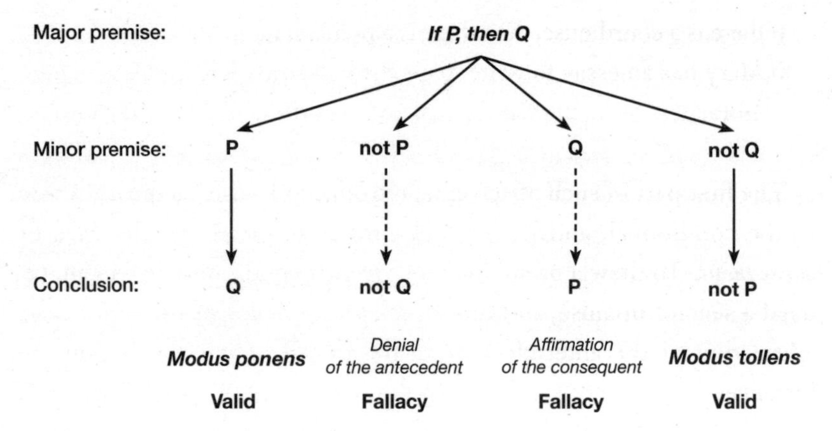

Figure 3. The four schemas of conditional inference.

The picture drawn by "mental logicians" and "mental modelers" is not quite rosy, however. Both approaches recognize that all except the simplest reasoning tasks can trip people and cause them to come to unwarranted conclusions. As they become more complex, reasoning tasks rapidly become forbiddingly difficult and performance collapses. But what makes a reasoning task complex? This is where the two theories differ. For mental logicians, it is the number of steps that must be taken and rules that must be followed. For mental modelers, it is the number of models that should be constructed and integrated to arrive at a certain conclusion.

The defense of reason would want these two schools to downplay their disagreements and to focus on a shared positive message: humans are equipped with general mechanisms for logical reasoning. Alas, the prosecution would find in the very work inspired by these two approaches much evidence to cast doubt on this positive message.

If there is one elementary pattern of reasoning that stands out as the most ubiquitous, the most important both in everyday and in scholarly reasoning, it is what is known as conditional reasoning—reasoning with "if . . . , then . . ." (see Figure 3). Such reasoning involves a *major premise* of the form "if *P*, then *Q*." For instance:

If you lost the key, then you owe us five dollars.
If pure silver is heated to 961°C, then it melts.

If there is a courthouse, then there is a police station.
If Mary has an essay to write, then she will study late in the library.

The first part of such statements, introduced by "if," is the *antecedent* of the conditional, and the second part, introduced by "then," is the *consequent.* To draw a useful inference from a conditional statement, you need a second premise, and this *minor premise* can consist either in the affirmation of the antecedent or in the denial of the consequent. For instance:

If there is a courthouse, then there is a police station. *(major premise: the conditional statement)*
There is a courthouse. *(minor premise: affirmation of the antecedent)*

There is a police station. *(conclusion)*

Or:

If there is a courthouse, then there is a police station. *(major premise: the conditional statement)*
There is no police station. *(minor premise: denial of the consequent)*

There is no courthouse. *(conclusion)*

These two inference patterns, the one based on the affirmation of the antecedent (known under its Latin name, *modus ponens*) and the one based on the denial of the consequent *(modus tollens),* are both logically valid: when the premises are true, the conclusion is necessarily true also.

But what about using as the minor premise the denial of the antecedent (rather than its affirmation) or the affirmation of the consequent (rather than its denial)? For instance:

If there is a courthouse, then there is a police station. *(major premise: the conditional statement)*

There is no courthouse. *(minor premise: denial of the antecedent)*

There is no police station. *(conclusion?)*

Or:

If there is a courthouse, then there is a police station. *(major premise: the conditional statement)*
There is a police station. *(minor premise: affirmation of the consequent)*

There is a courthouse. *(conclusion?)*

These two inference patterns (known by the name of their minor premise as "denial of the antecedent" and "affirmation of the consequent") are invalid; they are fallacies. Even if both premises are true, the conclusion does not necessarily follow—you may well, for instance, have a police station but no courthouse.

Surely, the prosecution would exclaim, all this is simple enough. Shouldn't people, if the defense were right, reliably perform the two valid inferences of conditional reasoning and never commit the two fallacies? Alas, the expert witnesses of the defense have demonstrated in countless experiments with very simple problems that such is not the case—far from it. True, nearly everybody draws the valid *modus ponens* inference from the affirmation of the antecedent. Good news for the defense? Well, the rest is good news for the prosecution: only two-thirds of the people, on average, draw the other valid inference, *modus tollens,* and about half of the people commit the two fallacies.[18] And there is worse . . .

Will She Study Late in the Library?

In a famous 1989 study, Ruth Byrne demonstrated that even the valid *modus ponens* inference, the only apparently safe bit of logicality in conditional reasoning, could all too easily be made to crumble.[19] Byrne presented participants with the following pair of premises:

Major premise:
If Mary has an essay to write, then she will study late in the library.
Minor premise:
She has an essay to write.

Participants had no difficulty deducing:

Conclusion: Mary will study late in the library.

So far, so good. To another group of people, however, Byrne presented the same problem, but this time with an additional major premise:

First major premise:
If Mary has an essay to write, then she will study late in the library.
Second major premise:
If the library stays open, then Mary will study late in the library.
Minor premise:
She has an essay to write.

From a strictly logical point of view, the second major premise is of no relevance whatsoever. So, if people were logical, they should draw the same valid *modus ponens* conclusion as before. Actually, only 38 percent of them did.

What Byrne was trying to prove was not that humans are irrational—mental modelers don't believe that—but that mental logicians have the wrong theory of human rationality. If, as mental logicians claim, people had a mental *modus ponens* rule of inference, then that inference should be automatic, whatever the context. Participants are instructed to take the premises as true, so, given the premises "If Mary has an essay to write, then she will study late in the library" and "Mary has an essay to write," they should without hesitation conclude that she will study late in the library. What about the possibility that the library might be closed? Well, what about it? After all, for all you know, Mary might have a pass to work in the library even when it is closed. A logician would tell you, just don't go there. This is irrelevant to this logic task, just as the possibility that a bubble might burst would be irrelevant to the arithmetic task of adding three bubbles to two bubbles.

Did mental logicians recognize, in the light of Byrne's findings, that their approach was erroneous? Well, no; they didn't have to. What they did instead was propose alternative explanations.[20] People might, for instance, consolidate the two major premises presented by Byrne into a single one: "If Mary has an essay to write *and* if the library stays open, then Mary will study late in the library." This, after all, is a realistic way of understanding the situation. If this is how people interpret the major premises, then the minor premise, "She has an essay to write," is not sufficient to trigger a valid *modus ponens* inference, and Byrne's findings, however intrinsically interesting, are no evidence against mental logic.

Is There a Defendant at This Trial?

The prosecution of reason might enjoy watching mental logicians and mental modelers, all expert witnesses for the defense, fight among themselves, but surely, at this point, the jury might grow impatient. Isn't there something amiss, not with the reasoning of people who participate in these experiments, but rather with the demands of psychologists?

Experimentalists expect participants to accept the premises as true whether those premises are plausible or not, to report only what necessarily follows from the premises, and to completely ignore what is merely likely to follow from them—to ignore the real world, that is. When people fail to identify the logical implications of the premises, many psychologists see this as proof that their reasoning abilities are wanting. There is an alternative explanation, namely, that the artificial instructions given to people are hard or even, in many cases, impossible to follow.

It is not that people are bad at making logical deductions; it is that they are bad at separating these deductions from probabilistic inferences that are suggested by the very same premises. Is this, however, evidence of people's irrationality? Couldn't it be seen rather as evidence that psychologists are making irrational demands?

A comparison with the psychology of vision will help. Look at Figure 4, a famous visual illusion devised by Edward Adelson. Which of the two squares, A or B, is of a lighter shade of gray? Surely, B is lighter than A—this couldn't be an illusion! But an illusion it is. However surprising, A and B are of exactly the same shade.

Figure 4. Adelson's checkerboard illusion.

In broad outline, what happens is not mysterious. Your perception of the degree to which a surface is light or dark tracks not the *amount* of light that is reflected to your eyes by that surface but the *proportion* of the light falling on that surface that is reflected by it. The higher this "reflectance" (as this proportion is called), the lighter the surface; the lower this reflectance, the darker the surface:

$$\text{reflectance} = \frac{\text{light reflected by the surface}}{\text{light falling on the surface}}$$

The same gray surface may receive and therefore reflect more or less light to your eyes, but if the reflectance remains the same, you will perceive the same shade of gray. Your eyes, however, get information on just one of the two quantities—the light reflected to your eyes. How, then, can your brain track reflectance, that is, the proportion between the two quantities, only one of which you can sense, and estimate the lightness or darkness of the surface? To do so, it has to use contextual information and background knowledge and infer the other relevant quantity, that is, the amount of light that falls on the surface.

When you look at Figure 4, what you see is a picture of a checkerboard, part of which is in the shadow of a cylinder.

Moreover, you expect checkerboards to have alternating light and dark squares. You have therefore several sound reasons to judge that square B—one of the light squares in the shade—is lighter than square A—one of the dark squares receiving direct lighting. Or rather you would have good reasons if you were really looking at a checkerboard partly in the shadow of a cylinder and not at a mere picture. The illusion comes from your inability to treat this picture just as a two-dimensional pattern of various gray surfaces and to ignore the tridimensional scene that is being depicted.

Painters and graphic designers may learn to overcome this natural tendency to integrate all potentially relevant information. The rest of us are prey to the illusion. When discovering this illusion, should we be taken aback and feel that our visual perception is not as good as we had thought it to be, that it is betraying us? Quite the opposite! The ability to take into account not just the stimulation of our retina but what we intuitively grasp of the physics of light and of the structure of objects allows us to recognize and understand what we perceive. Even when we look at a picture rather than at the real thing, we are generally interested in the properties of what is being represented rather than in the physical properties of the representation itself. While the picture of square A on the paper or on the screen is of the same shade of gray as that of square B, square A would be quite darker than square B on the checkerboard that this picture represents. The visual illusion is evidence of the fact that our perception is well adapted to the task of making sense of the three-dimensional environment in which we live and also, given our familiarity with images, to the task of interpreting two-dimensional pictures of three-dimensional scenes.

Now back to Mary, who might study late in the library. In general, we interpret statements on the assumption that they are intended to be relevant.[21] So when given the second major premise, "If the library stays open, then Mary will study late in the library," people sensibly assume that they are intended to take this premise as relevant. For it to be relevant, it must be the case that the library might close and that this would thwart Mary's intention to study late in the library. So, yes, participants have been instructed to accept as absolutely true that "if Mary has an essay to write, then she will study late in the library," and they seem not to. However, being unable to follow such instructions is not at all the same thing as being unable to reason well. Treating information that has been intentionally given to you as relevant isn't irrational—quite the contrary.

It takes patience and training for a painter to see a color on the canvas as it is rather than as how it will be perceived by others in the context of the whole picture. Similarly, it takes patience and training for a student of logic to consider only the logical terms in a premise and to ignore contextual information and background knowledge that might at first blush be relevant. What painters do see and we don't is useful to them as painters. The inferences that logicians draw are useful to them as logicians. Are the visual skills of painters and the inferential skills of logicians of much use in ordinary life? Should those of us who do not aspire to become painters or logicians feel we are missing something important for not sharing their cognitive skills? Actually, no.

The exact manner in which people in Ruth Byrne's experiment are being reasonable is a matter for further research, but that they are being reasonable is reasonably obvious. That people fail to solve rudimentary logical problems does not show that they are unable to reason well when doing so is relevant to solving real-life problems. The relationship between logic on the one hand and reasoning on the other is far from being simple and straightforward.

At this point, the judge, the jury, and our readers may have become weary of the defense's and the prosecution's grandstanding. The trial conceit is ours, of course, but the controversy (of which we have given only a few snapshots) is a very real one, and it has been going on for a long time. While arguments on both sides have become ever sharper, the issue itself has become hazier and hazier. What is the debate really about? What is this capacity to reason that is both claimed to make humans superior to other animals and of such inferior quality? Do the experiments of Kahneman and Tversky on the one hand and those of "mental logicians" and "mental modelers" on the other hand address the same issue? For that matter, is the reasoning they talk about the same thing as the reason hailed by Descartes and despised by Luther? Is there, to use the conceit one last time, a defendant in this trial? And if there is, is it reason itself or some dummy mistaken for the real thing? Is reason really a thing?

2

Psychologists' Travails

The idea that reason is what distinguishes humans from other animals is generally traced back to the ancient Greek philosopher Aristotle.[1] Aristotle has been by far the most influential thinker in the history of Western thought, where for long he was simply called "the Philosopher," as if he were the only one worthy of the name. Among many other achievements, he is credited with having founded the science of logic. In so doing, he provided reason with the right partner for a would-be everlasting union—or so it seemed. Few unions indeed have lasted as long as that between logic and reason, but lately (meaning in the past hundred years or so), the marriage has been tottering.

Reason and Logic? It's Complicated

Until the end of the nineteenth century, it went almost without saying that logic and the study of reasoning, while not exactly the same thing, were two aspects of a single enterprise. Logic, it was thought, describes good or correct reasoning. Not all reasoning is good—as we saw, far from it—but all reasoning, so the story goes, ought to be, and aims to be, logical. Bad reasoning is reasoning that tried to be logical but failed (or else it is sophistry merely pretending to be logical). Hence logic defines what reasoning is, just as a grammar defines a language, even if we often express ourselves ungrammatically.

Typical textbook examples of reasoning begin with a bit of simple logic and often end there—without a word on what goes on in the mind of the reasoner. They generally involve a pair of premises and a conclusion. For instance (to use one of Aristotle's best-known examples of so-called categorical syllogism):

Premises:	1. All humans are mortal.
	2. All Greeks are humans.
Conclusion:	All Greeks are mortal.

From the propositions that all humans are mortal and that all Greeks are humans, it logically follows that all Greeks are mortal. Similarly, from the propositions that Jack lent his umbrella either to Jill or to Susan and that he did not lend it to Jill, it logically follows that he lent it to Susan (this being an example of a "disjunctive syllogism"). One of the achievements of Aristotelian logic was to take such clear cases of valid deductions and to schematize them.

Forget about humans, mortals, and Greeks. Take any three categories whatsoever, and call them A, B, and C. Then you can generalize and say that all syllogisms that have the form of the following schema are valid:

Premises:	1. All As are Bs.
	2. All Cs are As.
Conclusion:	All Cs are Bs.

Forget about umbrellas, Jack, Jill, and Susan. Take any two propositions whatsoever, and call them *P* and *Q*. Then you can generalize and again say that all syllogisms that have the form of the following schema (corresponding to the "or" rule we talked about in Chapter 1) are valid:

Premises:	1. *P* or *Q*
	2. not *P*
Conclusion:	*Q*

What is the point of identifying such schemas? It is to go from the intuition that some particular deductions happen to be valid—deductions, for instance, about the Greeks being mortal or about Jack having lent his umbrella to Susan—to a formal account of what makes valid not just these particular deductions but all deductions of the same form. By replacing concrete contents (that are taken to be irrelevant to deduction) with arbitrary symbols such as capital letters—a device invented by Aristotle—you end up with a "logical form" that contains just terms such as "all," "or," or "not" that are relevant to premise-conclusion relationships. Deduction schemas display logical forms that stand in such relationships.

For more than two thousand years, scholars felt no need to go beyond Aristotelian logic. The author of the *Critique of Pure Reason,* Immanuel Kant, could, at the end of the eighteenth century, maintain that since Aristotle, logic "has been unable to take a single step forward, and therefore seems to all appearance to be finished and complete."[2] He couldn't have been more mistaken.

In the past two hundred years, logic has developed well beyond and away from its Aristotelian origins both in scope and in sophistication. It has diversified in many different subfields and approaches, and even into a plurality of logics. Modern deductive logic provides a formal account of a much greater variety of valid deductions than did classical logic. It does so not by means of a catalogue of deduction schemas but by deriving such schemas from first principles with elaborate methods. Many of the deductions studied in modern logic, however, even relatively simple ones, are no more part of ordinary people's repertoire than are advanced theorems in mathematics. True, there are some research programs in logic that aim at being relevant to psychology, but modern logic as a whole does not.

The experimental study of reasoning started in the twentieth century.[3] By then, many logicians saw logic as a purely formal system closely related to mathematics. Gottlob Frege, the German founder of modern logic, had denounced the very idea that logic is about human reasoning as a fallacy, the fallacy of "psychologism": logic is no more about human reasoning than arithmetic is about people's understanding and use of quantities. This is now the dominant view.

And yet, while most logicians were turning their backs on psychology, most psychologists of reasoning were still looking to logic in order to define their domain, divide it into subdomains, and decide what constitutes good and bad reasoning. Until recently, it rarely crossed their minds that this could amount to a fallacy of "logicism" in psychology symmetrical to the fallacy of psychologism in logic.[4]

True, thinking of reasoning as a "logical" process can seem quite natural. When people reason, some thoughts occur first in their mind, and have to occur first for other thoughts to occur afterward. It may be tempting to equate this temporal and causal sequence of thoughts with a logical sequence of propositions in a deduction. The very words "consequence" and "follows" used in logic evoke a time sequence. But no, these words do not, in logic, refer to temporal relationships. The order of propositions in a logical sequence is no more a genuine temporal order than is the order of the positive integers, 1, 2, 3, . . . , in arithmetic. Psychological processes have duration and involve effort. Logical sequences have not and do not.

In logic, the word "argument" describes a timeless and abstract sequence of propositions from premises to conclusion. In ordinary usage, on the other hand, an argument is the production, in one's mind or in conversation, of one or several reasons one after the other in order to justify some conclusion. What can we do here to avoid confusion? Since the psychology of reasoning has focused on classical deductive arguments, also known as "syllogisms," this is the term we will use in our critical discussion. We will always use "argument," on the other hand, in the ordinary, nontechnical sense.

Couldn't series of reasons given to convince an audience match logical sequences from premises to conclusion? Well, this is not what usually happens. Often, when you argue, you start by stating the conclusion you want your audience to accept—think of a lawyer pleading her client's innocence, or think of political discussions—and then you give reasons that support this conclusion. It is commonly assumed, all the same, that most, if not all, ordinary reasoning arguments must, to be arguments at all, correspond to syllogisms; if the correspondence is not manifest, then it must be implicit; some premises must have been left out for the sake of brevity. Most ordinary arguments are, according to this view, "enthymemes," that is, truncated syllogisms. This, we will argue, is just old dogma, so much taken for granted that little or no effort is made to justify it empirically.

Logic and the psychology of reasoning, which had been so close to one another, have moved in different directions. They still seem to have many concepts in common, but what they actually share are labels, words that have taken on different meanings in each discipline, creating much confusion.[5] "Argument" is not the only word used to describe both an abstract logical thing and a concrete psychological phenomenon. Many other words, such as "inference," "premise," "conclusion," "valid," or "sound," have been borrowed from one domain to the other and are used in both cases with little attention to the fact that they are used differently. Even the word "reasoning" has been used by logicians to talk about syllogisms, logical derivations, and proofs, and the word "logical" is commonly used as a psychological term (as in "Be logical!"). We will try to avoid the fallacies that may result from such equivocations.

Some of the Bakers Are Athletes

The unrequited love of psychology of reasoning for logic has had costly consequences. Many eminent psychologists chose, for instance, to investigate how people perform with Aristotelian categorical syllogisms. Why? Well, these syllogisms had been at the center of classical logic for more than two thousand years. Surely, then, they had to play a major role in psychology.

When all splittable hairs have been split, there turn out to be 256 possible forms of categorical syllogisms that could each be experimentally tested (twice as many when notational variants are included). To this end, many researchers invested years of work. Pity, too, the thousands of participants in these experiments who were given long series of dull and repetitive problems to solve, one after the other, in the style of the following:

Some of the bakers are athletes.
None of the bakers is a canoeist.
What, if anything, follows?

Only 24 of these 256 syllogistic forms are logically valid. It is not clear which of these valid and invalid syllogisms ever occur in human actual reasoning and how often they do.

Of course, in science the study of marginal or practically unimportant phenomena can be of major scientific relevance—think of the common fruit fly, also known as *Drosophila melanogaster,* and its place in modern biology—but this is hardly a case in point. In a review article published in 2012, after half a century of intensive studies, Sangeet Khemlani and Philip Johnson-Laird identified twelve competing theories of syllogistic reasoning, none of which, they say, "provides an adequate account." "The existence of 12 theories of any scientific domain," they add, "is a small disaster."[6] This indictment is made all the more powerful and even poignant by the fact that one of the twelve theories, and arguably the most influential, is Johnson-Laird's own mental-model account of syllogisms.

Proponents of different approaches to reasoning (mental logic, mental models, more recently Bayesian inference, and so on) have used the study of categorical syllogisms as evidence that their own approach is the best—evidence, however, that only the already-converted have found convincing.

There is another group of scholars, apart from psychologists, committed to the idea that classical syllogisms are still highly relevant: theologians, who have been teaching and using "syllogistics" since the Middle Ages. To give just one example, here is how Father Wojciech Giertych, the household theologian of Pope Benedict XVI, explained why women are not suited for priesthood: "Men are more likely to think of God in terms of philosophical definitions and logical syllogisms."[7] Not convinced? The relevance of the whole battery of Aristotelian syllogisms to psychology is, we are tempted to quip, equally mysterious.

"Never Do an Experiment If You Know Why You're Doing It!"

Few psychologists of reasoning, if any, had a greater impact on the field than Peter Wason. "Wason's way of doing research," Johnson-Laird told us,[8] "was pretty eccentric, e.g., never do an experiment if you know why you're doing it!"

In 1966, Wason introduced a new experimental design, the four-card selection task, which became—and remains to this day—a main tool and focus of research in the discipline. It wouldn't be completely wrong—just exaggerated and unfair to the few researchers who have resisted its lure—to say that the psychology of reasoning has to a large extent become the psychology of the Wason task. If Wason invented this experiment without knowing what purpose it would serve, then, it must be reckoned, this turned out to be an amazingly successful shot in the dark.

Figure 5. The four cards of the Wason selection task.

Here is how Wason's experiment goes. "In front of you are four cards," the experimenter tells you. "Each card has a letter on one side and a number on the other. Two cards (with an E and a K) have the letter side up; the two others (with a 2 and a 7) have the number side up" (see Figure 5).

"Your task is to answer the following question: Which of these four cards *must* be turned over to find out whether the following rule is true or false of these four cards: '*If there is an E on one side of a card, then there is a 2 on the other side*'?"

Which cards would *you* select?

The structure of the experiment derives from a standard type of inference in classical logic, conditional syllogisms, which we encountered in Chapter 1. Figure 3 in Chapter 1 laid out the four schemas of conditional syllogism; Figure 6 in this chapter shows how the selection task is built on these four schemas.

The "rule" of the selection task is the major premise of a conditional syllogism of the form "if *P*, then *Q*" (in our example, *if there is an E on one side of a card, then there is a 2 on the other side*). Each card provides one of the four possible minor premises (in our example, the E card represents the minor premise *P, there is an E;* the K card represents not-*P, there isn't an E*; the 2 card represents *Q*, *there is a 2;* and the 7 card represents not-*Q*, *there isn't a 2*). As you may remember, only two of these minor premises, *P* and not-*Q*, allow valid deductions (called *modus ponens* and *modus tollens,* respectively); trying to make a similar deduction from the two other possible minor premises, not-*P* and *Q*, yields the fallacies of "denial of the antecedent" and of "affirmation of the consequent."

The correct answer, then, is to select just the E and the 7 cards. The rule entails a prediction about what should be on the other side of these two cards,

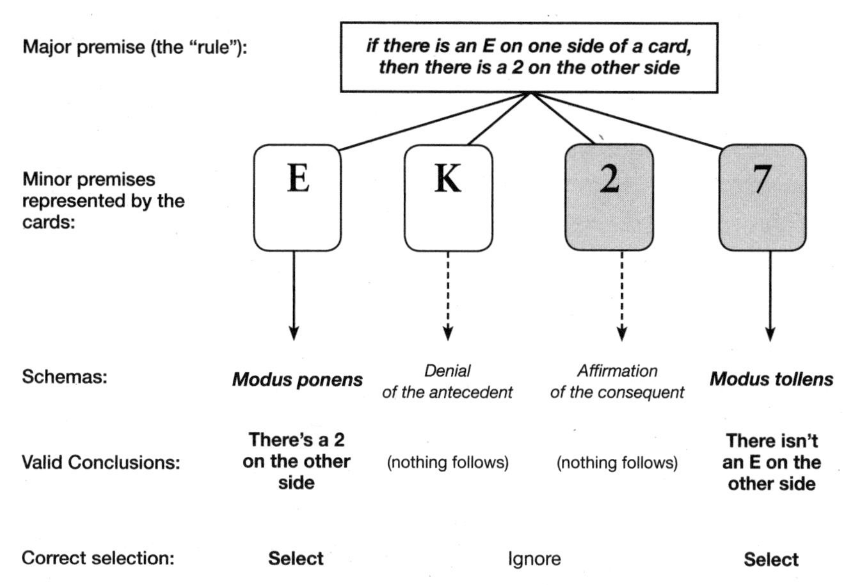

Figure 6. The selection task and the four schema of conditional inference.

a prediction that could be tested by turning these two cards over. Should the other side of either of these two cards fail to be as predicted, the rule would be falsified. The rule, on the other hand, doesn't entail any prediction as to what should be on the hidden side of the K and the 2 cards: they are therefore irrelevant. In particular, contrary to a common intuition, turning over the 2 card is useless. Suppose there *isn't* an E on the other side. So what? All the rule says is that an E must be paired with a 2; it does not say that *only* an E can be paired with a 2.

You selected the E and the 7 cards? Congratulations! You made another selection? Don't feel too bad. Only about 10 percent of participants make the right choice anyhow.

Once psychologists start experimenting with the Wason task, it is hard for them to stop. Many have become addicts. Why? Well, the design of the selection task lends itself to endless variations. You can alter the instructions, modify the content of cards, or invent a variety of contexts. You can then observe what happens and see in particular whether more participants give the correct answer than with Wason's original version of the task. If this happens, write an article. If it doesn't, try again. Moreover, not just psychologists but also philosophers, students, and sometimes your roommate or your cousin easily come up with conjectures to explain why participants respond the way they do and with suggestions for new variations. The selection task has proved an everlasting topic of conversation where many people, pros and amateurs, can have a go.

Ideally, of course, the selection task should owe its success to being, like the microscope for biology (to which we have heard it being compared), a superior tool that provides crucial evidence and helps answer fundamental questions. Has any theoretical breakthrough been made thanks to the selection task? No, whenever experimental evidence has been claimed to provide crucial support for a genuine theoretical claim, alternative interpretations have been proposed. As a result, much of the work done with the task has had as its goal to explain the task itself,[9] with the psychology of human reasoning serving just as a dull backdrop to colorful debates about experiments.

Much of the early research aimed at improving people's poor performance with the selection task. Would training help? Not much. Feedback? Hardly. Changing the wording of the rule? Try again. Monetary rewards for good performance? Forget it. Then could variations be introduced in the nonlogical content of the selection task (replacing numbers and letters with more interesting stuff) that would cause people to perform better? Yes, sometimes, but explanations proved elusive.

So a lot of noise has been produced, but what about light? A few findings relevant not just to understanding the task but to understanding the mind were generally stumbled upon rather than first predicted and then confirmed. What the story of the selection task mainly illustrates is how good scientists can go on and on exploring one blind alley after the other.

Ironically, the most important finding ever to come out of fifty years of work with the task is that people don't even use reasoning to resolve a task that was meant to reveal how they reason.

In the early 1970s, Jonathan Evans made a puzzling discovery by testing a simple variation on the standard selection task.[10] Take the usual problem with the rule, "*If there is an E on one side of a card, then there is a 2 on the other side*" and the four cards E, K, 2, and 7. As we saw, only about 10 percent select the E and the 7 cards, even though this is the logically correct solution.

Now just add a "not" in the rule, like this: *"If there is an E on one side of a card, then there is* not *a 2 on the other side."* Show the same cards. Ask the same question. Now, a majority of participants give the right answer.

Don't jump to the startling conclusion that a negation in the rule turns participants into good logical reasoners. Actually, in both conditions (with and without the negation in the rule), most participants make exactly the same selection, that of the E and the 2 cards, as if the presence of the negation made no difference whatsoever. It so happens that this selection is incorrect in the standard case but correct with the negated rule. (How so? Well, the affirmative rule makes no prediction on the letter to be found on the hidden side of the 2 card, but the negative version of the rule does: an E on the hidden side of the 2 card would falsify the negated rule. So with the negated rule, the 2 card *should* be selected.)

This shows, Evans argued, that people's answers to the Wason task are based not on logical reasoning but on intuitions of relevance: they turn over the cards that seem intuitively relevant. And why do the E and the 2 seem intuitively relevant? Because, explains Evans, they are mentioned in the rule, whereas other letters and numbers are not, and that's almost all there is to it.[11]

The long and convoluted story of the selection task well explains how and why the psychology of human reasoning ended up pivoting away from its early obsession with classical logic to new challenges.

Dual Process?

Look at work on the selection task and look more generally at experimental psychology of reasoning, and you will see psychologists at pains to be as thorough as possible. This makes it even more puzzling and disheartening to see how modest the progress, how uninspiring the overall state of the art—and this in a period where the study of cognition has undergone extraordinary developments. In many domains—vision, infant psychology, and social cognition, to name but three—there have been major discoveries, novel experimental methods, and clear theoretical advances at a more and more rapid pace. Every month, journals publish new and exciting results. There are intense debates, but with a clear common sense of purpose and the strong feeling of shared achievement, nothing of the sort in the psychology of reasoning. True, there are schools of thought that each claim major breakthroughs, but, for good or bad reasons, none of these claims has been widely accepted.

Still, if a survey was made and psychologists of reasoning were asked to mention what has been the most important recent theoretical development in the field, a majority—with a minority strongly dissenting—would name "dual process theory": the idea that there are two quite distinct basic types of processes involved in inference and more generally in human psychology.

A basic insight of dual process theory is that much of what people do in order to resolve a reasoning task isn't reasoning at all but some other kind of process, faster than reasoning, more automatic, less conscious, and less rule-governed. In the past twenty years, different versions of the approach have been developed. Talk of "system 1" and "system 2" is becoming almost as common and, we fear, often as vacuous as talk of "right brain" and "left brain" has been for a while.

Actually, an early sketch of dual process theory had been spelled out by Jonathan Evans and Peter Wason in a couple of articles published in 1975 and 1976 and quickly forgotten. As we saw, just by adding a *"not"* in the rule of the selection task, Evans had demonstrated that people make their selection without actually reasoning. They merely select the cards that they intuitively see as relevant (which happens to yield an incorrect response with the original rule and the correct response with the negated rule). Selection, then, is based on a type 1 intuitive process.

Evans and Wason redid the experiment, this time asking people to explain their selection, and then they did reason, no question about it. They reasoned not to resolve the problem—that they had done intuitively—but to justify their intuitive solution. When their solution happened to be logically correct (which typically occurred with the negated rule), they provided a sensible logical justification. When their solution happened to be incorrect, people gave, with equal confidence, a justification that made no logical sense. What conscious reasoning—a type 2 process—seemed to do was just provide a "rationalization" for a choice that had been made prior to actual reasoning.

There were three notable ideas in this first sketch of the dual process approach. The first was a revival of an old contrast, stressed by—among many others—the eighteenth-century Scottish philosopher David Hume and the nineteenth-century American philosopher William James, between two modes of inference, one occurring spontaneously and effortlessly and the other—reasoning proper—being on the contrary deliberate and effortful. A second, more novel idea was that people may and often do approach the same inferential task in the two modes. In the selection task, for instance, most participants produce both a spontaneous selection of cards and a reasoned explanation of their selection. The third idea was the most provocative: what type 2 deliberative processes typically do is just rationalize a conclusion that had been arrived at through intuitive type 1 processes. This idea so demeans the role of reasoning proper that Evans and Wason's dual process approach was met with reticence or incredulity.[12]

This early dual process approach to reasoning was not often mentioned, let alone discussed, in the next twenty years. When it did reappear on the front of the scene, gone were the youthful excesses; written off was the idea that reasoning just rationalizes conclusions that had been arrived at by other means. And so, in 1996, Evans and the philosopher David Over published a book, *Rationality and Reasoning*,[13] where they advocated a "dual process theory of thinking" but with type 1 processes seen as rational after all and type 2 processes "upgraded from a purely rationalizing role to form the basis of the logical component of performance." Moreover, the original assumption that the two types of processes occur in a rigid sequence—first the spontaneous decision, and then the deliberate rationalization—was definitely given up in favor of an alternative that had been suggested in passing in 1976, namely, that the two types of processes interact. Whereas the earlier Evans-and-Wason version of the dual process approach undermined humans' claims of rationality, the later Evans-and-Over version vindicates and might even be said to expand these claims.

Apparently, the time was ripe. That same year, 1996, the American psychologist Steven Sloman published "The Empirical Case for Two Systems of Reasoning" where, drawing on his expertise in artificial intelligence, he proposed a somewhat different dual process (or as he called it, "dual system" approach).[14] In 1999, the Canadian psychologist Keith Stanovich, in his book *Who Is Rational?*, drew on his expertise on individual differences in reasoning to propose another dual process approach.[15] In his Nobel Prize acceptance speech in 2002, Daniel Kahneman endorsed his own version of an approach that had been in many respects anticipated in his earlier work with Amos

Tversky.[16] Many others have contributed to this work, some with their own version of the approach, others with criticisms.

A typical device found in most accounts of a dual process approach is a table layout of contrasting features. Here are examples of contrasts typically found in such tables:

Type 1 processes	*Type 2 processes*
Fast	Slow
Effortless	Effortful
Parallel	Serial
Unconscious	Conscious
Automatic	Controlled
Associative	Rule-based
Contextualized	Decontextualized
Heuristic	Analytic
Intuitive	Reflective
Implicit	Explicit
Nonverbal	Linked to language
Independent of general intelligence	Linked to general intelligence
Independent of working memory	Involving working memory
Shared with nonhuman animals	Specifically human

The gist of these contrasts is clear enough: on the one side, features that are commonly associated with instincts in animals and intuition in humans; on the other side, features that are associated with higher-order conscious mental activity, in other terms with "thinking" as the term is generally understood. At first blush, such a distinction looks highly relevant to understanding human psychology in general and inference in particular: yes, we humans are capable both of spontaneous intuition and of deliberate reasoning. So, dual process approaches seem to be a welcome, important, and, if anything, long overdue development. How could one object to such an approach?

Well, one might object to the vagueness of the proposal. Aren't these features nicely partitioned on the two-column table somewhat intermingled in reality? For instance, we all perform simple arithmetic inferences automatically (a type 1 feature), but they are rule-based (a type 2 feature). So, are these simple arithmetic inferences a type 1 or a type 2 process? Moreover, many of the contrasts in such tables—between conscious and unconscious processing, for instance—may involve a difference of degree rather than a dichotomy of kinds. These and other similar examples undermine the idea of a clear dichotomy between two types of processes.

What, anyhow, is the explanatory import of the whole scheme? There are many more than two mechanisms involved in human inference. The key to explaining human inference is, we will argue, to properly identify the common properties of all these mechanisms, the specific features of each one of them, and their articulation, rather than classifying them into two loose categories on thin theoretical ground.

Still, for a while, dual process theory seemed to help resolve what we have called the enigma of reason by explaining why reasoning so often fails to perform its function. True reasoning (type 2 processes), the theory claimed, is indeed "logical," but it is quite costly in terms of cognitive resources. If people's judgments are not systematically rational, it is because they are commonly based on cheaper type 1 processes. Type 1 processes are heuristic shortcuts that, in most ordinary circumstances, do lead to the right judgment. In nonstandard situations, however, they produce biased and mistaken answers. All the same, using type 1 processes makes sense: the lack of high reliability is a price rationally paid for day-to-day speed and ease of inference. Moreover, type 2 reasoning remains available to double-check the output of type 1 intuitions. Intellectual alertness—intelligence, if you prefer—goes together with a greater readiness to let type 2 reasoning take over when needed. Enigma resolved? Not really.

The more dual process approaches were being developed, the more they inspired experimental research, the less this simple and happy picture could be maintained. Evans and Stanovich now call it a fallacy to interpret dual process theory as committed to seeing type 2 processes as necessarily "better" than type 1 processes. In fact, they acknowledge, type 2 reasoning can itself be a source of biases and even introduce errors where type 1 intuition had produced a correct judgment. We are not quite back to the early approach of Evans and Wason in the 1970s, if only because the picture is now so much richer, but the problems that dual process approaches seemed to solve are just posed in new and somewhat better terms. The enigma of reason still stands.

We won't discuss dual process theory in any detail: it is too much of a scattered, moving, and in part blurry target.[17] Our ambition, anyhow, is to offer something clearly better. More relevant to us than the varieties of dual process theories is the way the whole approach has shaken and in some sense shattered psychology of reasoning. For decades, the central question of the field had been: What is *the* mechanism by means of which humans reason? "Mental logic!" argued some psychologists; "mental models!" argued others. Some still see this as the central question and have offered novel answers, drawing on new ideas in logic or in probabilities. But with the dual process approach, doubt has been sown.

First there was the idea that there are not one but two types of processes at work. Then several dual system theorists came to see type 1 processes as carried out by a variety of different specialized mechanisms. More recently, even the homogeneity of type 2 processes has been questioned. The more it is recognized that human inference involves a variety of mechanisms at several levels, the less adequate become the labels "dual process" and "dual system theory." Reason and logic have split, and reason itself now seems to be broken into pieces. This is both a good end point for one kind of research and a good start for another.

II

Understanding Inference

The elephant trunk is a type of nose. However impressive it may be, it would not make sense to think of it as the epitome of noses. Similarly, reason is one type of inference mechanism; it is neither the best nor the model of all others. To understand reason, we must first understand inference in general, its diversity, and its powers. In Chapters 3 through 6, we show how widespread inference is in humans and other animals, and we look at its basic mechanisms and procedures. Just as the elephant trunk does things that other noses don't do, we explain how a category of higher-order human inferential mechanisms (that include reason) happens to have a uniquely wide reach.

3

From Unconscious Inferences to Intuitions

Animals don't think, Descartes had maintained. The eighteenth-century Scottish philosopher David Hume disagreed. He took it, on the contrary, as evident that animals think and are capable of drawing inferences in the same way humans do. He wrote, "Animals, therefore, are not guided in these inferences by reasoning: Neither are children: Neither are the generality of mankind, in their ordinary actions and conclusions: Neither are philosophers themselves, who, in all the active parts of life, are, in the main, the same with the vulgar." And he added: "Nature must have provided some other principle, of more ready, and more general use and application; nor can an operation of such immense consequence in life, as that of inferring effects from causes, be trusted to the uncertain process of reasoning and argumentation."[1]

Reasoning and argumentation had been viewed, not just by Descartes but by most philosophers, as the path to greater certainty and, moreover, as the only method for drawing inferences. Hume, unfazed, described reasoning as so unreliable that nature must have provided some other means for performing inferences. But if there are means better than reasoning that, moreover, are "of more ready, and more general use and application," why, then, bother to reason at all?

The two words "reasoning" and "inference" are often treated as synonyms. What Hume implied was that reasoning is only one way of performing inferences, and not such a reliable way at that. We agree.

The Why and the How

What is reasoning? Everybody has at least some working knowledge of it: it is, after all, something we all do and do consciously. But it is one thing to have a working knowledge of some mechanism, and another to really understand what exactly it does and how it does it. You know how to swallow, but do you know how swallowing works? To understand reasoning (rather than just make use of it), one needs, to begin with, an effective way of telling it apart from other psychological processes. And this is where the difficulty begins.

In the philosophical and psychological literature, reasoning is commonly defined in two ways, in terms of either its goal or its process. These two definitions, alas, fail to pick out the same phenomenon: the standard characterization of the goal picks out inference in general; the standard characterization of the process picks out reasoning proper.

Why do we reason? The goal of reasoning, so the story goes, is that of coming to new conclusions not through mere observation or through the testimony of others but by drawing these new conclusions from information already available to us.

How do we reason? The process of reasoning consists in attending to reasons for adopting new conclusions.

You hesitate, say, between spending the evening at home reading a novel and going to the cinema. You might, at some point, just find yourself reading the novel without having deliberated about what to do. Or you might think, "There is no very good film playing tonight and the weather is bad; I might have to walk back in the rain. On the other hand, there is this novel that Tomoko gave me and that looks really good . . ." If such was your train of thought, then your decision to stay at home involved reasoning (leaving open the question of whether you used reasoning to actually make the decision or just to rationalize it).

When we reason, conclusions do not just pop up in our mind as self-evident; we arrive at them by considering reasons to accept them. Or, if we already accept a given conclusion, we might still engage in reasoning in order to find reasons that justify our conclusion, and that should convince others to accept it too. This is quite a rudimentary sketch of the process of reasoning. Unlike most approaches, it does not even mention the role of logic in identifying reasons (which many would claim is essential). Because it eschews the logical framework, it speaks not of premises but, more broadly, of reasons. Still, it already raises a simple question: Is this process—attending to reasons—the *only* way to pursue the goal of extracting new information from information that we already possess? And Hume answers with force: Of course not! After all, even animals form expectations about the future. Their life depends on these expectations being on the whole correct. Since the future cannot be perceived, it is through inference that animals must form expectations. It is quite implausible, however, that, in so doing, animals attend to reasons.

Are we humans so different from other animals? Like them, we cannot perceive the future but only infer it. Like them, we base much of our everyday behavior on expectations that we arrive at unreflectively. You play tennis, for example, and have to adjust your position to return the ball; you infer the best position without reflection, in a fraction of a second. Or you call your father on the phone and, just from his tone of voice, you infer that he is in a bad mood. You refrain from telling him that you won't be able to come to the next family reunion. You don't have to reflect in order to understand your father's mood, to expect that he might be agitated if told that you won't come, and to avoid mentioning the issue: all this seems immediately obvious in the situation and you act accordingly. Just like other animals, humans are capable of forming expectations and drawing various kinds of inferences in a spontaneous and unreflective way.

Following Hume's example, we will use the term "inference" for the extraction of new information from information already available, whatever the process.[2] We will reserve the term "reasoning" for the particular process of pursuing this goal by attending to reasons. Humans, we will argue, cannot spend a minute of their waking life without making inferences. On the other hand, they can spend hours or even days without ever engaging in reasoning.

As long as attending to reasons was taken—or rather *mis*taken—for the only way of extracting new information from information already available, there was no need or incentive to distinguish "inference" from "reasoning." It is no surprise, then, that the two terms should have been, for so long, used as synonyms.

Even today, many philosophers, disagreeing with Hume, want to keep a narrow notion of inference as meaning more or less the same thing as reasoning in the traditional sense. This leaves them with two words, "inference" and "reasoning," for the same thing and no word for the many forms of inference in which reasons play no causal role. Apart from conservatism, there is no clear motivation for such an unhelpful terminological policy.

In some contexts, mind you, treating "inference" and "reasoning" as synonyms is innocuous and may even serve a purpose. Psychologists, for instance, often use the word "reasoning" to describe the remarkable inferences that animals and infants turn out to be capable of making. Unlike philosophers who insist on narrowing down the meaning of "inference" to that of "reasoning," what these comparative and developmental psychologists are doing is broadening the use of "reasoning" to cover all kinds of inference. This use of the superior-sounding word "reasoning" expresses well-deserved respect for the intellectual capacities of creatures—animals and infants—who had so often been described as quite dumb.

Here, however, we want to explore the very special place of reasoning among other forms of inference. For this, we had better sharply distinguish the two. When one looks for inference not just in reasoning but wherever it might occur, one sees it occurring everywhere, in lowly animals, in infants, and in human adults even when they are not reasoning at all.

Ants in the Desert

Charles Darwin read Hume's remarks on animal inference and commented in his diary that one should consider "the origin of reason as gradually developed," the first hint ever that reason should be approached in an evolutionary perspective.[3] It took more than a century for the evolution of human reason to become a serious topic of study, in particular in the new field of evolutionary psychology. Well before this, however, Darwin's ideas inspired the study of animal psychology. In the very process of investigating what distinguishes animal species from one another, "comparative psychology" (as this new discipline came to be called) has highlighted what all cognitive systems have in common—and, to begin with, the fact that they all perform inferences.

Take desert ants.[4] The University of Zurich biologist Rüdiger Wehner has devoted more than thirty years to their study. He explains:

> The salt pans of the Sahara Desert—vast expanses of flat, hot, and dry terrain—are inhabited by very few animals. *Cataglyphis fortis,* a skillful and vivacious species of ant, is certainly the most remarkable of these species. It dashes, leaps, and scrambles across the desert surface, and sweeps it for widely scattered food particles, mostly carcasses of other insects that have succumbed to the stress of the harsh desert environment. In searching, the ant leaves its underground colony and scavenges across the desert floor for distances of more than 200m, winding its way in a tortuous search for fodder. Once it has found its bit of food, the ant grasps it, turns around, and runs more or less directly back to the starting point of its foraging excursion, a tiny hole that leads to its subterranean nest [see Figure 7].[5]

How do these insects manage, on their way back, to orient themselves in precisely the right direction and to stop speeding forward when they are in the vicinity of their nest? Wehner and his collaborators have, through many careful experiments and observations, shown that ants have in their "navigational toolkit" a "celestial compass" that allows them to assess their changes of direction and an "odometer" that keeps track of the distance covered between two such changes of direction. Needless to say, both tasks involve much more than mere recording of sensory information.

The celestial compass uses the ants' sensitivity to the polarization of the sun's light to determine an axis and to record the angle of each segment of the ant's outbound run relative to this axis. The odometer infers the distance covered between two changes of direction on the basis of the number of steps used to cover it. Ants' brains then put together the information inferred by their compass and odometer to further infer the direction and the distance to their nest. This process of "path integration" is comparable to the technique of dead reckoning used by sailors and aviators in order to compute the position of a ship or an airplane in the absence of landmarks (and, nowadays, of GPS input). While human dead reckoning is a complex intellectual task done with the help of measuring instruments, ants' path integration is achieved within their minute brain by means of these automatic and unconscious computations.

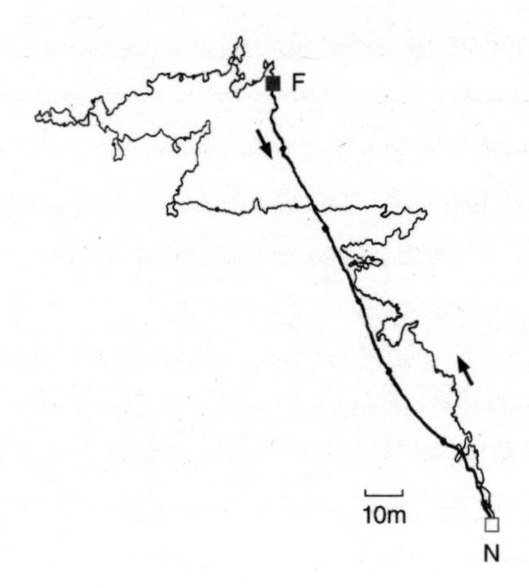

Figure 7. The trajectory of a desert ant from its nest (N) to food (F) and back.

For all its specificity, the desert ants' case well illustrates three of the basic properties of all cognitive systems:

1. Cognition is first and foremost a means for organisms that can move around to react appropriately to risks and opportunities presented by their environment. Cognition didn't evolve in plants, which stay put, but in animals that are capable of locomotion. Cognition without locomotion would be wasteful. Locomotion without cognition would be fatal. Desert ants in particular moving out of their nest would, without their cognitive skills, quickly fry in the sun.
2. Cognition involves going well beyond the information available to the senses. All that sensory organs get by way of information, be it in ants or in humans, are changes of energy at thousands or millions of nerve endings. To integrate this information, to identify the events in the environment that have caused these sensory stimulations, to respond in an appropriate manner to these events, cognition must, to a large extent, consist in drawing inferences about the way things are, about what to expect, and about what to do. Foraging ants draw inferences every second.
3. Inferences may be performed by specialized mechanisms that are each remarkably good at dealing with just one quite specific task:

inferring distance on the basis of number of steps, inferring angular changes of direction on the basis of the position of the sun in the sky, inferring a best path back to the nest on the basis of distance and direction, and so on.

Ptolemy on a Boat

Do humans have specialized mechanisms that each deal with one kind of cognitive task? Well, of course they do—to begin with, in perception. That perception is performed by specialized mechanisms—vision, audition, olfaction—is a truism. What is less immediately obvious, however, is that these mechanisms perform inferences. We experience ordinary perception as a mere registration of the way things are, as a "taking in" of facts rather than as a construction of mental representations. If there are inferences involved in perception, they are typically unconscious.

In cognitive psychology textbooks, the scholar usually credited with having discovered unconscious inference in perception is the nineteenth-century German scientist Herman von Helmholtz. The discovery, however, is much older. It is the Greco-Roman scientist Ptolemy, who in the second century CE first talked about the role of unconscious inference in vision. He was also the first to use perceptual illusions as providing crucial evidence on the workings of the mind.[6]

You may have had the experience of sitting on a train that actually had imperceptibly started moving, giving you for a brief instant the impression that it was not your train but rather the train on the other side of the platform that was in motion. There were no trains at the time of Ptolemy, but he described a similar illusion on a boat. Suppose, he wrote, "we sail in a boat along the shore . . . and we do not sense the motion of the [boat] carrying us, then we judge the trees and topographical features of the shoreline to be moving. The illusion stems from the fact that . . . we *infer* that the visible objects are moving" (emphasis added).

The best explanation of many illusions is that they arise from the inferences we automatically draw to make sense of our sensations. In most cases these inferences are correct: when our position relative to objects around us changes,

Figure 8. Monsters in a tunnel.

either we or they are moving. If it seems to us that we are not moving, it is reasonable—even if not fail-safe—to infer that they are.

In modern psychology, illusions provide crucial evidence for studying perception. We saw the remarkable example of Adelson's checkerboard illusion in Chapter 1. Roger Shepard created another striking visual illusion (Figure 8).[7] Look at this image of one monster chasing another in a tunnel. The chaser looks bigger than the chased, right? In fact, as you can measure yourself, their two images are exactly of the same size. They therefore project same-size images on the retina. Why, then, does one look bigger than the other? Because we don't just read the size of the object perceived off retinal stimulation; we automatically use contextual evidence relevant to assessing an object's distance in order to infer its size. In the picture, the images of the two monsters are at the same distance from our eyes. Still, because we spontaneously interpret the picture as representing a three-dimensional scene where the chaser is behind the chased, hence farther away from us, we see him as bigger. It so happens that, when we look at this particular picture, the information we use is misleading and our assumptions are false; we are prey to a visual illusion. The very fact, however, that this illusion is surprising tells us how confidently, in normal conditions, we rely on such unconscious inferences to provide us with true perceptions. In most cases, we are right to do so.

The inferential processes involved in perception are typically so fast—their duration is measured in milliseconds—that we are wholly unaware of them. Perception seems to be some kind of immediate and direct access to reality. Inference, which involves both the use of information about things distant in space or time and the risk of error, seems out of place in this direct relationship. Well, there *is* a risk of error in perception; misperceptions and illusions do occur; our perceptions *are* informed, and sometimes misinformed, by previous experience. All this goes to show that our "intuitions" about what it is to perceive shouldn't be given more authority than, say, our intuitions about what it is to digest or to breathe. The conscious intuitions—or "introspections"—we have about our mental and other biological processes do not provide a reliable description, let alone an explanation of these processes. Rather, these intuitions—like all intuitions—are themselves something to be explained.

Still, you might object, how useful can it be to put together under a single category of "inference" wholly unconscious, superfast processes in perception and the conscious and slow processes—sometime painfully slow—that occur in reasoning? Isn't this as contrived as it would be to insist that, say, jumping and flying should be studied as two examples of one and the same kind of process of "moving in the air"? Actually, this objection doesn't work too well. There are fundamental discontinuities between jumping and flying, whereas automatic inference in perception and deliberate inference in reasoning are at the two ends of a continuum. Between them, there is a great variety of inferential processes doing all kinds of jobs. Some are faster, some slower; they involve greater or lesser degrees of awareness of the fact that some thinking is taking place.

That all inferential mechanisms stand on a continuum doesn't mean that they are the same. What it suggests, rather, is that in spite of sharing the function of drawing inferences, they may well be quite different from one another. And reasoning? Reasoning is only one of these many mechanisms.

Inferences We Are Unaware Of

When you see the picture of the two monsters in the tunnel, you do not merely register visible features of the scene (which, as we pointed out, already involves some inference). You also interpret what you see. For instance, you assume that the two monsters are running (rather than, say, standing still on one foot). You assume that one is chasing the other (rather than trying to copy his movements). You assume that the chaser has hostile intentions and that the chased is afraid. Even though the two faces are identical, you interpret them differently.

Perception may involve some degree of freedom in interpreting what it is exactly that we perceive. While we just *see* one monster bigger than the other and we are not that easily persuaded that we are mistaken, we are more willing to entertain the idea, rather than one monster chasing the other—our first interpretation—that the two monsters might both be chased by a third even bigger monster who is off the picture. We came to our first interpretation spontaneously, but that this is an interpretation and not a mere registration of fact is something of which we can easily be made aware.

Memory, too, involves inference. The expression "stored in memory," evoking as it does a storage place where things can be safely kept to be taken out when needed, turns out to be quite misleading. The British psychologist Frederick Bartlett published in 1932 a still-influential book, *Remembering,* where he introduced a now classical distinction between *reproductive* and *reconstructive* memory.[8] If your task is to remember a random list of numbers, you learn them by rote and you indeed try to reproduce the list when you have to. But this is not at all typical of how memory works most of the time. As Bartlett wrote, we should get rid of the notion that "memory is primarily or literally reduplicative, or reproductive. In a world of a constantly changing environment, literal recall is extraordinarily unimportant."[9] So *how* do we remember?

Just as the mechanisms of perception are often best revealed by means of perceptual illusions, the normal mechanisms of memory are often revealed by tricking them into producing false recollections. Brent Strickland and Frank Keil, for instance, showed people short videos of someone kicking or throwing a ball.[10] In half of the videos, the moment of contact (or release) was omitted. Immediately after each video, participants were shown a series of still pictures and had to indicate whether the picture had appeared in the video. When the whole sequence of events in the video, and in particular the movement of the ball, had implied that contact must have taken place, a majority of participants "remembered" having seen the contact event that actually had not been shown. What must have happened is that people inferentially reconstructed and "remembered" the sequence of events that had to have taken place, rather than what they had actually seen.

In a similar vein, Michael Miller and Michael Gazzaniga presented to participants in an experiment detailed color pictures of characteristic scenes of American life such as a grocery store, a barnyard, or a beach scene.[11] The original pictures contained many typical items. In the beach scene, for instance, there were a beach ball, beach blankets, beach umbrellas, and the lifeguard's life preserver. The pictures that the participants actually saw were doctored: a pair of such typical objects had been removed (different pairs for different groups of participants). What Miller and Gazzaniga surmised was that people would "remember" items that they had not actually seen.

Half an hour after having seen the pictures, participants were read a list of items and asked whether these items had been in the pictures they saw. Indeed, they misremembered having seen, say, the umbrella or the life preserver, which had been deleted, almost as often as they remembered having seen a beach ball and blankets, which had actually been there. How can this be?

In all cases, true and false memory, recall involves inference—inference, for instance, about the kicking of a ball that explains its subsequent trajectory, or inference about what there "must have been" in that picture of a beach. Often the inference is wholly unconscious and recall seems immediate and effortless. Sometimes, however, there is a hesitation—was there really an umbrella in the picture?—which gets rapidly resolved one way or another. How? By means of inferences that are correct most of the time but not always.

In perception and memory, inference is always at work. Most of the time, we are wholly unaware of its role. It is as if what we perceived was immediately present to us, and as if what we remember was retrieved just as it had been stored. Still, not so rarely, we become aware of having interpreted what we see, or of having reconstructed what we remember. Perception and recall lose some of their apparent immediacy and transparency. In these cases, we are aware of the fact that our perceiving or our remembering involves some intuitive insight.

That inference can be more or less conscious—or is more or less likely to become conscious at some point—is even better illustrated by what happens in verbal comprehension. Suppose that you are sitting in a café and you overhear a woman at the next table say to the man sitting with her, "It's water." You have no problem decoding what this ordinary English sentence means, but still, you don't know what the woman meant. As the philosopher Paul Grice insisted, sentence meaning and speaker's meaning are two quite different things.[12]

The man may have pointed to a wet spot on his shirt, and she might be reassuring him that it is only water. She may be complaining that her tea is too weak by saying hyperbolically, "It's water." It could also be that her meaning has nothing to do with the immediate situation; they may have been discussing what poses the greatest problem to the planet, and she might be maintaining that it is the shortage of fresh water supplies; and so on.

The woman's interlocutor, unlike you, understands her meaning. Not, however, because of a superior command of English. What he has and you don't is relevant contextual knowledge, knowledge about what they had said before, about each other, and about whatever experiences and ideas they happen to share. From this contextual knowledge and from the fragmentary indication given by the linguistic meaning of the words she used, he is in a position to *infer* what she meant. For instance, if he knows she likes strong tea and sees her frown after having taken a fist sip, he will as a matter of course understand her to mean that the tea is too weak.

Most of the time, the inferences involved in comprehension are done as if effortlessly and without any awareness of the process. It is as if we just picked up our interlocutor's meaning from her words. At times, however, we hesitate. The man in our story may not have known that his companion liked her tea quite strong, and may have gone through a moment of puzzlement before grasping her meaning. He would have become aware, then, that he had to infer what she intended to convey. Comprehension always involves inference, even if, most of the time, we are not aware of it.

Intuitions

Intuitions contrast with ordinary perceptions, which we experience as mere recordings of what is out there without any awareness of the inferential work perception involves. In the illusion of the two monsters in a tunnel, for instance, seeing one as bigger than the other feels like a mere registration of a fact. On the other hand, interpreting the scene as one monster pursuing the other may be experienced as more active understanding. Asked why you believe one monster to be bigger than the other, you might answer, "I see it." Asked why you believe that one is chasing the other, you might answer, "It seems intuitively obvious."

Similarly, in the verbal exchange at the next table in the café, if the man interprets the woman's statement "It's water!" as meaning that the spot on his shirt is caused by a drop of water, it seems to him that he is merely picking her meaning from her words. If he furthermore interprets her to imply that, since it is merely water, he shouldn't worry, then his understanding of this implicit meaning may well feel like an intuition.

Intuitions also contrast with the conclusions of conscious reasoning where we know—or think we know—how and why we arrive at them. Suppose you are told that the pictures of the two monsters in the tunnel are actually the same size; you measure them and verify that such is the case. You then accept this unintuitive conclusion with knowledge of your reasons for doing so. This is reasoned knowledge rather than mere intuitive knowledge.

Or the man in the café could reason: "Why is she telling me 'It's water' with such a patronizing tone? Because she thinks I worry too much. Well, she is right—I was worrying about a mere drop of water! A drop of water doesn't matter. It dries without leaving any trace. I shouldn't worry so much." When he comes to the conclusion that he shouldn't worry so much, he pays attention to reasons in favor of this conclusion. Some kind of reasoning is involved.

A simple first-pass way to define intuitions is to say that they are judgments (or decisions, which can also be quite intuitive) that we make and take to be justified without knowledge of the reasons that justifies them. Intuition is often characterized as "knowing without knowing how one knows." Our conscious train of thought is, to a large extent, a "train of intuitions." Intuitions play a central role in our personal experience and also in the way we think and talk about the mind in general, our "folk psychology."

A common idea in folk psychology is that our many and varied intuitions are delivered by a general ability itself called "intuition" (in the singular). Intuition is viewed as a talent, a gift that people have to a greater or lesser degree. Some people are seen as more gifted in this respect, as having better intuition than others. It is a stereotype, for instance, that women are more intuitive than men. But is there really a general faculty or mechanism of intuition?

Perception is the work not of a single faculty but of several different perceptual mechanisms: vision, audition, and so on. That much is obvious. Ordinary experience, on the other hand, doesn't tell us whether, behind our sundry intuitions, there is a single general faculty. The idea of intuition as a kind of faculty, however, isn't supported by any scientific evidence. What the evidence suggests, rather, is that our intuitions are delivered by a variety of more or less specialized inferential mechanisms.

Say, then, that there are many inferential mechanisms that deliver intuitions. Have these mechanisms some basic features in common that differentiate them from inferential mechanisms of perception on one side, and from reasoning on the other side? Actually, intuitive inferences are generally defined by features they lack more than by features they possess. This comes out with characteristic clarity in Daniel Kahneman's figure of the "three cognitive systems," perception, intuition, and reasoning (the latter two being the two systems of dual process and dual system theory) (Figure 9).[13]

Reasoning, in this picture, is positively defined by properties of the process it uses: slow, serial, controlled, and so on. Perception is positively defined by properties of the contents it produces: percepts, current stimulation, stimulus bound.

Intuition, on the other hand, is described as using the same kind of processes as perception, and producing the same kind of content as rea-

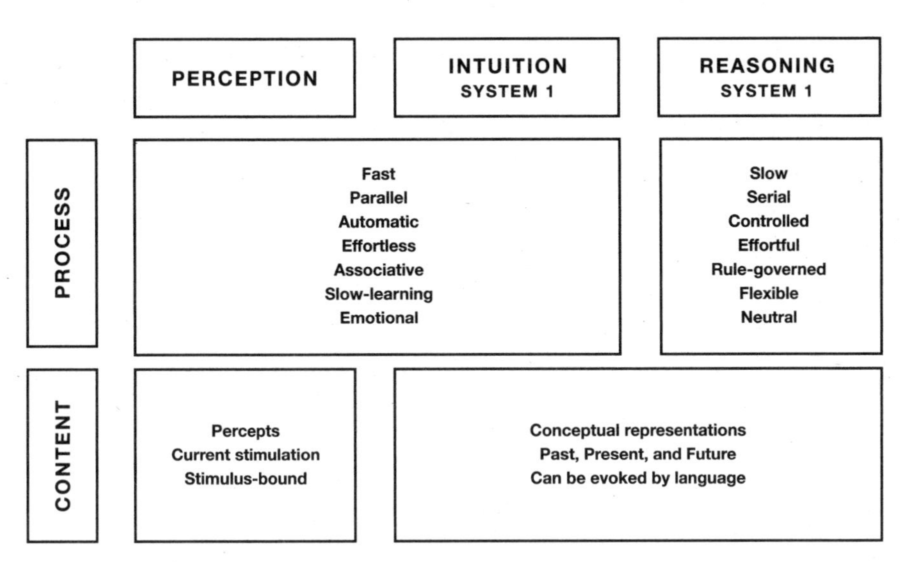

Figure 9. Daniel Kahneman's "Three cognitive systems."

soning. While it may be handy to classify under the label "intuition" or intuitive inference all inferences that count neither as perception nor as reasoning, the category so understood is a residual one, without positive features of its own. This should cast doubt on its theoretical significance. Still, this needn't be the end of the matter.

If intuitions stand apart at least in folk psychology, it is not because they are produced by a distinct kind of mechanism—this is something folk psychology knows little or nothing about—but because they are experienced in a distinctive way. When we have an intuition, we experience it as something our mind produced but without having any experience of the process of its production. Intuitions, in other terms, even if they are not a basic type of mechanism, may well be a distinctive "metacognitive" category.

"Metacognition," or "cognition about cognition," refers to the capacity humans have of evaluating their own mental states.[14] When you remember, say, where you left your keys, you also have a weaker or stronger feeling of knowing where they are. When you infer from your friend Molly's facial expression that she is upset, you are more or less confident that you are right. Your own cognitive states are the object of a "metacognitive" evaluation, which may take the form either of a mere metacognitive feeling or, in some cases, of an articulated thought about your own thinking.

As the Canadian psychologist Valerie Thompson has convincingly argued, intuitions have quite distinctive metacognitive features.[15] We want to make the even stronger claim that the only distinctive features that intuitions clearly have are metacognitive.

Intuitions are experienced as a distinct type of mental state. The content of an intuition is conscious. It would be paradoxical to say, "I have an intuition, but I am unaware of what it is about." There is no awareness, on the other hand, of the inferential processes that deliver an intuition. Actually, the fact that intuitions seem to pop up in consciousness is part of their metacognitive profile. Intuitions are not, however, experienced as mere ideas "in the air" or as pure guesses. They come with a sense of metacognitive self-confidence that can be more or less compelling: intuitions are experienced as weaker or stronger. One has little or no knowledge of reasons for one's intuitions, but it is taken for granted that there exist such reasons and that they are good enough to justify the intuition, at least to some degree. Intuitions also come with a sense of agency or authorship. While we are not the authors of our perception, we are, or so we feel, the authors of our intuitions; we may even feel proud of them.

So, rather than think of intuitions as mental representations produced by a special type of inferential process called "intuitive inference" or "intuition" (in Kahneman's sense of process rather than product), it makes better sense to think of intuitive inferences as inferences, the output of which happens to be experienced as intuitions. "Intuitive inference," in this perspective, stands between "unconscious inference" and "conscious inference." These inferences are not distinguished from one another by properties of the inferential mechanisms involved but by the way the process of inference and its conclusion are or are not metacognized.

The metacognitive experience of having an intuition is sometimes vivid and indeed salient, and sometimes little more than an elusive feeling of self-confidence in a judgment or a decision. There is a continuum from wholly unconscious inferences to inferences of which we have some partial awareness, and there is no clear way to draw a boundary on this continuum. Is there, on the other hand, a boundary between intuitions and the conclusions of reasoning? Is it that, in the case of reasoning, not just the conclusion but also the process of inference is conscious? Don't bet on it yet. In reasoning, we will argue, the opacity of the inferential processes that is typical of intuitions is not eliminated; it is merely displaced. The metacognitive properties of intuitions will thus help us look at reasoning in a new perspective.

4

Modularity

Hume was right: humans and other animals perform inferences all the time. But are they, in doing so, using the same kind of mechanisms? Animals are born with instincts, whereas humans, the old story goes, having few and quite limited instincts, compensate with higher intellectual abilities, and acquire knowledge and skills through learning. But is this really an "either . . . or . . ." kind of issue, either instincts or higher intelligence and learning? Or could learning actually rely on specialized inferential mechanisms that may, to a variable degree, have some instinctual basis?

Between Instinct and Expertise

Comparative psychologists have shown that many animals, such as songbirds, corvids, or apes, acquire complex skills by observing and emulating others or even by discovering on their own new ways of solving problems. Developmental psychologists, on their side, have shown that humans have strong evolved dispositions that influence their cognitive processes from birth. To mention again just one example, from the day they are born (if not already in utero), infants pay special attention to speech sounds and start working on acquiring their mother's tongue.[1] This would already be enough to talk of a "language instinct" (an instinct that, Steven Pinker has famously argued, may do much more than merely focus infants' attention on the sounds of speech).[2]

Bridging the gap between instinct and learning, the ethologist and songbird expert Peter Marler suggested that animals have "learning instincts."[3] Depending on how it is interpreted, the expression might be seen as a contradiction in terms—what is learned isn't instinctive; what is instinctive isn't learned—or on the contrary as an original way to make a rather trivial point: that some animals, humans in particular, have a biologically inherited disposition to learn. Marler, however, understood the expression he had coined in a more specific and interesting way. A learning instinct, as he meant it, isn't an indiscriminate disposition to learn anything; it is an evolved disposition to acquire a given type of knowledge, such as songs (for birds) or language (for humans). A learning instinct not only targets a specific learning goal, it also provides the instinctive learner with appropriate perceptual and inferential mechanisms to extract the right kind of knowledge from the right kind of evidence.

Instincts could be seen as "natural expertises." Expertises could be seen as "acquired instincts." What makes Marler's idea of a "learning instinct" a source of insight is that it suggests that, rather than a gap, there can be a continuum of cases between wholly evolved instincts at one end and wholly acquired expertises at the other end. Cognitive mechanisms may occupy various positions on this continuum. In particular, when psychologists study a human cognitive mechanism, the question shouldn't be: Is it innate or is it acquired? It should be: How much and how is the development of this mechanism in each individual prepared by evolved capacities and dispositions (whether these are present at birth or mature later)? To what extent do some evolved learning capacities target specific learning goals? To what extent do other such capacities, on the contrary, facilitate learning in several domains that in spite of their differences happen to be best understood by using one and the same "mode of construal"[4] (as when we spontaneously use psychological category to learn and think not only about people but also about groups and organizations)? How do learning instincts take advantage of experience to produce mature cognitive mechanisms?

Faces, Norms, and Written Words

There has been an ongoing debate on the mechanisms of face recognition. The ability to recognize other people's faces plays a major role in human social life. Does this competence result just from a lot of practice starting in infancy, a practice strongly motivated by the social benefits of recognizing others and the costs of failing to do so? Or are there also evolved predispositions that drive the attention of infants to faces and provide them with an innate "face template" and with procedures to exploit sensory input in a way uniquely efficient for processing faces?

While nobody denies that experience plays an important role in acquiring the adult competence in the matter, much evidence seems to point to the existence and essential role of evolved predispositions. A small brain area (in the inferior temporal lobe of both hemispheres) named by the American neuropsychologist Nancy Kanwisher the "fusiform face area" is crucially involved in face recognition.[5] Lesion to this area results in an inability to recognize faces. Face recognition has certain features not found (at least to the same degree) in the recognition of other visual stimuli of comparable complexity. For instance, face recognition is much less effective when the face is upside down, an effect that is much stronger for faces than for any other kind of stimuli. Still, some researchers have put forward evidence and arguments to suggest that there is no special skill dedicated to face recognition. It is just that we are much more experienced at visually recognizing faces than at recognizing almost anything else. We have in the matter, they suggest, an expertise that comes of habit.

It seems to us—but we are not specialists—that the case for an evolved basis that guides the individual development of a domain-specific face-recognition mechanism is strong and, over the years, has become compelling. Be that as it may, our point in evoking this example is not to take a stand on the issue but to illustrate how today, research on specific inferential mechanisms involves finding out where they might fit on a continuum of possibilities from specialized cognitive "learning instincts" to a proficiency in using general mechanisms to handle specific types of information based on an expert's level of experience. The very fact that different answers can be given implies that inference is a function that may well be carried out through quite diverse mechanisms.

Similar questions arise for types of learning and inference that are less closely linked to perceptual recognition and more conceptual in character. Parents, for instance, commonly note how eager their toddlers are to learn the "right way" of doing various things and how eager they are then to show their grasp of a norm they have just acquired. This raises a puzzle. When very young children observe actions performed around them, how are they to distinguish those that exemplify norms from other actions that are socially acceptable but neither positively nor negatively sanctioned? Perceptual cues are of very limited help here. Teaching of norms, moreover, varies greatly across cultures; it is much more often implicit than explicit; it never comes close to being exhaustive. So how do very young learners recognize which of the behaviors they observe exemplify norms?

The Hungarian psychologists Gergely Csibra and György Gergely have shown that infants are already disposed to treat information addressed to them "ostensively"—that is, addressed in an attention-arresting, manifestly intentional way—as information of general relevance in their community. When adults ostensively demonstrate some novel behavior, infants readily infer that such behavior exemplifies something like "the way we do things."

In a famous study, infants were shown a dome-shaped table lamp that could be switched on by pressing directly on the dome. This was demonstrated by an adult who switches on the lamp by pressing on it not with her hand as would have been normal, but with her forehead. When this uncommon action has been ostensively demonstrated to them (as if it were some kind of game or ritual), infants readily imitate it. If, on the other hand, infants witness the same head-touch action but this time not performed ostensively, then they don't imitate it; rather, when it is their turn to manipulate the lamp, they switch it on by pressing it with their hand. This research[6] reveals a disposition to selectively imitate actions that because they have been ostensively demonstrated are understood to exemplify the "proper way" to perform them.

Toddlers are able to infer that a way to act is normative from the fact that it is ostensively demonstrated. At a later age, they become able to also use characteristic properties of the action itself to infer its normative character. Psychologists Hannes Rakoczy, Marco Schmidt, Michael Tomasello, and Felix Warneken have shown how two- or three-year-olds spontaneously demonstrate their newly acquired understanding of a norm by trying to enforce it when another individual fails to obey it.[7] All this suggests that there may well exist in humans a quasi-instinctive disposition to identify and acquire social norms.

Face recognition and norm-obeying behavior are universal features of human social life. It is quite plausible therefore that, in both cases, the cognitive mechanisms involved have been in good part shaped by biological evolution. On the other hand, natural selection is much too slow to have evolved the specialized competencies involved in recent practices such as surfing or computer programming. Even more ancient cultural skills such as reading or chess playing are still much too recent for something like a learning instinct to have evolved in order to help with their individual acquisition.

Until recently, reading was a skill possessed by a minority of experts: scholars and scribes. Even though reading is now quite widespread, it still is, from a cognitive point of view, an expertise, that is, a skill acquired through intense practice guided by organized teaching. Such expert skills stand in stark contrast to ordinary instincts, which need little or no learning at all. And yet neuropsychological studies of reading have shown that its neural basis is quite similar to that of more "instinctive" competencies like face recognition.

The French neuroscientists Stanislas Dehaene and Laurent Cohen have established that the recognition of written words in reading recruits a small and precise brain area they named the "visual word form area" that is next to the fusiform face area in the left hemisphere of the brain. This area recognizes letters and words in the script acquired by the individual independently of whether they are in upper- and lowercase, in handwriting or in printing fonts. There is evidence that the same area is involved in blind people reading in Braille with their fingers. Clearly, the information that this brain mechanism extracts is much more abstract than the visual or tactile stimuli from which it is inferred.

How can it be that a small brain area, the same across individuals, societies, and writing systems, should be recruited for reading? Dehaene and Cohen's hypothesis is that the development of a dedicated area in readers is the result of a process of "neuronal recycling":

> On the one hand, reading acquisition should "encroach" on particular areas of the cortex—those that possess the appropriate receptive fields to recognize the small contrasted shapes that are used as characters, and the appropriate connections to send this information to temporal lobe language areas. On the other hand, the cultural form of writing systems must have evolved in accordance with the brain's learnability constraints,

> converging progressively on a small set of symbol shapes that can be optimally learned by these particular visual areas.[8]

The visual word form area is situated, as this hypothesis suggests it might be, in a zone that includes several mechanisms dedicated to perceptual recognition of specific kinds of input (including faces). Moreover, the left hemisphere location is close and well connected to the language areas where, once read, written words must be interpreted.

As the cases of face recognition and reading jointly illustrate, both evolved and expert cognitive skills exploit quite specific brain areas, and when these areas are injured, these skills are impaired. In important respects, evolved and expert skills work in a similar manner: much of their operations are fast, quasi-automatic, and tailored to their specific task. Evolved and expert mechanisms are remarkably efficient in their specific domain of competence. When these mechanisms are presented with inputs that do not belong to their proper domain but that nevertheless stimulate them (the way a pattern in a cloud or a smiley may stimulate face recognition), then their performance may be poor or result in "cognitive illusions."

Modules

All these mechanisms on the instinct-expertise continuum are what in biology (or in engineering) might typically be called *modules:* they are autonomous mechanisms with a history, a function, and procedures appropriate to this function. They should be viewed as components of larger systems to which they each make a distinct contribution. Conversely, the capacities of a modular system cannot be well explained without identifying its modular components and the way they work together.

The biological notion of a module is a broad one. It allows for big and small modules, for sub- and sub-submodules (such as the various components of the visual system), and for teams of modules that are themselves modular (such as the combination of nervous system submechanisms and features of the human hand that permits both the "power grip" and the "precision grip"). The whole brain is a biological module, and so is a single neuron. A biological module may be an anatomical feature of all the members of a species such as the elephant's proboscis, or a behavior such as rumination in cows. A biological module may also be an anatomical trait or a behavioral disposition that manifests itself only in certain environmental circumstances such as a callus growing on skin exposed to friction or the collective behavior known as stampede in herd animals.

There are deep reasons why organisms are, to a large extent, modular systems, that is, articulations of relatively autonomous mechanisms that may have distinct evolutionary and developmental trajectories. Individual modules are each relatively rigid, but the articulation of different modules provides complex organisms with adaptive flexibility. Could a nonmodular organism achieve a comparable kind of flexibility? It is not quite clear how. Modular systems, moreover, are more likely to overcome a local malfunction or to adjust to a changing environment. Most importantly, modular systems have a greater—some have argued a unique—ability to evolve.[9]

In psychology, the mind had long been viewed as a unitary general intelligence with an integrated memory, and connected to the world through various sensory and motor organs. Today, evidence and arguments from neuroscience, developmental psychology, and evolutionary psychology has favored a view of the mind as an articulation of a much greater variety of autonomous mechanisms. Identifying and describing these mechanisms have become a central task of cognitive science.

In philosophy of mind and in psychology, however, talk of modules or of modularity is, for historical reasons, controversial. The notion that the mind might include several autonomous modules was famously defended by the philosopher Jerry Fodor in his groundbreaking 1983 book *The Modularity of Mind.*[10] Fodor argued that the mind's input systems (perception and language) are modular, while its central processes, reasoning in particular, are not. His stringent and, in retrospect, somewhat arbitrary definition of a module and his ideas about the limited modularity of the mind were a source of inspiration but also of endless polemics. To avoid these polemics, a number of psychologists have resorted to talking of "mechanisms" or, like Stanislas Dehaene, of "processors." The terminological price they pay in giving up "module" is that they thereby forsake the use of "modular," "modularization," and "modularity," notions that are useful in asking, for instance, how modular is a given mechanism, under what conditions an explicitly taught skill may modularize, or what is the role of modularity in evolution.

Here, we are wholly eschewing the somewhat stale debates raised by the idiosyncratic Fodorian notion of modularity by simply defining cognitive modules as biological modules having a cognitive function. The notion of a biological module and that of a cognitive function are understood, if not perfectly, at least well enough to allow us to combine the two and move ahead. Contrary to a common misinterpretation of the idea of biological modularity, this does not imply that cognitive modules must be "innate." Reading, we saw, is a perfect illustration of a cognitive module realized in brain tissues, the biologically evolved properties of which lent themselves to being recycled for a novel, culturally evolved function.

Another common misinterpretation of modularity is to assume that a modular mind must be quite rigid. It makes no sense to compare the relative rigidity of individual modules to the flexibility of cognitive systems as a whole (or to the plasticity of the brain as a whole). In biology, flexibility and plasticity are typically explained in terms of a modular organization. This, we suggest, should also be the case in psychology.

We acknowledge that, given the way research is fast progressing in the domain and how much remains to be done, our current understanding of the organization of the human mind is likely to improve and change considerably in the near future. The idea that the mind consists in an articulation of modules provides, when properly understood, a challenging working hypothesis to contribute to this improvement.[11] To move ahead, it is crucial to improve our grasp not just of what inferential modules do but of how they do it. How can individually fairly dumb micromodules combine into smarter but still quite limited larger modules that jointly provide for the kind of superior intelligence that we humans feel confident in attributing to ourselves?

5

Cognitive Opportunism

A large army moving as a unit ignores, when it can, irregularities of the terrain, or else it treats them as obstacles to be overcome. Autonomous guerrilla groups, on the other hand, approach such local features as opportunities and try, when possible, to use them to their advantage. Steering a motorboat involves making minor adjustments to take into account the effect of winds on the boat's course. Sailing, on the other hand, involves treating winds and wind changes as opportunities to be exploited. The general contrast is clear: similar goals may be achieved sometimes by planning a course of action and using enough power to be able to stick to it, and sometimes by exploiting opportunities along the way and moving forward in a more frugal manner.

The classical view of inference assumes a powerful logic engine that, whatever the peculiarities of the task at hand, steers the mind on a straight and principled path. The view we favor is that inference, and cognition more generally, are achieved by a coalition of relatively autonomous modules that have evolved in the species and that develop in individuals so as to solve problems and exploit opportunities as they appear. Just as guerrilla warfare or sailing, cognition is opportunistic.

Without Darwin's idea of evolution by natural selection—the paradigm of an opportunistic process—would the idea that mental processes are opportunistic ever have emerged? The fact is that it emerged only when Darwin's ideas started influencing psychology.

Still, already well before Darwin, the discovery of unconscious inference presented a challenge to the classical view of the mind as unitary and principled. The first to properly understand and address the challenge was the Arab scientist Ibn Al-Haytham (also known as Alhacen), born in Basra in 965 CE, who took up the study of unconscious inference in visual perception where Ptolemy had left it eight centuries before and who developed it much further.[1]

Ibn Al-Haytham's Conjecture

How does unconscious inference proceed? Ibn Al-Haytham wondered. Does it use the same method as conscious inference? At first sight, there is little in common between the immediate and automatic inferences involved in, say, perception and the deliberate and often painstakingly slow inferences of conscious reasoning. Ibn Al-Haytham realized that there is, as we have argued in Chapter 4, a continuum of cases between conscious and unconscious inference. He conjectured that in spite of their apparent differences, conscious and unconscious inference make use of the same tools. What tools? Aristotelian syllogisms, he thought. In his days, there were no real alternatives.

Today, there are several quite different accounts of how inference may proceed. There are many different systems of logic. In psychology, there are several "mental logic" accounts, and there is the theory developed by Johnson-Laird and Byrne that all genuine inference is achieved by constructing and manipulating mental models. Probabilistic models of inference—in particular, those based on the ideas of the eighteenth-century English cleric and scholar Thomas Bayes—have recently inspired much novel research.[2] It could be that several of these approaches each provide a good account of some specific type of inference while none of them offer an adequate account of inference in general. Most proponents of these approaches, however, tend to agree with Ibn Al-Haytham that there must exist one general method that guides inference in all its forms. They disagree with him and among themselves as to what this true method might be.

Assuming that all inferences use the same general method, whichever it might be, raises, Ibn Al-Haytham realized, a deep puzzle. How can it be that one and the same method is sometimes deployed in a slow and effortful manner, and sometimes without any conscious expenditure of time or effort? Why not use the fast mode all of the time? His answer was that all inferences must initially be performed through conscious and effortful reasoning. Some of these inferences, having been done again and again, cease to present any difficulty; they can be performed so fast that one isn't even aware of them. So, he argued, degrees of consciousness do not correspond to different types of inference but only to different levels of difficulty, with the most routine inferences being the easiest and least conscious. From sophisticated reasoning on philosophical issues (a rare occurrence) down to automatic inference in perceiving relative size (that occurs all the time), all inference proceeds, Ibn Al-Haytham maintained, in one and the same way.

Arguing that, initially, all inferences are conscious and that some of them become unconscious by the force of habit is quite ingenious, but is it true? Most probably not, since it entails blatantly wrong predictions. If fast and unconscious inferences were so because they have become wholly routinized, one should, for instance, expect infants to draw inferences in a slow and conscious manner. They should reach the automaticity of routine only at a later age and through extended practice. Developmental psychologists have shown, however, that infants perform automatically a variety of ordinary inferences, years before they start engaging in deliberate, conscious reasoning, contrary to what Ibn Al-Haytham's explanation would lead us to predict.

Here is one example among many. Psychologists Amy Needham and Renée Baillargeon showed 4.5-month-old infants either a possible or an impossible event (see Figure 10).[3] In the "possible event" condition, infants saw a hand put a box on a platform. In the "impossible event" condition, they saw the hand release the box beyond the platform in midair. In both cases, the box stayed where the hand had released it. Infants looked longer at the impossible event of the box staying in midair without falling. This difference in looking time provides good evidence that the infants expected the box to fall, just as adults would have.

Let's assume with Ibn Al-Haytham that all inferences are made by following a logical schema. One should, then, conclude that infants had expected the unsupported box to fall because they had performed something like the following conditional syllogism:

Premises:	1. If an object is unsupported, it will fall.
	2. The object is unsupported.
Conclusion:	The object will fall.

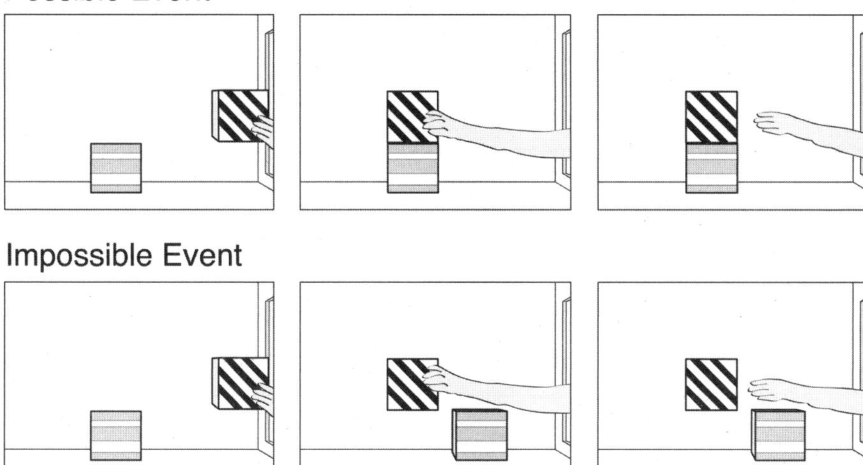

Figure 10. 4.5-month-old infants shown the physically impossible event are surprised.

One should expect, moreover, infants to make this inference in a slow, effortful, and conscious manner (until, with age and experience, it happens so fast as to be unconscious). This is not, however, what psychologists observe.

The evidence shows that experience does matter, but not in the way Ibn Al-Haytham might have predicted. At 4.5 months of age, infants don't pay attention to the amount of support the box gets. Even if only 15 percent of its bottom surface is supported by the platform, they expect it to remain stable. By 6.5 months of age, they have learned better and expect the box to fall when it is not sufficiently supported.[4] There is no evidence or argument, however, that this progression from the age of 4.5 to that of 6.5 months is achieved through slow, conscious, effortful reasoning becoming progressively routinized. What is much more plausible is that infants, using procedures that are adjusted to the task, automatically and unconsciously extract statistical regularities in a way that ends up enriching the procedures themselves.

With all the extraordinary work done in the past fifty years on infant cognition, it is no longer controversial that babies are able to take account of basic properties of physical objects in their inferences and to do so with increasing competence. What is dubious is the idea that in expecting an unsupported object to fall, infants actually make use of a general conditional premise about the fall of objects. Do infants really have such general knowledge? Do they, between 4.5 and 6.5 months of age, correct this knowledge by representing the amount of support an object needs not to fall? For Ibn Al-Haytham and many modern authors, the answer would have had to be yes: an inference must be based on a logical schema and on mentally represented premises. No logic, no inference.

If Ibn Al-Haytham had been right that, without logic, there can be no inference, shouldn't this claim be true not just of human but also of animal inference? The philosopher Jerry Fodor has argued it is: "Darwinian selection guarantees that organisms either know the elements of logic or become posthumous."[5]

Well, there is another way.

Representations and Procedures

All inference, whether made by ants, humans, or robots, involves *representations* and *procedures*. This distinction has played an important role in the development of artificial intelligence (under labels such as "data" versus "procedures" or "declarative" versus "procedural").[6] It is also highly relevant to our understanding of the evolution of the modular mind.

A word, first, about "representation," a notion that causes a lot of confusion. It is quite common to understand the notion of a representation on the model of an image or of a verbal statement. Pictures and utterances are familiar objects in our environment, which we produce and use to communicate with one another. We also use them as cognitive tools. We use written numerals to calculate; maps to plan a trip; shopping lists as external memory props; and so on.

Unlike pictures and spoken or written utterances, however, most of the representations we use are located not in our environment but in our brains; we use them not to communicate with others but to process information on our own. All the same, it is tempting to assume that mental representations are somehow structured like pictures or like utterances. Don't we, after all, have mental images? Don't we silently talk to ourselves in our mind? Couldn't all of our thinking be done with a mixture of images and inner speech? Such considerations, however, fall quite short of demonstrating that all or even most of our mental representations must be structured like public representations or, for that matter, must be structured at all.

So, you might ask, what else could representations be?

Representations, as we will use the term,[7] are material things, such as activation of groups of neurons in a brain, magnetic patterns in an electronic storage medium, or ink patterns on a piece of paper. They can be inside an organism or in its environment. What makes such a material thing a representation is not its location, its shape, or its structure; it is its function. A representation has the function of providing an organism (or, more generally, any information-processing device) with information about some state of affairs. The information provided may be about actual or about desirable states of affairs, that is, about facts or about goals.

As a very simple example, consider motion detectors used in alarm systems. Better-quality motion detectors use simultaneously two types of sensors, such as (1) a microwave sensor that emits microwaves and detects, in the reflected waves, changes typically caused by moving bodies, and (2) an infrared sensor that detects radiations emitted by a warm body. The joint use of two types of sensors lowers the risk of false alarms. When activated, each of the sensors emits an electric signal that has the function of informing the next device in the system that a sensor-activating event has happened. This next device is the same for both sensors and is known as an "AND-gate" (Figure 11). Its function is inferential: when it is informed by two electric inputs that both the first *and* the second sensor are being activated, it triggers an acoustic signal. This signal informs human agents that the probability of an intrusion has reached a threshold such that action is called for.

Although they are neither picture-like nor statement-like and although they need no internal structure to perform their function, the electric signals of the two sensors that serve as inputs to the AND-gate and the acoustic signal that is the output of the AND-gate have the function of providing an electronic device or a human agent with information about a specific type of occurrence, and can therefore be described as "representations" in the sense in which we use the term.

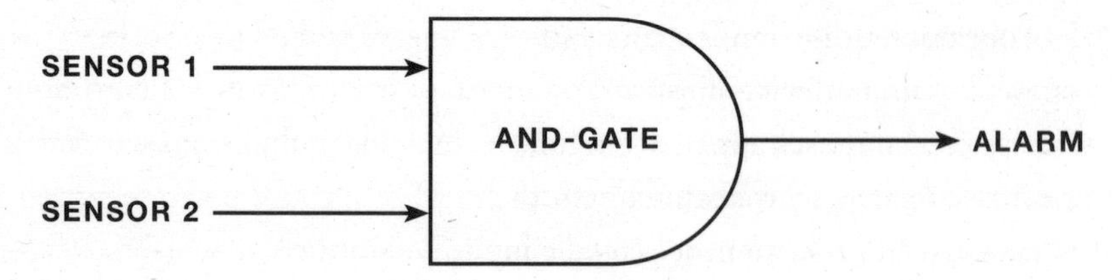

Figure 11. The AND-gate used in a dual-technology motion detector.

Of course, the whole process could be described in physical terms, without talk of information, function, or representation. Still, to understand why people build, sell, and buy motion detectors, going beyond a purely physical account and describing what the device does in terms of information and function is perspicuous. Here, we exploit the case of motion detectors to introduce in the simplest possible way the notion of representation, which we need to tell our story. Our use of "representation" is quite pragmatic:[8] we know of no sensible way to talk about inference and reasoning without using some such notion (whether one uses the term or not).

Inferential procedures apply to representations. They take representations as input and may erase or modify them, or they may produce new ones (as does the AND-gate of a motion detector when it gets the proper pair of inputs). Just as representations are defined by their function, so are inferential procedures. What makes a procedure *inferential* is that it has the function of making more information available for processing, or of making information that is already available more reliable. An inferential procedure may, for instance, erase a representation when new evidence implies it was a *mis*representation; it may modify a representation in order to correct or update it; it may produce new representations that follow from other representations already available; it may increase or decrease the cognitive system's reliance on a representation. A successful inferential procedure results in richer or more reliable relevant information.

Cognitive procedures are implemented in mental modules (in the way programs can be implemented in computers or apps in smartphones). Very simple procedures, like reflexes, may implement a single procedure, whereas more complex modules may implement and combine several of them (and modules with submodules may articulate many procedures).

A mental module implements and uses one or several procedures (just as an electronic device implements and uses programs or a smartphone implements and uses apps). A module, through its connections with other modules, feeds its procedures with the kind of input they are equipped to process. In order to process their input, procedures may need to have access to some special data. The procedures used by a reading module, for instance, need information about the shape of letters. Such data are made available to the procedures by the module, which may store it in a proprietary database or be able to request it from other modules. Modules make their output available to other modules to which they are connected.[9]

In the brain of the desert ant, for example, the odometer and the compass feed their output to an integrative module that computes and updates a representation of the direction and the distance at which the ant's nest is located, a representation that in turn is used by a motor control module to direct the ant's movements on its way back.

Several modules may process the same inputs but submit them to different procedures. A main benefit of having a modular system with many modules working mostly in parallel is to simultaneously achieve a plurality of outcomes. This, after all, is the kind of inferential ability an animal would need to monitor its complex environment and to detect in time different threats and opportunities.

In the history of philosophy and psychology, the focus has been on conscious reasoning and the explicit procedures that it uses sequentially, in the slow and concentrated manner of a scholar—picture a gentleman of leisure with scholarly interests, living a well-ordered life, having entrusted the daily chores and vicissitudes of daily life to servants and womenfolk. When, starting with Ibn Al-Haytham, scholars paid attention to the mechanisms of unconscious inference as it occurs, for instance, in visual perception, they generally assumed that the procedures involved were identical or quite similar to those involved in their own conscious reasoning and that they operated on statements or statement-like representations. This, however, is neither a necessary truth nor an empirically well-supported hypothesis. What had been a daring conjecture in the work of Ibn Al-Haytham has become an old dogma.

Beyond the Dogma

For a long time, the dogma that all inference, conscious or unconscious, uses the same Aristotelian logical procedures and applies them to statement-like representations profited from a lack of alternatives. How else could inference proceed? Until recently, this would have been a mere rhetorical question, but not anymore. The dogma has been undermined both by formal and by empirical research.

On the formal side, the progressive emergence of the theory of probabilities since the seventeenth century as well as the growth and diversification of modern logic since the nineteenth century have rendered the Aristotelian model of inference obsolete. The effect of these formal developments, however, has hardly been to question the idea that inference must be based on the same small repertoire of general procedures across domains; it has been rather to open a debate on what these general procedures might be.

On the empirical side, work on cognition, its evolution, its diversity across species, its development in children, and its implementation in the brain, as well as advances in artificial intelligence and mathematical modeling of cognitive and brain processes, has demonstrated that inference can proceed in many different ways. A great variety of procedures may be involved, many of them specialized in extracting information from one specific empirical domain or in performing just one specific type of inferential task. Some of these procedures have little in common over and above their being inferential. Whatever their differences, they are all procedures that find in the information already available a basis to revise or expand it.

There may well exist important commonalities across some procedures. Transitive inference (of the type "*A* is bigger than *B; B* is bigger than *C;* therefore *A* is bigger than *C*") is, for instance, relevant in a variety of domains, from the physical to the social. It is quite plausible also that a great many inferential procedures are in the same business of updating probabilities of future events—making and revising probabilistic predictions, if you prefer—while doing so each in a way fine-tuned to the regularities of its specific domain.[10]

Some extremely specialized inferential modules are little more than cognitive reflexes. The Russian psychologist Pavlov famously conditioned dogs to salivate at the ring of a bell that, in the dogs' experience, had been repeatedly followed by food. The study of such conditioned reflexes played a major role in the development of behaviorism, an approach to psychology that denied or at least ignored mental states. From a postbehaviorist, cognitive perspective, the conditioned reflex of Pavlov's dogs is both cognitive and behavioral.[11] It causes the dog to expect food—a cognitive response— and to salivate—a behavioral response.

Here is what, presumably, is *not* happening in the dog's mind. There is no general inferential procedure, no Aristotelian syllogism, which uses as premises two statement-like representations present in the dog's mind and that we could paraphrase as "If the bell is ringing, food is coming" and "The bell is ringing." There is no "if . . . then . . ." major premise in the dog's mind to which a *modus ponens* rule of conditional inference could be applied. Rather, Pavlov's conditioning has produced in the dog a wholly specialized module that exploits this bell–food regularity but doesn't represent it.

What gets represented in the dog's mind each and every time the bell rings is the event of the bell ringing. This representation informs the dog's cognitive system and, more specifically, the conditioned reflex module of the fact that the bell has been ringing, thus launching the procedure. This module has just one cognitive effect, which is that it produces an expectation of food, and one behavioral effect, which is salivating. In a cognitivist perspective, this is an inferential module all the same: its job is to derive a relevant conclusion (that food is coming) from an observation (that the bell is ringing). This reflex inference is cognitively sound as long as the bell–food regularity is maintained in the environment.

About events in a wholly chaotic world, no relevant inference could ever be drawn. Logic would be pointless. Probabilities would be of no help. What makes relevant inferences possible—be they those of a physicist or those of a dog—is the existence in the world of dependable regularities. Some regularities, like the laws of physics, are quite general. Others, like the bell–food regularity in Pavlov's lab, are quite transient and local. It is these regularities—the general and the local ones—that allow us, nonhuman and human animals, to make sense of our sensory stimulation and of past information we have stored. It allows us, most importantly, to form expectations on what may happen next and to act appropriately. No regularities, no inference. No inference, no action.

Animals, including humans, have evolved to take advantage of regularities in their environment. They have not evolved to attend to all regularities or to regularities in general. Attempting to do so would be an absurd waste of time and energy. Rather, animals take into account only regularities that, sometimes directly and more often indirectly, matter to their reproductive success.

Animals that move around exploit, to begin with, physical features of their environment that can help or hinder their locomotion. Foraging animals exploit regularities relevant to their finding food. Preys exploit regularities in the behavior of their predators; predators, in that of their preys. Sexually reproducing animals exploit regularities in the behavior of potential mates. Members of social species exploit regularities in the behavior of conspecifics. And so on. Even humans, whose curiosity may seem boundless and who hoard vast amounts of information that may never turn out to be of any use, ignore many regularities in their environment. You are likely, for instance, to be aware of more regularities in the behavior of mosquitoes than in that of dust bunnies even if there are more dust bunnies than mosquitoes near you. If you were immune to mosquito bites and allergic to dust bunnies, it might be the other way around.

The fact that relevant inferences must exploit empirical regularities is, of course, compatible with the classical approach to inference. The classical method relies on formal procedures, or general inference rules, that apply to representations. The way to exploit empirical regularities in this framework is to represent them and to use these representations of regularities as major premises in inferences. "If . . . then . . ." statements (such as "If it is a snake, it is dangerous") have a simple format for representing many regularities and for combining these general representations with representations of particular facts (such as "It is a snake"). From such a combination of general and particular (or major and minor) premises, formal rules may derive relevant conclusions (for instance, a so-called *modus ponens* rule would derive, "it is dangerous"). Alternatively, not just some but all regularities can be represented in probabilistic terms, and rules of probabilistic inferences can then be applied to these representations.

Exploiting a large database of representations of regularities and of particular facts by means of a small set of formal inference rules makes for a formally powerful inferential system. Arguably, anything that can be inferred at all can be inferred in that way. Don't assume, however, that such power and generality make for an optimal—or even a superior—inferential system, one that natural selection should have favored.

The alternative to drawing inferences by means of logical or probabilistic methods working across the board in the same way is to use many specialized modules, each taking advantage of a given regularity by means of a procedure adjusted to the task.[12] This is what presumably happens, for instance, in species that have an automatic fear of snakes (whether innate or acquired). A specialized inferential procedure takes as input the perception of a snake in the environment and produces as output a fear response (with its cognitive and behavioral aspects). Such a procedure relies neither on a premise describing the regularity that snakes are dangerous nor on a formal rule of conditional inference. It directly produces the right response when a snake has been detected, and otherwise it does nothing.

Procedures that exploit a regularity don't appear in evolution or in cognitive development by magic. They are biological or cognitive adaptations to the existence and relevance of the regularity they exploit. They contain, in that sense, information about the regularity (just as a key contains information about the lock it opens, or antibodies contain information about the antigens they neutralize).

What, then, is the difference between the representation of a regularity and a procedure that directly exploits it if both the representation and the procedure somehow contain information about the regularity? Here is the answer. The representation of a regularity doesn't *do* anything by itself, but it provides a premise that may be exploited by a variety of inferential procedures. A dedicated procedure *does something:* given an appropriate input, it produces an inferential output. What a dedicated procedure does not do is make the information it exploits available for other procedures. So, for instance, if you have two representations, "If it's a snake, it is dangerous" and "If it is a scorpion, it is dangerous," then formal rules may allow you to infer "Snakes and scorpions are dangerous" or "There are at least two species of dangerous animals." On the other hand, you might have two danger-detecting procedures, one for snakes and the other for scorpions, and be unable to put the two together and make such simple inferences.

Note that a cognitive system can contain the same information twice: in a procedure that directly exploits the information, and in a representation that serves as a premise for other kinds of procedures—you may have both a reflex fear of snakes and the knowledge that snakes are dangerous.

Which of the two methods, exploiting regularities through specific procedures or through representations, is better? So put, the question is meaningless. What is better depends on costs and benefits that may vary across organisms, environment, situations, and purposes. When the purpose of an organism is to avoid being harmed by snakes, then a fast, reflex-like specialized module is likely to be the best option. When its purpose is to gain general knowledge about snakes, general statement-like representations and more formal argument patterns might be the way to go.

There is no evidence that other animals are interested in any form of general knowledge (but let's keep an open mind about the possibility). In the case of humans, all of them are definitely interested in not being harmed by snakes (and in other types of specific knowledge with practical import), and most are also interested in some general knowledge about snakes without immediate concern for its practical import. They want not just to exploit regularities but also to represent them. Does this means that humans are better off using just the classical method? Or both methods? Or is, as we will suggest, something merely resembling the classical method itself modularized in the human cognitive system?

There are many relevant arguments in the controversies about modularity purporting to show that human inference is basically classical or basically modular. While we are more swayed by arguments in favor of a modular view (and have contributed arguments of our own),[13] we strongly feel that the debate suffers by pitting one against another mere sketches of two alternative accounts.

The classical approach has been around a much longer time and, as a result, it is much more developed both from a formal and from an experimental point of view. What remains quite sketchy, not to say problematic, however, in the classical picture is the way it explains or fails to explain how human reasoning may have evolved in the history of species, how it develops in individuals, how it succeeds in producing just the inferences that are relevant in a given situation rather than starting to produce all the mostly irrelevant inferences it is capable of producing. (This is the so-called frame-problem that doesn't arise, or at least not to the same degree in a modular system.) What remains sketchy at best is also the way the classical picture tries to explain why people who reason from the same premises commonly arrive at divergent or even contradictory conclusions.

The way we aim here to contribute to the debate is not by rehashing it but by fleshing out the modular picture and in particular by explaining how human reason fits into it.

6

Metarepresentations

Is the mind really just an articulation of many modules? Animal minds, perhaps, but, critics argue, surely not the human mind! Animal inferences might be exclusively performed by modules that exploit regularities without ever representing them. Humans, on the other hand, are capable not just of *exploiting* but also of *representing* many empirical regularities. Regularities in the world aren't just something humans take advantage of, as do other animals; they are also something that humans think and talk about. Humans, moreover, are capable of consciously using representations of empirical regularities to discover even more general regularities. We are not disputing this. How could we? After all, it is by exercising this capacity that we scientists make a living.

More generally, doesn't the very existence of reasoning demonstrate that humans are capable of going well beyond module-based intuitive inference? Doesn't reason stand apart, above all, from these specialized inference modules? Don't be so sure. Reasoning, we will argue, is a form of intuitive inference.

The classical contrast between intuition and reasoning isn't better justified than the old hackneyed contrast between animals and humans beings (and its invocation of reason as something humans possess and beasts don't). To contrast humans not with *other* animals but simply with animals is to deprive oneself of a fundamental resource to understand what it is to be human and how indeed humans stand out among other animals. Similarly, to contrast reason with intuitive inference in general rather than with *other* forms of intuitive inference is to deprive oneself of the means to understand how and why humans reason.

Folk Ontology

If reason is based on intuitive inference, what, you may ask, are the intuitions about? The answer we will develop in Chapters 7, 8, and 9 is that intuitions involved in the use of reason are ***intuitions about reasons.*** But first, we need to set the stage.

Intuitions about reasons belong to a wider category: intuitions about representations. The ability to represent representations with ease and to draw a variety of intuitive inferences about them may well be the most original and characteristic features of the human mind. In this chapter, we look at these intuitions about representations.

Humans have a very rich "folk ontology." That is, they recognize and distinguish many different basic kinds of things in the world, and they do so intuitively, as a matter of common sense. Folk ontology contrasts with scientific ontology, much of which is neither intuitive nor commonsensical at all. As humans grow up, their folk ontology is enriched and modified under the influence of both direct experience and cultural inputs. It may even be influenced by scientific or philosophical theories. Still, the most basic ontological distinctions humans make are common to all cultures (and some of these distinctions are, no doubt, also made by other animals).

Everywhere, humans recognize inanimate physical objects like rocks and animate objects like birds; substances like water and flesh; physical qualities like color and weight; events like storms and births; actions like eating and running; moral qualities like courage and patience; abstract properties like quantity or similarity. Typically, humans have distinct intuitions about the various kinds of things they distinguish in their folk ontologies. This suggests—and there is ample supporting evidence—that they have distinct inferential mechanisms that to some extent correspond to different ontological categories.[1]

Modules may evolve or develop, we have argued, when there is a regularity to be exploited in inference—and, needless to say, when it is adaptive to exploit it. Many of these regularities correspond to ontological categories. For instance, animate and inanimate objects move in quite different ways, and their movements typically present humans and other animals with very different risks and opportunities. There is a corresponding evolved capacity to recognize these two types of movements and treat them differently.

Some relevant regularities, however, have to do less with basic properties of an ontological category than with a practical interest of humans (or of other animals). Various omnivorous animals, including humans, may have special modules for making inference about the edibility of plants, for example, although *edible plant* is not a proper ontological category. Actually, modules are task specific, problem specific, or opportunity specific as often as domain specific, if not more often. Still, ontology is a terrain that inferential modules typically exploit.

Not only do humans represent many kinds of things in their thoughts and in their utterances, they also recognize that they are doing so. In their basic ontology—and here humans seem quite exceptional—there are not only things but also representations of things. In fact, for most things humans can represent, they can also represent its representation. They can represent rocks and the idea of a rock, colors and color words, numbers and numerals, states of affairs (say, that it is raining) and representations of these states of affairs (the thought or the statement that it is raining).

Representations of things are themselves a very special kind of things in the world. Representations constitute a special ontological category (with subcategories), for which humans have specialized inferential mechanisms. Representations of representations, also known as *higher-order representations* or as *metarepresentations,* play a unique role in human cognition and social life.[2] Apart from philosophers and psychologists, however, people rarely think or talk about representations as such. They talk, rather, about specific types of representations.

People talk about beliefs, opinions, hopes, doubts, fears, desires, or intentions—all these occur in people's minds and brains; they are *mental representations.* Or they talk about the public expression of such mental representations, spoken or written utterances as well as gestures or pictures—they are *public representations.*

Mental and public representations are concrete objects that are differently located in time and space. A belief is entertained at a given time in someone's head; a spoken statement is an acoustic event that occurs in the shared environment of interlocutors. A written statement or a picture is not an event but an object in the environment. What makes these mental and public representations *representations* isn't, however, their location, duration, or other concrete features. It is a more abstract property that in commonsense psychology is recognized as "meaning" or "content." When we say that we share someone's belief, what we mean is that we have beliefs of closely similar content. When we say of someone that she expressed her thoughts, what we mean is that the meaning of what she said matched the content of what she thought.

Often, when people think or talk about a representation, they consider only its content, and they abstract away from the representation's more concrete properties. They may say of an idea that it is true, contradictory, confused, profound, or poetic without attributing it to anyone in particular either as a thought or as a statement. When they do so, what they talk about are *representations considered in the abstract* (or "abstract representations" for short). Cultural representations such as Little Red Riding Hood, the Golden Rule, or multiplication tables are, most of the time, considered in the abstract, even though they must be instantiated in mental and public representations in order to play a role in human affairs.

Since representations are recognized in our commonsense ontology, the question arises: What cognitive mechanisms do we have, if any, for drawing inferences about them? What kinds of intuitions do we have about representations? As we saw, there are several kinds of representations, each with distinct properties. There is no a priori reason to assume that humans have a module for drawing inferences about representations in general. It is not clear what regularities such a module might exploit. On the other hand, various types of representations present regularities that can be exploited to produce different kinds of highly relevant inferences.

The Infant, the Caterpillar, and the Hidden Piece of Cheese

Metarepresentations, that is, representations of representations, became a major topic of study in psychology after David Premack and Guy Woodruff asked in a famous 1978 article, "Does the Chimpanzee Have a Theory of Mind?"[3] The phrase "theory of mind" made for an attention-grabbing title, but it also became a source of theoretical misunderstandings. Premack and Woodruff's question was not really whether chimpanzees have theoretical beliefs about minds (or, in our terms, use representations of psychological regularities as premises in inference). It was rather whether chimpanzees are capable of attributing specific beliefs or intentions to each other (or to humans).

Some authors, such as Alison Gopnik and Henry Wellman or Josef Perner, believe that you need some theoretical understanding of mental states to attribute mental states to others.[4] Others authors, such as Renée Baillargeon and Alan Leslie, don't—and we agree with them.[5] To avoid the confusion caused by the phrase "theory of mind," we will use the metaphor *mindreading* to describe the cognitive ability involved.

That humans are capable of mindreading is all too obvious. We attribute mental representations to one another all the time. We are often aware of what people around us think, and even of what they think we think. Such thoughts about the thoughts of others come to us quite naturally.

There is no evidence, on the other hand, that most animals, say, desert ants, snakes, or cows, attribute mental states to others. Cows, presumably, don't have mental states in their ontology. They see other cows as living bodies behaving in ways that make biological sense—eating, ruminating, sleeping, walking, and so on—rather than as agents carrying out decisions based on their desires and beliefs. For a few other particularly clever social species such as chimpanzees, dogs, crows, or dolphins, the question raised by Premack and Woodruff remains controversial: yes, these animals may well be capable of some rudimentary mindreading, but nothing approaching the virtuosity of humans in the matter.

Premack and Woodruff's article had a huge impact on the study not only of animal psychology but also of children's. At what age do children start reading minds? A whole field of research initiated in the early 1980s showed that around the age of four, children readily attribute false beliefs to others (which, for good or bad reasons, has become the litmus test of genuine mindreading).[6] And then, in 2005, a groundbreaking study by Onishi and Baillargeon,[7] followed by many more studies confirming their findings, showed that not just four-year-olds but even infants are paying some attention to what others around them have in mind and even expect an agent's actions to be consistent with its beliefs, whether true or false.

Luca Surian, Stefana Caldi, and Dan Sperber, for instance, showed thirteen-month-old infants a video of a hungry caterpillar.[8] The caterpillar, having

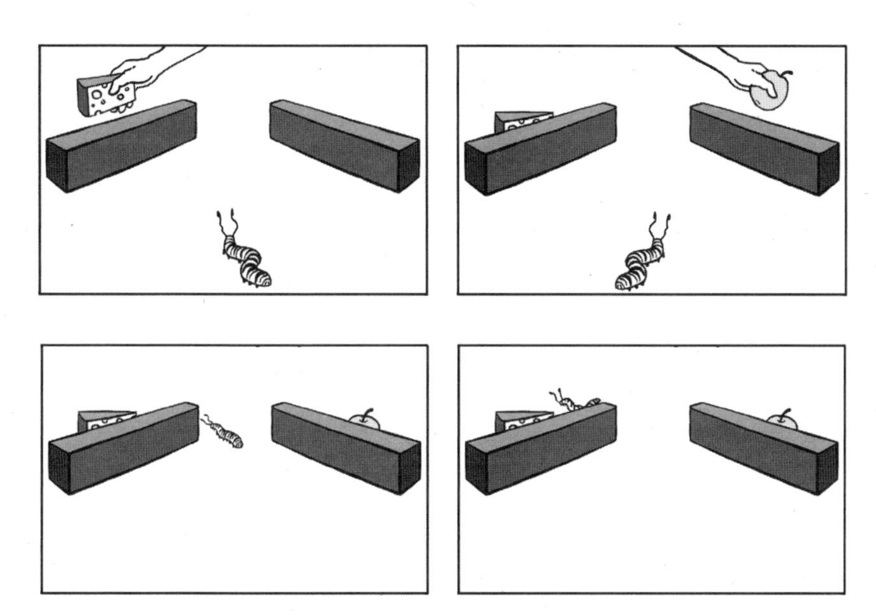

Figure 12. In the familiarization phase, infants see a caterpillar go and nibble at a piece of cheese.

seen a hand put a piece of cheese behind a screen on its left and an apple behind another screen on its right, went around the left screen and nibbled at the cheese (Figure 12). The infants saw this video several times so that it would be clear to them that the caterpillar had a preference for the cheese. Then came the crucial test phase. This time, infants saw the hand put the cheese behind the right screen and the apple behind the left screen. They saw the caterpillar arriving on the scene only after the two food items had been hidden, so that it wouldn't know that their positions had been switched.

What would, in such conditions, the infants expect the caterpillar to do? To go and look behind the left screen where it had repeatedly seen the cheese being hidden, or to go to the right screen behind which the cheese had, this time, been put? To us adults, it is obvious that the caterpillar would go where it had grounds to believe the cheese to be, that is, behind the left screen, even though this belief would, in this case, be false. But are infants able to draw this kind of inference and in particular to attribute false beliefs to others? Remarkably, infants behaved like adults would. They expected the caterpillar to go to the left, where it had grounds to believe (falsely) it would find the cheese. They looked longer when the caterpillar went straight to the cheese on the right side, where it had no way to know that the cheese was. In other words, infants expected an agent to behave according to its beliefs, whether true or false.

Should we conclude that the infants in this study have a mental representation of a general psychological fact—that agents form beliefs and intentions rationally—and that the agents use this psychological fact as a premise in inference? Here again, it makes better sense to assume that what infants have is a specialized procedure that exploits some regularity in the way agents form beliefs and intentions. Infants need not represent this regularity.

How can mindreading exploit the fact that agents such as humans and caterpillars tend to be rational without actually using a mental representation of this fact as a premise? What makes agents rational, we have suggested in Chapters 3 through 5, isn't a general mechanism or disposition to think and act rationally, but a variety of inferential mechanisms with different inferential specializations. These mechanisms, notwithstanding their diversity, have all been shaped by a selective pressure for efficiency—in this case, for performing in an efficient way the kind of inferences it is their function to perform.

Arguably, "rationality" in a most basic sense is synonymous with inferential efficiency. The degree to which rationality is achieved depends to a large extent on the way many inferential modules each perform their functions and on the way these modules are articulated. In order to take advantage of agents' rationality, mindreading must then turn to specific inferential mechanisms and exploit their tendency to deliver, each in its domain, efficient inferences.

To better understand how this multifaceted rationality of real people can be exploited in mindreading, we had better step out of the laboratory and away from the narrow focus on the "false belief task" and a few related experimental paradigms, however well designed they may be.

Here is a simple example of everyday life mindreading. You enter the waiting room at your doctor's office. There is already another patient. You both exchange glances and say, "Hello!" You sit down. She is intermittently typing on her smartphone and staring at the screen. You take a magazine. She looks at her watch and sighs. You exchange another glance. No doubt, you each have a train of personal thoughts that is opaque to the other, but still, you both do some light mutual mindreading as a matter of course.

When you arrived in the waiting room, she knew that you would understand that she would see the doctor before you. You were disappointed that there was already someone waiting, but you tried not to show it, not to let her read your mind on this, but she probably did all the same. You understood that her alternately tapping on her smartphone and staring at it were part of an ongoing interaction with someone else with whom she was chatting at a distance (even if you had no idea what they were chatting about). You guessed that she was looking at her watch because it might already be past the time of her appointment with the doctor, and that she was sighing because she was unhappy to have to wait. You understood from the exchange of glances that she had understood that you had understood her sigh. And so on. All this mindreading that occurred in this minimal interaction was done spontaneously and without effort.

The same kind of spontaneous mindreading occurs even in the absence of interaction, when one is just watching another agent that is unaware of being watched. The thirteen-month-old infants understand that the caterpillar, having seen the cheese repeatedly placed behind the left screen, assumes that it is again in the same place. The infants, who, unlike the caterpillar, have witnessed the cheese been placed on the right side this time nevertheless expect the caterpillar to search for the cheese on the left side, as before.

Such mindreading is so obvious! What is not obvious, however, is what makes it so obvious. Here is a possible sketch of what happens. We humans tend to constantly monitor our social environment (as in the waiting room example). We open, maintain, and update "mental files"[9] on all the people we know (including people we only know of, like kings or famous actors, not to mention fictional characters whose thought we also know quite a bit about). In these files about people (and other kinds of agents such as caterpillars and gods), there may be all kinds of information: information about their names, family, history, appearance, dispositions, doings; information also about what is in *their* mental files where they store information about other people, us included. Our mental files about other people contain information about the contents of their mental files, and that information is provided by mindreading (and is, of course, metarepresentational).

Some of your mental files about people are very thin and short-lived, such as the file you opened about the other patient in the doctor's waiting room. Other files are thick and permanent, such as files about members of your close family. Some of the mindreading information in these files is provided by your spontaneous interpretation of what you observe others doing. Some information is provided by the people themselves who, in communicating, help you read their minds. Further information is provided through people talking about people: gossip. The point is that we read minds on the basis of a great variety of evidence and in a great variety of ways.

There must be a mindreading module—actually a *minds*-reading module, with "minds" in the plural—that has the job of managing, in our mental files about other people, what these people have in *their* mental files. No such module, however, could do the job on its own. In order to perform mindreading inferences about the inferences that are performed in other people's mind, the mindreading module must be linked to a great variety of other inferential modules and use them for updating the information represented in individual files.

Take a very simple example. Tim asks you for candies. You give him some. He looks at them, says, "Three candies, eh?" and puts them in an empty paper bag. In the file you keep about Tim, there is now a metarepresentation of the belief you assume he has that there are three candies in the bag. He says, "Some more, please!" You give him another handful of candies, and he looks at them and says, "Five candies—thanks—with the first three, that's enough," and puts them in the bag. In your Tim file, there is now a metarepresentation of his belief that there are five more candies in his bag. As it happens, you counted not five but six more candies, so you believe that what Tim believes is wrong.

On the basis of Tim's beliefs that he first put in his bag three candies and that he then added five more, you might attribute to him the further (false) belief that there are eight candies in the bag. But how would you carry out this mindreading inference? Performing arithmetic operations—for instance the addition $3+5=8$—is not something your mindreading module is equipped to do on its own. For this, your mindreading module has to share the content of the two relevant beliefs of Tim (that he first put three candies in the bag and that he then added five more) with your arithmetic module and copy back the output of its operation into Tim's file, thus coming to metarepresent Tim's belief that there are eight candies in the bag. Your arithmetic module, presented with your own belief that six candies were added to the initial three (rather than five, as Tim wrongly believes), drew the conclusion that there are now nine candies in the bag. This, however, goes in the mental file you have opened for the bag—we have mental files not just for people, but for all kinds of things—and not in the file you have opened for Tim.

There is, actually, growing evidence that our highly social minds track and anticipate all the time not only what happens in our physical environment but also what happens in the minds of others around us.[10] To achieve this double kind of tracking, many of our inferential modules routinely perform inferences not just to update our own beliefs about the world but also to update our metarepresentations of the beliefs of other people around us.

Does this mean that we understand what others think by taking their point of view? Actually, only whole persons have a point of view and may attempt to see things from another person's point of view. Mental modules in the individual's brain, on the other hand, are "subpersonal"[11] mechanisms and don't have points of view.

Is it that modules are occasionally used "offline" to simulate the mental processes of other people? We suggest, rather, that tracking the mental processes of others is part of their regular "online" job. Our arithmetic module, for instance, computes quantities in a perspective-neutral manner, and the output of these computations may update our representations of the way things are as well as our metarepresentations of the way things are represented in other people's minds.[12]

In the caterpillar experiment, infants see the cheese being repeatedly placed behind the left screen and see the caterpillar seeing the same thing. They draw, both for the caterpillar and for themselves, the inference that this is where the cheese regularly goes. When, in the absence of the caterpillar, the infants now see the cheese being placed behind the right screen, they update their own representation of the location of the cheese, but not their metarepresentation of the caterpillar's representation. Later, when the caterpillar arrives on the scene, it is the belief metarepresented in its file (a belief that now is false) that the infants' mindreading module uses to predict where it will look for the cheese.

How, then, do infants form sensible expectations about the future actions of the caterpillar? We suggest that when the caterpillar arrives on the scene, the infants' mindreading module (1) transfers information from the caterpillar's file to a goal-directed-movement module, the job of which is to compute a rational path to a goal in space; and (2) uses the result of this computation to update the caterpillar's file and to anticipate its movements.

In the waiting-room example, you exploit, among other modules, a modularized competence informed by social conventions that guides you in the kind of situation where you happen to be for a while physically close to strangers because of goals that are parallel but not shared (as, for instance, in a waiting room, in an elevator, or on a plane). There, in the waiting room, you register the events relevant to your interaction simultaneously from your perspective and from that of the other patient. You interpret your brief salutations and your exchange of only short glances as a means to maintain the social distance you probably both feel most comfortable with. In order to do this mindreading, you do not have to actively decide to take the other person's perspective. You were, from the start, automatically updating the file you opened about her mental states just as you were, in a much more fine-grained manner, updating your own files about the physical objects and events around you.

This kind of automatic, permanent tracking of the mental states of others around us routinely involves, we are assuming, the attribution of beliefs, intentions, decisions, and other contentful mental states. It develops from infancy to adulthood (with individual and cultural variations) into a fairly elaborate capacity to understand others, so to speak, on the fly.[13]

Cultural traditions diversely enhance, hinder, and otherwise influence the way we understand each other, and so do social roles and professional occupations. Across cultures and historical times and even within cultures, there is a great variety of explicit ideas about the human mind, which can be expressed in proverbs as well as in elaborate folk or scholarly theories. The mindreading we do every day in interacting with others (or in hearing or reading about them) remains, however, quite spontaneous and intuitive.

It doesn't use these cultural ideas as premises in spontaneous inference aimed at recognizing the mental states of others. These ideas, rather, are used, when need be, to help explain and justify conclusions that were arrived at quite intuitively.

Virtual Domains

Mindreading, which provides us with intuitions about people's beliefs, desires, and intentions and about how these relate to what people perceive and do, by no means exhausts our ability to draw inferences about representations. Numerical cognition, for instance, provides us with a sharply different kind of metarepresentational intuitions.

As much recent work on numerical cognition has shown,[14] human infants share with other animals the ability to mentally represent quantities of discrete objects. These representations are exact for very small quantities—one, two, and three—and are approximate for larger quantities. The acquisition of a language with names for numbers provides humans with lexical tools to represent in an exact manner quantities much larger than three. Linked to language and to writing, the cultural emergence of numeral systems, that is, of symbols used to produce public representations of quantities, has led in some cultures to the development of a whole new branch of knowledge, arithmetic. Anybody who has learned some arithmetic (and even, it seems, anybody who just uses numerals to count)[15] has intuitions not only about concrete quantities but also about formal relationships among numbers apprehended through the numerical symbols that represent them.

To give just one example, it is intuitively obvious to you—you don't have to make any computation—that 900 is three times 300. This intuition exploits your knowledge of relations among very small numbers (in this case, that $3 \times 3 = 9$) and a property not of quantities but of a particular way of representing them, the decimal system (with Arabic numerals). If you used a base nine system, it wouldn't be as immediately obvious that 1,210 is three times 363 (even though these numerals represent the same two numbers as do 900 and 300 in the decimal system). On the other hand, if you used a base nine system, it would be quite obvious to you that 1,000 is three times 300. It is with the decimal system this time that it is less immediately obvious that 729 is three times 243 (even though these numerals represent in base ten the same two numbers that 1,000 and 300 represent in base nine). We have better intuitions about rounded numbers; however, rounding isn't a property of numbers but of the system of numerals used to represent them.

This example of numbers and numerals illustrates three important points:

1. Our intuitions about things (here, numbers) are not the same as our intuitions about their representations (here, numerals).
2. Our intuitions about representations exploit properties of the representations that need not match properties of the things represented (such as roundedness).
3. Our intuitions about representations of things may nevertheless be a source of insight about the things represented themselves. (For instance, that 900 is three times 300 is a fact about the numbers themselves; this fact is intuitively grasped because of the intuitive relationship between the numerals used in the decimal system to represent these two numbers.)

Numbers are a very special kind of thing, and numerals are a very special kind of representation. All the same, the three general points we have just made about their relationship readily extend to other kinds of metarepresentational abilities.

Take explanations. Explanations are a certain kind of representation. Children, well before being themselves capable of providing anything resembling a genuine explanation, start asking all kinds of why-questions. That is, they start requesting explanations for a wide variety of things. Soon enough, they start themselves providing explanations. More generally, asking for and providing explanations is a common aspect of conversation across culture.

We have clear intuitions about the "goodness" of various explanations. As the psychologist Frank Keil and his colleagues have shown, these intuitions may not be very reliable when they concern our own ability to provide explanations.[16] We often greatly overestimate, for instance, our ability to explain how domestic appliances we use every day actually work. We are, however, better at evaluating the explanations given by others. Even children are typically quite adept at recognizing the expertise of others in specific domains of explanation, and at taking advantage of it. Keil describes a division of cognitive labor between seekers and providers of explanation in different domains that is quite similar to the division of cognitive labor that is often at work in the exchange of arguments (as we will show in Chapter 15).

Quite parallel to the case of numbers and numerals,

1. Our intuitions about good and bad explanations are not the same as our intuitions about the things explained.
2. Our intuitions about explanations exploit properties such as cogency, generality, or coherence that are properties of the explanations themselves and not of the things explained.
3. Our intuitions about explanations (which make us prefer good explanations) is nevertheless a major source of insight about the things explained.

More generally, representations are a very special kind of thing in the universe. They are found only inside and in the vicinity of beings with minds. By any sensible criterion, metarepresentational modules and the inferences they perform about representations are very specialized, domain- or task-specific devices. At the same time, our inferences about representations are quite relevant to our understanding of the things represented.

If you intuitively grasp that 900 is three times 300, your intuition is driven by properties of the numerals, but the relevant information you gain is about the numbers these numerals represent. If you recognize the cogency of a good explanation of, say, how a dual-technology motion detector works, then you have learned something not just about the explanation but also, and more importantly, about motion detectors.

Mindreading, arguably the most important of our metarepresentational modules, is no exception. It informs you, through your intuitions about what others believe, about the subject matter of their beliefs. When the woman in the waiting room looks at her watch and sighs, you guess not only that she *believes* that the doctor is keeping her waiting but also that indeed the doctor *is* keeping her waiting. From this, you may draw further inferences of your own, such as about how well the doctor keeps her appointments. We learn a lot about the world by discovering what other people think about it. Mind-reading provides a window on the world at large.

Metarepresentational modules provide information not only about the representations metarepresented but also, indirectly, about the things these representations represent. So while, as we insisted, these modules have very specific domains, namely, specific aspects of specific kinds of representations, they may nevertheless have a different and much wider *virtual domain* corresponding to the things represented.

A standard objection to modularist views of the human mind is that reasoning couldn't possibly be modular since it is not specialized, not restricted to a single domain. Indeed, humans can and do reason about everything and anything. They bring together in their reasoning evidence pertaining to quite different domains.

To compute the circumference of the earth, the Alexandrian scholar Eratosthenes, for instance, used a mixture of astronomy, geography, geometry, and local observation of shadows. Or, to take a more recent debate, consider the ancient piece of cloth known as the Shroud of Turin. Is it, as some claim and others deny, the very burial shroud of Jesus and hence a sacred relic? Arguments on the issue have drawn on history, theology, various branches of chemistry, and physics (radiocarbon dating methods in particular). Aren't these two examples—Eratosthenes's discovery and the debate on the Shroud of Turin—perfect illustration of the fact that human reasoning isn't domain specific, let alone modular?

Actually, while such illustrations raise interesting issues, they do not provide strong objections to the modularist approach. They ignore the fact that metarepresentational modules may have virtual domains that extend way beyond their real domain.

Reasoning, we will argue, is based on a metarepresentational module that provides intuitions not about the world in general but about *reasons*. Reasons are a kind of representation. The real domain of the reason module—reasons—is rather narrow. Still, reasons themselves can be about anything or any combinations of things in the world, bringing together, for instance, the pace of camels in the desert and the position of the sun in the sky of Alexandria at noon on the summer solstice, or the story of the crucifixion of Jesus and the rate of radioactive decay of radiocarbon. Reasoning, therefore, can be both quite specialized in its operations and quite general in its import. In fact, its universal import is best explained by the specific properties of its specialized domain. Inferences about reasons that are themselves about anything result in a kind of *virtual domain-generality*.

III

Rethinking Reason

What is reason? How does it work? What is it for? How could it evolve? In Chapters 7 through 10, we develop a novel interactionist approach that answers these questions. Reason, we argue, is a mechanism of intuitive inferences about reasons in which logic plays at best a marginal role. Humans use reasons to justify themselves and to convince others, two activities that play an essential role in their cooperation and communication. Just as echolocation evolved as an adaptation to the ecological niche inhabited by bats, reason evolved as an adaptation to a very special ecological niche, a niche that humans built and maintain for themselves with their intense social relationships, powerful languages, and rich culture.

7

How We Use Reasons

Humans appeal to reasons not just in reasoning but also in explaining and justifying themselves.[1] Nevertheless, explanation and justification on the one hand and reasoning on the other have been studied independently of one another, as if two quite different psychological mechanisms were involved. We believe that the difference is less one of mechanism than of function, and we want to articulate, in this third part of the book, an integrated approach to the psychology of reasons.

Why do you think this? Why did you do that? We answer such questions by giving reasons, as if it went without saying that reasons guide our thoughts and actions and hence explain them. These reasons are open to evaluation: they may be good or bad. Good reasons justify the thoughts or actions that they explain. This picture of the role of reasons in explanation and in justification may seem self-evident. It is based, however, on a convenient fiction: most reasons are after-the-fact rationalizations. Still, this fictional use of reasons plays a central role in human interactions, from the most trivial to the most dramatic.

The Commonsense Picture

Start with the dramatic: in the middle of the night of November 2, 2013, Theodore Wafer, a middle-aged man living in Dearborn Heights, Michigan, a white suburb of Detroit, was awakened by loud banging on his front door. Earlier that night, Renisha McBride, a young African American woman, had crashed her car, walked out, and wandered for hours in a state of confusion. She ended up at Wafer's door, probably looking for help. Wafer thought his house was being attacked, took his shotgun, opened the door, and fired, killing McBride. At the trial, he explained his action by saying that he had been afraid for his life and had shot in what he believed was self-defense against several attackers. An unreasonable fear, the prosecution argued. In any case, even if the fear had been reasonable, the reasonable response would have been not to open the door but to lock himself in and call the police. Wafer was found guilty of second-degree murder.

Why did Wafer act the way he did? No doubt the reasons he gave at the trial were the best he could muster. We will never know for sure, and it may well be that he himself didn't know what exactly had gone on in his mind at the time. But say we accept that he acted for the reasons he himself invoked: he felt under attack and wanted to defend himself. Even so, most of us would agree with the prosecution and the jury that he didn't have good enough reasons to believe he was in great danger or to act the way he did.

As the case of Wafer illustrates, some reasons may seem good enough to explain but not good enough to justify. We can accept an explanation and, at the same time, be critical of the reasons it invokes. For the purpose of explanation, it is enough that these reasons should have seemed adequate to the person we are trying to understand. On the other hand, to judge that what the person thinks or does is justified, the same reasons must seem adequate *to us.*

Much less dramatically, we all invoke minor or even minute reasons at every turn in daily social interactions. Rob, for instance, asks Ji-Eun, "Do you want a milkshake?" and she answers, "Thank you, but, you know, most of us Koreans are lactose intolerant." Why doesn't Ji-Eun just say no? Why does she bother to give a reason for her refusal? By mentioning a reason for declining Rob's offer of a milkshake, Ji-Eun suggests that she is appreciative of the offer and might have accepted it otherwise. Rob is unlikely to question the reason invoked by Ji-Eun. What should matter to him is that by giving a reason rather than just saying, "No!" Ji-Eun has been considerate toward him. In such ordinary interactions, our giving reasons manifests the kind of consideration others can expect of us and we might expect of them.

How must people understand reasons to be able to use them as they do in their thinking and interactions? To answer this question, it would be of limited use to look at the way people understand the word "reason" itself (in the sense of a reason for something). This understanding varies across people, across circumstances, and, obviously, across cultures. The English word "reason" in that sense doesn't have straightforward translations in all languages. This fuzziness and this variability in the way people think and talk about reasons in general are worth studying in their own right. What we are investigating here, though, are not folk notions or folk theories of reasons but the way people make use of particular reasons and do so whether or not they categorize them as "reasons."

To make better sense of the way reasons are produced and used, it will be helpful to make a distinction between objective and psychological reasons. Such a distinction (with subtle differences and various terminological choices) is common in the rich philosophical literature on reasons. For our present purpose we will, when useful, draw inspiration from this literature without getting involved in its controversies.[2]

An *objective reason,* as we will use the phrase, is a fact that objectively supports some conclusion. This conclusion may be descriptive (about what is the case) or practical (about what is desirable). For instance, the fact that today is Friday is an objective reason to conclude that tomorrow will be a Saturday; the fact that the plums are ripe is an objective reason to pick them from the tree without waiting.

Facts, as we are using the term, are true propositions, abstract objects without causal powers. It is not, strictly speaking, the fact that some plums are ripe that causes them to fall from the tree; it is their ripeness itself or, to be more precise, it is the chemical and physical changes that define ripeness. Objective reasons, being facts, are themselves without causal powers. From our empirical point of view, it doesn't matter whether objective reasons exist in the world independently of human interests and what well-defined criterion of objectivity they meet, if any. If we talk about objective reasons at all, it is because they are represented both in what people think and in what they say and, unlike facts, representations of facts do have causal powers.

A *psychological reason* is a mental representation of an objective reason. Like all mental representations, it may be a misrepresentation; that is, it may represent as an objective reason for a conclusion a false proposition rather than a fact, or it may represent a genuine fact as supporting a conclusion it actually does not support. Still, in our cognitive and evolutionary perspective, we assume that in general, people tend to represent as objective reasons for a given conclusion facts that do indeed support it: cognition is imperfect but not random. Psychological reasons are mental representations in the brain and, as such, play a causal role in people's lives. (When we use "reasons" without qualifying the term, we are talking about psychological reasons.)

It is generally thought that the main role psychological reasons play is to motivate and guide people's actions and beliefs (guidance being little more than a fine-grained form of motivation and motivation a coarse-grained form of guidance). We disagree. The main role of reasons is not to motivate or guide us in reaching conclusions but to explain and justify after the fact the conclusions we have reached.

The uses of reasons to explain and to justify are not just related; they are intertwined. To explain people's beliefs or decisions, psychological reasons must at least point in the direction of a conceivable justification, that is, of a good, objective reason. Wafer, for instance, could, perhaps not with sufficient good sense but not absurdly either, see his reasons as objectively justifying his shooting the person at the door. Had he, on the other hand, given as his reason for having fired a shot that Elvis Presley was dead and that therefore life was meaningless, we would see this not as a genuine reason-based explanation, not even a defective one, but as an admission (or a claim) of temporary insanity.

Similarly, when we invoke reasons to justify other people's thoughts or actions, the normal implication is that they had these reasons in mind and were guided by them. How could people be justified by reasons they knew nothing about? This seems to be a commonsensical constraint on justification, but is it really? Actually, there are interesting exceptions described in the philosophical literature as cases of "moral" or "epistemic luck."[3]

Imagine, for instance, what would have happened if the person Wafer shot and killed had turned out to be not a young woman looking for help but a dangerous criminal on the "Most Wanted" list. In that case, Wafer's action might well have been commended even though he would have acted without knowing that there happened to be good, objective reasons to do what he did—a typical case of moral luck. Reasons must be in people's minds to explain their behavior, but in some cases at least, they need not be in their minds to make their behavior either blameworthy or praiseworthy.

Just as there are cases of moral luck—an action undertaken with inadequate reasons in mind but that turns out to have had, unbeknownst to the agent, an excellent objective justification—there are cases of epistemic luck—a belief held for inadequate reasons but that turns out to be justified all the same.

The Alexandrian scholar Eratosthenes, as we saw in Chapter 1, was the first person ever to measure the circumference of the earth. The result of his calculation, 252,000 stades, was just 1 percent shy of the modern measurement of 24,859 miles, or 40,008 kilometers. What makes this precision even more astonishing is that Eratosthenes had made two mistakes in his assumptions. It was essential to his calculation that the town of Syene should be right on the Tropic of Cancer and due south of Alexandria, and he was convinced that such was indeed the case. Actually, Syene is 1° north of the Tropic of Cancer and 3° east of Alexandria. So how could he come so close? By pure chance, the effects of these two errors practically canceled one another out. This bit of epistemic luck has not been used to downplay Eratosthenes's accomplishment. It is as if his overall method was so brilliant and his result so impressive that the inadequacy of some of his reasons was irrelevant.

Moral and epistemic luck are puzzling phenomena: they suggest that we may find people justified by reasons that actually didn't motivate or guide them and that therefore cannot explain their thoughts or actions. Could the commonsense picture of reasons not be as clear and coherent as it seems at first sight?

The Commonsense Picture Challenged

Contrary to the commonsense picture, much experimental evidence suggests that people quite often arrive at their beliefs and decisions with little or no attention to reasons. Reasons are used primarily not to guide oneself but to justify oneself in the eyes of others, and to evaluate the justifications of others (often critically). When we do produce reasons for guidance, most of the time it is to guide others rather than ourselves. While we would like others to be guided by the reasons we give them, we tend to think that we ourselves are best guided by our own intuitions (which are based, we are sure, on good reasons, even if we cannot spell them out).

Whether or not it would be better to be guided by reasons, the fact is that in order to believe or decide something, we do not need to pay any attention to reasons. Purely intuitive inference, which generates so many of our beliefs and decisions, operates in a way that is opaque to us. You look at your friend Molly and somehow intuit that she is upset. What are your reasons for this intuition? Or you check what films are playing tonight at the Odeon: *Star Wars 12* and *Superman 8*. You decide to go and see *Superman 8*. What are your reasons for this choice? If asked, sure, you would produce reasons, but the fact is that at the moment of intuiting that Molly was upset or of choosing *Superman 8*, you were not consciously entertaining, let alone pondering, reasons. The opinion and the choice came to you intuitively.

Still, one might object, reasons may well have guided us unconsciously. Moreover, we are generally able to introspect and to become conscious of our unconscious reasons. But is this really what is happening? When we explain ourselves, do we really bring to consciousness reasons that have guided us unconsciously? Let us first look at some challenging evidence and, in the next section, propose an even more radical challenge.

The commonsense confidence in one's ability to know one's mind was, of course, undermined by the work of Sigmund Freud and its focus on what he called "the Unconscious." The existence of unconscious mental processes had been recognized long ago, by Ptolemy or Ibn Al-Haytham, but until Freud, these processes were seen as relatively peripheral. Mental life was regarded, for the most part, as typically conscious, or at least open to introspection. However, Freud made a compelling case that we are quite commonly mistaken about our real motivations. A century later, in a cognitive psychology perspective, the once radically challenging idea of the "Unconscious" seems outdated. Not some, but all mental processes, affective and cognitive, are now seen as largely or even wholly unconscious. The problem has become, if anything, to understand why and how there is something like consciousness at all.[4] Freud's challenge to the idea that we know our reasons has been, if anything, expanded.

In a very influential 1977 article, two American psychologists, Richard Nisbett and Timothy Wilson, reviewed a rich range of evidence showing that we have little or no introspective access to our own mental processes and that our verbal reports of these processes are often confabulations.[5] Actually, they argued, the way we explain our own behavior isn't that different from the way we would explain that of others. To explain the behavior of others, we take into account what we know of them and of the situation, and we look for plausible causes (influenced by the type of causal accounts that are accepted in our culture). To know our own mind and to explain our own behavior, we do the same (drawing on richer but not radically different evidence). In his book *The Opacity of Mind,*[6] philosopher Peter Carruthers showed how much recent research had confirmed and enriched Nisbett and Wilson's approach (which he expands both empirically and philosophically).

Our inferences about others are often quite insightful; our inferences about ourselves needn't be worse. We may often succeed in identifying bits of information that did play a role in our beliefs and decisions. Where we are systematically mistaken is in assuming that we have direct introspective knowledge of our mental states and of the processes through which they are produced.

How much does the existence of pervasive unconscious processes to which we have no introspective access challenge our commonsense view of ourselves? The long-established fact that the operations of perception, memory, or motor control are inaccessible to consciousness isn't really the problem. Much more unsettling is the discovery that even in the case of seemingly conscious choices, our true motives may be unconscious and not even open to introspection; the reasons we give in good faith may, in many cases, be little more than rationalizations after the fact.

We have already encountered (in Chapter 2) a clear example of such rationalization from the psychology of reasoning. In the Wason four-card selection task, participants, before they even start reasoning, make an intuitive selection of cards. Their selection is typically correct in some versions of the task and incorrect in others, even though the problem is logically the same in all versions. Asked to explain their selection, participants have no problem providing reasons. When their choice happens to be correct, the reasons they come up with are sound. When their choice happens to be mistaken, the reasons they come up with are spurious. In both cases—sound and spurious reasons—these are demonstrably rationalizations after the fact. In neither case are participants aware of the factors that, experimental evidence shows, actually drove their initial selection (and which are the same factors whether their answer is correct or mistaken). Still, such experimental findings, however robust, smack of the laboratory.

Fortunately, not all experimental research is disconnected from real-life concerns. The brutal murder of Kitty Genovese in New York, on March 13, 1964, with dozens of neighbors who heard at least some of her cries for help but didn't intervene, prompted social psychologists John Darley and Bibb Latané to study the conditions under which people are likely or unlikely to help.[7] They discovered that when there are more people in a position to be helpful, the probability that any of them will help may actually decrease. The presence of bystanders causes people to ignore a person's distress (a phenomenon Darley and Latané dubbed "the bystander effect"), but this is a causal factor of which people are typically unaware.

In one study (by Latané and Judith Rodin),[8] people were told that they would participate in a market study on games. Participants were individually welcomed at the door of the lab by a friendly assistant who took them to a room connected to her office, gave them a questionnaire to fill out, and went back to her office, where she could be heard shuffling paper, opening drawers, and so on. A while later, the participant heard her climb on a chair and then heard a loud crash and a scream, "Oh, my God, my foot . . . I . . . I . . . can't move it. Oh . . . my ankle! . . . I . . . can't get this . . . thing . . . off me."

In one condition, the participant was alone in the room when all this happened. In another condition, there was a man in the room who acted as if he were a participant too (but who was, in fact, a confederate of the experimenter). This man hardly reacted to the crash and the scream. He just shrugged and went on filling out the questionnaire. When real participants were on their own in the room, 70 percent of them intervened to help. When they were together with this apparently callous participant, only 7 percent of them did.

Immediately after all this happened, participants were interviewed about their reactions. Most of those who had taken steps to help said something like: "I wasn't quite sure what had happened; I thought I should at least find out." Most of those who didn't intervene reported having thought that whatever had happened was not too serious and that, moreover, other people working in nearby offices would help if needed. They didn't feel particularly guilty or ill at ease. Had it been a real emergency, of course, they would have helped (or so they claimed).

When asked whether the presence of another person in the room had had an effect on their decision not to help, most were adamant that it had had no influence at all. Well, we know that they had been ten times more likely to intervene when they were alone in the room than when they were not. In other terms, the presence of that other person had a massive influence on their decision. Various factors can help explain this "bystander effect": when there are other people in a position to help, one's responsibility is diluted, the fact that other people are not reacting creates a risk of appearing silly if one does, and so on. What is relevant here is the forceful demonstration of how badly mistaken one can be about what moves one to act or not to act (expect more striking examples in Chapter 14).

What is happening? Do we form beliefs and make decisions for psychological reasons that are often unconscious, that we are not able to introspect, and that we reconstruct with a serious risk of mistake? Or is what generally happens even more at odds with the commonsense view of the role of reasons in our mental life?

Modules Don't Have Reasons

The evidence we have considered so far suggests that humans have limited knowledge of the reasons that guide them and are often mistaken about these reasons. We want to present an even more radical challenge to the commonsense picture. It is not that we commonly misidentify our true reasons. It is, rather, that we are mistaken in assuming that all our inferences are guided by reasons in the first place. Reasons, we want to argue, play a central role in the after-the-fact explanation and justification of our intuitions, not in the process of intuitive inference itself.

Of course, few philosophers or psychologists would deny the obvious fact that we often form beliefs and make decisions without being conscious of reasons for doing so. Still, they would argue that whether or not we are conscious of our reasons, we are guided by reasons all the same. It is just that these reasons are "implicit." This is what happens in intuitive inference. But what are implicit reasons? How can they play their alleged guiding role without being consciously represented?

The word "implicit" is borrowed from the study of linguistic communication, where it has a relatively clear sense. On the other hand, when psychologists or philosophers talk of implicit reasons, they might mean either that these reasons are represented unconsciously or that they aren't represented at all (while somehow still being relevant). Often, the ambiguity is left unresolved, and talk of "implicit reasons" is little more than a way to endorse the commonsense view that people's thought and action must in some way be based on reasons without committing to any positive view of the psychological reality and actual role of such reasons.[9]

We believe that the explicit-implicit distinction is a clear and useful one only in the study of verbal communication and that the only clearly useful sense in which one may talk of "implicit reasons" is when reasons are implicitly communicated. When, for example, Ji-Eun answers Rob's offer of a milkshake by saying, "Thank you, but, you know, most of us Koreans are lactose intolerant," the reason she gives for her refusal is an implicit reason, in the linguistic sense of "implicit": it is not explicit—she doesn't say, for instance, that she herself is lactose intolerant—but it can and should be inferred from her utterance.

Still, one might maintain that psychological reasons can be conscious or unconscious, that at least some unconscious reasons can be made conscious, and that it makes sense to call unconscious reasons that can be made conscious "implicit reasons." But are there really implicit reasons in this sense? We doubt it.

Unconscious and intuitive inferences are carried out, we argued, by specialized modules. Modules take as input representations of particular facts and use specialized procedures to draw conclusions from them. When a module functions well and produces sound inferences, the facts represented in the input to the module do indeed objectively support the conclusion the module produces. This, however, is quite far from the claim that the representations of particular facts are unconscious reasons that guide the work of the module.

Here is why not. A fact—any fact—is an objective reason not for one conclusion but for an unbounded variety of conclusions. The fact, for instance, that today is Friday is an objective reason to conclude not only that tomorrow will be a Saturday but also that the day after tomorrow will be a Sunday; that tonight begins the Jewish Shabbat; that, in many offices, employees are dressed more casually than on other working days of the week; and so on endlessly.

The same fact, moreover, may be a strong objective reason for one conclusion and a weak one for another. For instance, the fact that the plums are ripe is a strong reason to conclude that if they are not picked they will soon fall and a weaker reason to conclude that they are about to be picked. The same fact may even be an objective reason for two incompatible conclusions. For instance, the fact that it has been snowing may be a reason to stay at home and also a reason to go skiing.

It follows from all this that the mental representation of a mere fact is not by itself a psychological reason. The representation of a fact is a psychological reason only if this fact is represented as supporting some specific conclusion. I may know that you know that it has been snowing and not know whether this fact is for you a reason to stay home, a reason to go skiing, a reason for something else, or just a mere fact and not a reason for any particular conclusion at all. We cannot attribute reasons to others without knowing what their reasons are reasons for. Well, you might think, so what? Surely, if the representation of a fact (real or presumed) is used as a premise to derive a conclusion, then it is a psychological reason for this very conclusion. This, however, is mistaken. A belief used as a premise to derive a conclusion is not necessarily a psychological reason for this conclusion.

A long time ago, Ibn Al-Haytham argued that the mind performs unconscious inferences by going through the steps of a syllogism. If this were truly the case, then the representation of a particular fact would serve as the minor premise of such a syllogism, the regularity that justifies inferring the conclusion from this particular fact would serve as the major premise, and these two premises taken together could be seen as a psychological reason for the conclusion of the syllogism (each premise being a partial reason in the context of the other premise). This logicist understanding of all inferences, conscious or unconscious, is still commonly accepted, and this may be why it may seem self-evident that whatever information is used as an input (or a "premise") to an inference has to be a reason for the conclusion of this inference. The view, however, that unconscious inferences are produced by going through the steps of a syllogism or more generally through the steps of a logical derivation has been completely undermined by modern research in comparative psychology and in psychology of perception (as we argued in Chapter 5).

Unconscious inferences, we argued, are produced by modules; these modules exploit the regular relationship that exists between the particular facts they use as input and the conclusion they produce as output, but they don't represent this relationship either as the major premise of a syllogism or in any other way. Representations of particular facts, we pointed out, are not by themselves psychological reasons for any particular conclusion. Modules, in any case, don't need reasons to guide them. They can use representations of facts as input without having to represent, either as a reason or in any other way, the relationship between these facts and the conclusions they derive from them. Modules don't need motivation or guidance to churn out their output.

Consider, as a first illustration, the case of a rudimentary inference that, even though it takes place in the nervous system, doesn't, properly speaking, take place in the mind. Perspiration is a well-developed mechanism in humans (and in horses).[10] When body temperature rises too much, the hypothalamus, a brain structure, triggers the production of sweat that cools the body. This is obviously something that happens in us and to us rather than something that we intend. Still, the perspiration mechanism performs a rudimentary practical inference: it takes as input information about body temperature provided by various neural detectors and, when appropriate, it yields as output instructions to the sweat glands. In doing so, it exploits the general fact that, above some temperature threshold, sweating is appropriate, but it does not *represent* this general fact. Information about current body temperature, which *is* represented in the module, isn't by itself a reason for anything in particular. The module clearly does its job without being guided by any psychological reason, and it doesn't need any such reason to perform its job.

As a second example, take desert ants. Once they have found some piece of food, they speed back to their nest in an almost straight line (as we saw in Chapter 3). Ants know where to go thanks to what Wehner calls a "navigational toolkit," a complex cognitive module with specialized submodules. The procedures used by the submodules (counting steps, taking into account angular changes of trajectory, and so on) each evolved to take advantage of a reliable regularity without, however, representing this regularity. On each foray the ants make outside the nest, relevant information inferred by these submodules is automatically integrated and contributes to determining the ants' return path. In explaining this process, there is no ground to assume that ants are guided by explicit reasons or that the modules involved are guided by implicit reasons.

Why do we bother, you may ask, to make the obvious point that human perspiration and desert ant orientation are not guided by reasons, either explicit or implicit? Because considerations that are relevant in these two cases extend quite naturally to less straightforward cases and, to begin with, to the case of inference in perception.

Remember the Adelson checkerboard illusion we talked about in Chapter 1 (see Figure 4)? Participants are asked which of two squares in the picture of a checkerboard is darker when, actually, both are exactly the same shade of gray. They see—here is the illusion—one of the two squares as much lighter than the other. They do so because they infer the relative darkness of a surface not just from the light it reflects to their eyes but also from the light they assume it receives. One of the two squares is depicted as being in the shadow of a cylinder and therefore as receiving less light than the other. Participants automatically compensate for this difference in light received, and see this square as lighter than it is.

In this textbook example of the role of inference in perception, no one would argue that we have conscious reasons to see one square as darker than the other. The module involved computes the relative darkness of each square as the ratio of the light the square reflects to the light it receives because it evolved to do so; in nonillusory cases, there are objective reasons why it should do so; these reasons, however, are not represented in the mechanism. The module is not guided by unconscious psychological reasons.

Some basic perceptual inferences, we hope you now agree, are not guided by psychological reasons. But does the argument extend to less automatic, more interpretive aspects of perception and to intuitive inference generally? When participants in the Latané and Rodin experiment heard various noises coming from the next room, they recognized some of these as noises of impact, others as words spoken in English. This recognition was, presumably, quite automatic: mentally represented reasons had no role to play in the process (even though, of course, it was an inferential process). Putting these perceptions together and inferring that the woman next door had fallen and hurt herself involved more interpretation, understanding that she might need help even more so. Does this mean that the participants had to have explicit reasons to come to these further more interpretive conclusions or that the mechanisms involved had to have unconscious reasons?

Any socially competent person familiar with the kind of modern environment where the experiment was taking place would have developed a modularized capacity to make the relevant inferences. They would spontaneously recognize some noises not just as noises of impact but, using subtle acoustic features, as noises of a crash involving both hard pieces of furniture and a softer body such as a human body falling on a hard surface. We suggest that the inferential perception procedure that permits such recognition exploits correlations between features and probable causes of noises without representing these correlations, let alone representing them as reasons.

When the participants heard the words, "Oh, my God, my foot . . . I . . . I . . . can't move it. Oh . . . my ankle! . . . I . . . can't get this . . . thing . . . off me," they spontaneously understood "it" to refer to the speaker's foot, they understood "this . . . thing" to refer to some large solid object like a piece of office furniture that had fallen on the person's foot, and in so doing they were going well beyond what "it" and "this thing" linguistically encode. To perform these pragmatic inferences, participants used a modular comprehension procedure that exploits, without representing them, relevance-based regularities in verbal communication.[11] They homed in on one of many linguistically possible interpretations without representing, either consciously or unconsciously, reasons to do so.

Having understood in this intuitive manner that the woman next door had hurt herself and was communicating about her predicament, most participants would spontaneously conclude that she needed help. Again, we suggest that they could come to that conclusion without having to mentally represent (consciously or unconsciously) a reason for the inference. In fact, if someone had to mentally represent such a reason in order to realize that a person in pain and complaining might need help, this absence of spontaneous empathy might be taken as a symptom of impaired cognitive and social competence.

Still, there is no doubt that reasons did play a role in Latané and Rodin's experiment. The question is when reasons appeared in the course of events. After the crash in the next office, some participants stood up and went to see whether they could help. Others remained seated. Was their decision either to try to help or to do nothing a spontaneous decision arrived at without considering reasons? We would suggest that unless they showed some hesitation, there is no strong ground to assume that they had been guided by a conscious or unconscious reason represented in their mind. Be that as it may, a few minutes later, when the experimenter asked them what had gone through their mind, they readily provided reasons that, they claimed, had motivated their action or inaction.

Some participants, especially those who had been helpful, may have come up with reasons just when they were asked. For them, finding reasons was effortless: in circumstances where it is obvious that someone may need you to help, wanting to help is an obviously good reason. Those who had remained seated and had done nothing had had more time to think about reasons they might invoke to justify themselves. They may also have felt a greater need to think of such a justification. Still, even for them, devising a plausible justification was not that difficult. In both cases, thinking about personal reasons for one's decision typically came after having made the decision itself.

If, as we suggest, the point of reasons isn't to guide the formation of beliefs and the making of decisions, then what are reasons for?

Reasons Are for Social Consumption

Whatever humans do is likely to contribute for better or worse to the way they are seen by others—in other words, to their reputation. These indirect reputational effects may turn out to be no less important than the direct goal of their action, whatever it is. Socially competent people are hardly ever indifferent to the way their behavior might be interpreted. By explaining and justifying themselves, people may defend or even improve their own reputation. By failing to do so, they may jeopardize it.

Thinking about good reasons for their actions is something that people often do proactively, anticipating that they may be called upon to explain or justify themselves. The minute you have engaged in a course of action that may have reputational costs—and sometimes even before, when you are merely considering it—a different mental mechanism may start working. Its function is to manage your reputation and for this, to provide an explanation that will justify your behavior. Participants in the Latané and Rodin experiment who ignored the events taking place in the next room may have been aware that their passivity might be open to criticism. This may have prompted them to think of reasons they might provide to justify themselves.

The reputation management mechanism acts like a lawyer defending you, whatever you have done. Still, given the opportunity, a lawyer may advise a regular client against a course of action that would be hard or impossible to defend. What happens when the reputational mechanism cannot come up with a good narrative? Then the course of action considered (and possibly already undertaken) turns out to have costs that weren't initially taken into account in the decision process. In such a case, the failure of the reputational mechanism to produce an adequate narrative may have a feedback effect and cause the initial decision to be rescinded, or at least revised. It is possible, for instance, that a few of the participants in the Latané and Rodin experiment who had first dismissed the idea of helping the person in distress in the next room then realized that they would be hard put to justify their passivity; hesitated, that is, compared the ease with which they could justify their two possible responses; and chose to do something after all.

There is some fascinating experimental evidence that the search for reasons aimed at justification may, in fact, influence action (more about this in Chapter 14). Still, living up to the story you want to be able to tell about yourself isn't quite the same thing as telling a true story.

So, when people produce reasons to explain and justify their beliefs or actions, the narrative they come up with may be at odds with what really happened in their mind in three different ways, each interesting in its own right.

First, people commonly present themselves as having considered reasons and been guided by them in the process of reaching a belief or a decision that, in fact, was arrived at intuitively. The error we all make here is to falsely assume that we have direct knowledge of what happens in our mind when we draw intuitive inferences and that we are guided by reasons of which we are conscious or that we can easily introspect. Contrary to such an assumption, in coming to an intuitive decision or belief, we are not guided by reasons (not even "implicit" ones).

Even so, we often do correctly identify a piece of information that served as input to our inferential processes. When this happens, our mistake is just to describe this piece of information as a personal reason. Participants who went to see if they could help the person in the next room had mentally represented the fact that that person was in pain and needed help; what they hadn't mentally represented at the moment was the higher-order consideration that the person needing help was a reason to go and help. This appropriateness of helping people in need (in certain circumstances at least) is so obvious that in socially competent people, it is exploited by a dedicated procedure, which works without representing it. For our social interactions, such a mistaken self-attribution of reasons doesn't matter at all. In fact, it is less an individual mistake than a socially encouraged use of commonsense psychology. From a scientific point of view, on the other hand, this should be recognized as a misrepresentation: acting spontaneously to help others need not be guided by conscious or unconscious reasons any more than smiling at someone who smiles at you.

Second, people may be mistaken about the information that served as input to their inferences. Participants who chose not to help were first and foremost influenced by the presence and passivity of the other person in the room, but they failed to acknowledge this crucial factor. In all likelihood, when they claimed that their decision had been motivated by the thought that nothing too serious had happened and that their help was not needed anyhow, they were just reporting thoughts that occurred to them only after they had made the intuitive decision to align their behavior to that of the other person in the room. In giving these reasons to the experimenter, they may have been sincere, even if self-deluded: people who think of themselves as nice and helpful and who realize they didn't help when they should have may be puzzled by their own behavior. The easy, too easy, solution to this puzzle is to assume that they must have had what looked at the time like good reasons to think that their help wasn't required.

Third, there are people who fail to find good enough reasons for what they are about to do and who, as a result, waver and change or at least readjust their course of action. In such cases, the personal reasons people invoke have truly played a causal role in their final decision. These reasons didn't directly shape the decision that was finally taken, but they inhibited the carrying out of the initial decision. Still, contrary to what these people may believe and claim, the true causal process didn't go from reasons to decision, but from tentative decision to search for justification, failure of search, and revised decision aimed at a more easily justifiable course of action.

What we have said so far implies that there is no role for unconscious reasons in human cognition. If unconscious reasons have no role to play, then they probably don't exist, and if they don't exist, then implicit reasons don't exist either. When we attribute to ourselves an implicit reason, we are just interpreting our thoughts or actions in terms of newly constructed conscious reasons that we fictitiously represent as having been implicit before.

Does all this suggest that psychological reasons, whether "implicit" or conscious, have no reality whatsoever, that they are a pure construction? No, reasons are indeed constructed, but under two constraints that ensure that they have some degree of both psychological and social reality. The reasons we invoke for justification have to make psychological sense. Talk of reasons need not—in fact, we have argued, cannot—provide an accurate account of what happens in our minds, but it tends to highlight factors that did play a causal role. Reasons are typically constructed out of bits of psychological insight.

In order to fulfill their justificatory function, moreover, psychological reasons have to represent objective reasons socially recognized as good reasons. This social recognition may be biased by many social and cultural factors, but if it were a purely arbitrary affair, how would it hold any sway? For many beliefs and practices—and in particular for the use of reasons as justification or argument—genuine objectivity is, we maintain, a major factor in cultural success (a fact well explained by our account of the evaluation of reasons, as discussed in Chapter 13).

To attribute reasons to oneself or to others is not so much to formulate a hypothesis about how things are as to construct a tool for social action: justification, evaluation, or, we will see in Chapters 8 through 10, argumentation. But what makes reasons a useful tool?

When we give reasons for our actions, we not only justify ourselves, we also commit ourselves. In the first place, by invoking reasons, we take personal responsibility for our opinions and actions as described by us, that is, as attitudes and behavior that we had reasons to adopt. We thereby indicate that we expect others to either accept that we are entitled to think what we think and do what we do or be ready to challenge our reasons. When what we thought or did is unlikely to be approved, by giving reasons, we may indicate a line of defense: we had, if not good reasons, at least reasons that *seemed* good at the time. A defense based on reasons typically allows us to accept responsibility while denying guilt.

By giving reasons, we also commit ourselves to a future line of thought and conduct. Invoking reasons as motivations of one's past views and actions expresses a recognition of the normative aptness of these reasons and a commitment to being guided by similar reasons in the future. For our audience, this commitment to accepting responsibility and to being guided in the future by the type of reasons we invoked to explain the past is much more relevant than the accuracy of our would-be introspections. This is why we all pay attention to the reasons of others and why we produce our own.

To put it in more sociological terms: Reasons are social constructs. They are constructed by distorting and simplifying our understanding of mental states and of their causal role and by injecting into it a strong dose of normativity. Invocations and evaluations of reasons are contributions to a negotiated record of individuals' ideas, actions, responsibilities, and commitments. This partly consensual, partly contested social record of who thinks what and who did what for which reasons plays a central role in guiding cooperative or antagonistic interactions, in influencing reputations, and in stabilizing social norms. Reasons are primarily for social consumption.

8

Could Reason Be a Module?

On a sunny day, an elder from the Dorzé tribe in southern Ethiopia was exhorting a group of idle young men: they shouldn't smoke cigarettes—it was against their religion! The young men pointed out that the young French anthropologist (Dan) who had been living among the Dorzé for the past few months was sitting nearby, doing what? Smoking cigarettes! The elder turned to the foreigner and demanded an explanation: "Why do you smoke?" The anthropologist didn't quite know what to say: "Well," he mumbled, "my father smokes, my grandfather smoked . . ."

The elder turned back to the young men: "See," he exulted, "the foreigner shows respect for his father and his forefathers. They smoked, so he smokes. Follow his example, show respect for your forefathers: they never smoked a cigarette; you shouldn't either!"

As this example illustrates, the same reasons that can be used to justify oneself can also be used as an argument to convince others. (The extra twist here is that when combined with other considerations, the same reasons can be used to justify one individual's practice and to argue against others adopting it: the import of reasons depends on the context.)

Retrospective and Prospective Uses of Reasons

Sometimes people use reasons to explain or to justify decisions already taken and beliefs already held. This is a retrospective use of reasons. Sometimes people use reasons as arguments in favor of new decisions or new beliefs. This is a prospective use of reasons. When reasons are used prospectively, it may

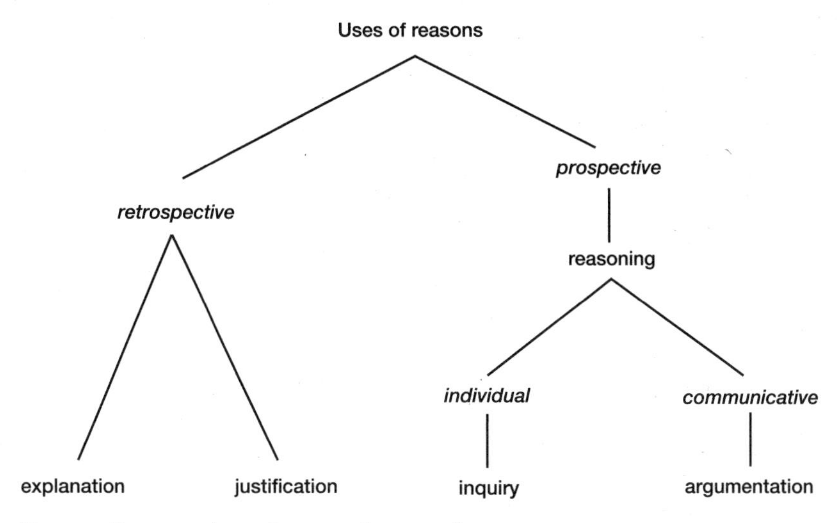

Figure 13. Retrospective and prospective uses of reasons.

be to answer a question of one's own to which one doesn't yet know the answer or to convince others of an opinion one already has. In the first case, it is inquisitive reasoning, and in the second case, it is argumentative reasoning. Figure 13 represents these uses of reason.

The philosophical and psychological literature treats retrospective and prospective uses of reasons as if they were altogether different topics, with little or no attention to what they have in common. We, on the contrary, will argue that there is no dividing line between retrospective and prospective uses of reasons, and that there are many cases of overlap. The use of reasons in explanation is often treated as more important than their role in justification (or, at least, as equally important). We, on the contrary, will argue that the justificatory role of reasons is more important than their explanatory role. The individual, inquisitive form of reasoning, aimed at answering questions on one's own, is considered reasoning par excellence, while the communicative, argumentative form is considered secondary. We, on the contrary, will argue that argumentation is primary.

Reasoning is a major topic in philosophy and psychology in its own right. The dominant view is that reasoning is aimed at truth and at good decisions and should be impartial and objective. The reasons used in reasoning should be impersonal "arguments" that owe their force to their formal properties (studied in logic and probability theory). The use of reasons to explain and justify ideas already adopted or decisions already made and acted upon have, on the other hand, a personal component and cannot be really impartial.

How warranted is this contrast between impersonal reasoning and personal justification? Are personal considerations always irrelevant in good reasoning? In practical reasoning, at least, the relevance of personal considerations is generally recognized. Not surprisingly, philosophers working on practical reason and moral philosophy, in particular Joseph Raz,[1] have been more open to a unified approach to what we call retrospective and prospective uses of reasons.

There is, anyhow, a simple argument in favor of approaching in a unified framework retrospective and prospective reasons. The argument is that all reasons that can be used in justification can be used in reasoning and conversely. If you doubt this, imagine, for instance, a reason that would be good enough to retrospectively justify the opinion that no one should smoke. Then, surely, this reason could be used prospectively to convince oneself or others that no one should smoke. Conversely, imagine that a math teacher asks her pupils whether there is a greatest prime number and helps them reason and rediscover the proof that there is not. Then, surely, that very same proof, which caused the pupils in the first place to come to the conclusion that there is no greatest prime, becomes thereafter an impeccable justification for their belief, as it already was for the teacher.

In standard cases of argumentation, that is, in the production of reasons to convince others, the same reasons have both retrospective and prospective relevance. The arguer presents herself as trying to convince the addressee of an opinion she already holds. Olav and Livia, for example, are about to order raw oysters, and Olav wonders which wine they should have. He asks Livia, and she answers: "A Muscadet! It has just the right acidity and minerality to go with oysters." Olav has heard bad things about Muscadet, but given that Livia is more knowledgeable than he is about wine, her argument convinces him that they should have Muscadet and that he should revise his negative opinion. Livia's arguments to convince Olav are at the same time reasons that justify her own opinions. A sincere arguer uses as arguments to convince her audience reasons that she thinks provide good, retrospective justifications of her own views.

Not only do retrospective justification and prospective reasoning overlap in many ways, not only do they draw on the same pool of reasons; they also rely, we want to argue, on one and the same mechanism, a module that delivers intuitions about reasons.

Reasons Themselves Must Be Inferred

It would be quite surprising (and interesting) to find animals other than humans that think about reasons. Reasons occupy an important place in human thinking because, we have suggested, of the unique role they play in humans' very rich and complex social interactions. Reasons help establish personal accountability, mutual expectations, and norms. Saying this, however, doesn't tell us how humans are capable of knowing their reasons (even if this is a quite imperfect knowledge, as we have seen).

As we pointed out in Chapter 7, it takes more to have a reason than to just recognize some fact. You can walk out and see that the pavement is wet, but you cannot just *see* that this is a reason to believe that it has been raining. That the pavement is wet may be an objective reason for concluding that the pavement is slippery, that the outside temperature is not below freezing point, that one's shoes will get dirty, and so on. To think, "The pavement is wet" is not by itself to entertain a reason. You may, moreover, intuitively infer that it has been raining from the fact that the pavement is wet without this relationship between premise and conclusion being mentally represented in the process. Only if you were to entertain a thought like "From the fact that the pavement is wet it follows that it must have been raining" would you be recognizing the reason for your conclusion.

Suppose you do entertain a reason for inferring that it has been raining. The question still arises: *How* did you come to know that the fact that the pavement is wet is a reason for your conclusion? Reasons do not appear in our head by magic. Recognizing that some fact is a reason for inferring a given conclusion can only be achieved through—what else?—another, higher-order inference.

So, how are reasons inferred? By finding further reasons for our reasons? Sometimes, yes; most of the time, no. Assuming that the recognition of a reason must itself always be based on a higher-order reason would lead to an infinite regress: to infer a conclusion *A,* you would need some reason *B;* to infer that *B* is a reason to infer *A,* you would need a reason *C;* to infer that *C* is a reason to infer that *B* is a reason to infer *A,* you would need a reason *D;* and so on, without end. Hence, the recognition of a reason must ultimately be grounded not in further reasoning but in intuitive inference. This infinite regress argument is an old one: in 1895 Lewis Carroll (of *Alice in Wonderland* fame) published an early version of it in a short and witty note entitled "What the Tortoise Said to Achilles."[2] But there is a question that had not been addressed: What implications, if any, does all this have for psychology?

Inferences, we have argued, are made possible by the existence of regularities in the world (general regularities like the laws of physics or local regularities like the bell–food association in the lab where Pavlov kept his dogs). A regularity that makes an inference possible need not be represented as a premise in the inferential process; it can instead be incorporated in a dedicated procedure. Intuitive inferences are produced by autonomous modules using such procedures. In this perspective, the fact that the recognition of reasons is grounded in intuitive inference suggests that there must be some regularity that a module can exploit in order to recognize reasons. Is there really such a module? If so, what is the regularity involved? Is there some better alternative account to explain how reasons are identified? In the psychology of reasoning, these issues are not even discussed.

Psychologists have studied the format in which people represent premises and the method by which they infer conclusions from these premises (this is what the debate between "mental modelers" and "mental logicians" that we evoked in Chapter 1 was all about). Much less studied is when and how people infer that specific premises provide a reason for a specific conclusion. The fact that reasons must ultimately be grounded in intuitive inference has been either ignored or deemed irrelevant, as if this ultimate intuitive grounding of reasons were too far removed from the actual processes of reasoning to be of consequence to psychology. Actually, just the opposite is the case.

In everyday reasoning, higher-order reasoning about reasons is quite rare. Most of the reasons people use are directly grounded in intuitive inference. It is intuitively obvious, for instance, that the pavement being wet is a reason to infer (with a risk of error) that it has been raining. When the intuitive grounding of reasons is indirect, the chain is short. A fastidious reasoner might add one extra step: intuitively, the most likely explanation of the pavement being wet is that it has been raining; hence the pavement being wet is a reason to infer that it has been raining. Even in more formal reasoning, where people do reason about reasons, the intuitive ground is never very far. What makes reasoning possible, not just in principle but in practice, is the human capacity to intuitively recognize reasons.

The whole dual process approach of Evans, Kahneman, Stanovich, and others that we considered in Chapter 2 has at its core the assumption that intuitive inference and reasoning are achieved through two quite distinct types of mechanisms. We disagree. One of the main claims of this book is that *reasoning is not an alternative to intuitive inference; reasoning is a use of intuitive inferences about reasons.*

What makes humans capable of inferring their reasons is, we claim, their capacity for metarepresentational intuitive inference. To articulate this claim, we revisit and expand ideas introduced in Chapters 3 through 5 about three topics essential to understanding the human mind: intuitions, modules, and metarepresentations.

Intuitions about Reasons

Intuitions, we suggested in Chapter 3, are produced neither by a general faculty of intuition nor by distinct types of inferential process. They are, rather, the output of a great variety of inferential modules, the output of which is to some degree conscious while their operations remain unconscious. Our question now is: Is there a module that draws intuitive inferences about reasons? To answer, we must first sharpen our understanding of intuitions generally.

The inferential mechanisms that produce intuitions, that is, conscious conclusions arrived at through unconscious processes, are quite diverse. There is no intrinsic feature of intuitive inference that they would all share among themselves but not with other types of inference. By way of illustration, here are two cases of inference that are clearly both intuitive but that otherwise have very little in common.

Drawing on earlier work by Wolfgang Köhler (one of the founding fathers of Gestalt psychology), the neuroscientist Vilayanur Ramachandran and the

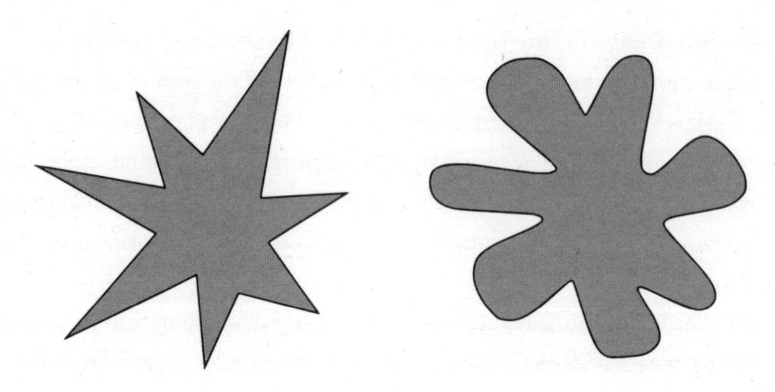

Figure 14. Kiki and Bouba.

psychologist Edward Hubbard presented people with two shapes from Figure 14.

They told them, "In Martian language, one of these two figures is a *bouba* and the other is a *kiki.* Try to guess which is which." Ninety-five percent of people picked the left figure as *kiki* and the right as *bouba.*[3] The strong intuition that it should be so is based, it seems, on a synesthetic association between sounds and shapes. Here intuition is close to perception.

The *bouba–kiki* intuition is quite concrete. Other intuitions are quite abstract. The English philosopher G. E. Moore noted in 1942 that it would be absurd to make a statement of the form "*P,* but I don't believe that *P*" such as, "It is Monday, but I don't believe that it is Monday." This observation is intuitively obvious, but it is not that easily explained. Contrary to what may seem, the proposition isn't self-contradictory: it could very well be true that it is Monday and that I don't believe that it is. The absurdity isn't in what is stated but in its being stated in the first person and in the present tense. That much is clear, but it isn't enough to explain the intuition of absurdity. In fact, while the intuition of "Moore's paradox" is uncontroversial, its explanation remains to this day a topic of controversy among philosophers.[4]

As the *bouba–kiki* and the Moore paradox examples illustrate, what renders some conclusions intuitive is neither their content nor the way in which they are produced. It is the kind of confidence we have in these conclusions. Intuitions are distinguished not by their cognitive features, but by their *meta*-cognitive features.[5]

Confidence in our intuitions is unlike confidence in our perception or in our memory. Perception is experienced as a direct registration of how things are. Correctness of perception is generally taken for granted. Similarly, the way we use memory at every moment of our life is experienced as the direct recall of information that had been mentally stored. When perception and memory work fluently and unhampered, we are wholly unaware of the inferential work they involve.

We experience intuitions, on the other hand, as something our mind comes up with rather than as information that we just pick up from the environment or from memory. Our confidence in our intuitions is confidence in our mind's ability to go beyond the information given, in other terms, to draw inferences. It is not just that our intuitions feel right; we feel that *we* are right in coming up with them. The conclusions of intuitive inferences are experienced as personal thoughts. When we think of objective reasons to justify our intuitions, we readily assume that we must have had these objective reasons in mind to produce these intuitions. This sense of cognitive competence needn't be conceptually articulated. It may be just a metacognitive feeling that is activated only to the degree to which we pay attention to our intuition. Still, we claim, it is this distinctive kind of self-confidence that characterizes intuitions.

We have rejected the old idea that intuitions are the outputs of a distinct faculty of intuition and the currently fashionable idea that they are the outputs of a system 1 type of mechanisms of inference. This raises a puzzle. If, in and of themselves, intuitions are neither similar to one another nor distinct from other types of inference, why should we group them together and distinguish them from other mental states at all? Why do we have this special form of self-confidence that makes us set apart as intuitions some inferences that do not have that much in common otherwise?

Why indeed do we have intuitions at all? Here, as an aside, is a speculative answer. To distinguish a thought of your own as an intuition is to take a stance of personal authority on the content of that thought. This stance, we suggest, is less relevant to your own individual thinking than it is to the way in which you might communicate that thought to others. An intuition is a thought that, you feel, you may assert on your own authority, without an argument or an appeal to the authority of a third party. To make an assertion (or propose a course of action) on the basis of your intuition is a social move that puts others in the situation of having either to accept it or to express distrust not just in what you are saying but in your authority for saying it. By expressing an intuition as such, you are raising the stakes: you stand to gain in authority if it is accepted, and to lose if it is not. Even if your assertion is rejected, however, putting it forward as an intuition of yours may help you withstand the authority or arguments of others. Intuition may license stubbornness, which sometimes is a sensible social strategy.[6]

Metacognition can take not only the simpler form of evaluative feelings but also the more elaborate form of metarepresentational judgments.[7] Say you are asked how long it will take you to finish writing the paper you have been working on. You reply, "A couple of days," but your feeling of self-confidence isn't very strong. Thinking about it, you are more confident that finishing the paper will take you *at least* a couple of days. What is involved now is more elaborate than a mere metacognitive feeling. You are metarepresenting two representations (*I will have finished writing the paper in a couple of days,* and . . . *in at least a couple of days*) and comparing your relative confidence in them. This time, metacognition is also metarepresentational. Distinguishing mere metacognitive feelings and more elaborate metacognitive metarepresentations is, we will show, essential to understanding reasons.

Our intuitions result from inferences about an indefinite variety of topics: *bouba* and *kiki,* Moore's paradox, the mood of a friend, or what film we might enjoy. Our metarepresentational abilities make us, moreover, capable of having intuitions about our intuitions.

Metarepresentational intuitions about our first-order intuitions may be focused on various aspects of these intuitions: they may be, for instance, about the reliability of first-order intuitions or about their acceptability to other people with whom we would like to share them. Some of our metarepresentational intuitions are not just about our degree of confidence in our first-order intuitions but—and this is crucial to the present argument—they are about the *reasons* for these intuitions.

You arrive at the party and are pleased to see that your friend Molly is there too. She seems, however, to be upset. When you have a chance to talk to her, you say, "You seem to be upset tonight." She replies, "I am not upset.

Why do you say that?" Just as you had intuited that she was upset, you now intuit reasons for your initial intuition. Here are what your two intuitions might be:

> *First-order intuition*: Molly is upset.
> *Metarepresentational intuition about your reasons for your first-order intuition:* the fact that Molly isn't smiling and that her voice is strained is what gives me reasons to believe that she is upset.

You want to go to the cinema and hesitate between *Superman 8* and *Star Wars 12*. You intuitively decide to go and see *Superman 8*. The film turns out to be rather disappointing, and you ask yourself why you had made that choice. An answer comes intuitively:

> *First-order intuitive decision*: To go to see *Superman 8*.
> *Metarepresentational intuition about your reasons for your first-order intuitive decision*: The fact that you had enjoyed *Superman 7* more than *Star Wars 11* was your reason for deciding to go to see *Superman 8* rather than *Star Wars 12*.

We typically care about our reasons when our intuitions are challenged by other people or by further experience.

We have intuitions not only about our reasons for our own intuitions but also about other people's reasons for their intuitions. You have spent a few hours inside, in a windowless conference room; you are now walking out of the building with your colleague Lin. The sky is blue, as it was when you walked in, and yet Lin says, "It has been raining." What reasons does he have to say that? Yes, it is a bit cooler than you might have expected, but is this enough to justify Lin's intuition? You look around and see a few puddles. Their presence, you intuit, provides a reason for Lin's assertion:

> *Lin's intuition:* It has been raining.
> *Metarepresentational intuition about Lin's reasons for his intuition:* The fact that there are puddles is Lin's reason to assume that it has been raining.

If, moreover, you think that the reason you attribute to Lin is a good reason, you don't have to accept his assertion just on his authority; you now share a reason to accept it.

The Reason Module

What is this mechanism by means of which we intuitively infer reasons? What is the empirical regularity that makes reasons identifiable as such in the first place? While the attribution of reasons is hardly discussed, the attribution of beliefs and desires has been a central topic of debates in philosophy and psychology. As we saw, a common view—about which we expressed reservations—is that the regularity that helps us attribute beliefs and intentions to people is the presumption that they are rational beings. Couldn't such a presumption of rationality help us identify people's reasons? Should one attribute to others reasons that make their beliefs and intentions rational?

And what about our own reasons? How do we come to know them—or think we know them? Are the identification of other people's reasons and that of our own reasons based on one and the same mechanism, or on two distinct mechanisms operating in quite different ways?

As we saw in Chapter 7, there are good grounds to reject the idea that there is a power or a faculty of introspection that allows us to directly read our own mind. Moreover, it is not that we had reasons in mind when reaching our intuitive conclusions, reasons that we might then introspect if there were such a thing as introspection. Rather, we typically construct our reasons after having reached the conclusions they support. In order to attribute reasons to ourselves, then, we have to infer them, just as we have to infer the reason we attribute to others. Of course, we typically have much richer evidence about ourselves. We have some degree of direct access to our sensations and feelings. We can talk to ourselves in inner speech. Notwithstanding these differences in the evidence available, the way we draw inferences about our own reasons should be, in essential respects, similar to the way we draw inferences about the reasons of others.

Does this mean that, when we attribute reasons to ourselves or to others, we must be using *as a premise* the presumption that we (or they) are rational and tend to come to conclusions for which we (or they) have good reasons?

No, from a modularist point of view, regularities—such as the regularity that humans tend to think and act rationally—can be exploited by modular procedures without being represented as premises at all. We may, in other terms, exploit properties of human rationality in order to attribute reasons to others or to ourselves without having to entertain any general thought about human rationality itself.

There is both something appealing and something quite problematic about giving a central role to rationality in the discovery of reasons (whether in a standard or in a modularist perspective). What is appealing is the highlighting of a close link between reasons and rationality. It seems commonsensical that if we weren't rational, we wouldn't understand reasons, let alone care for them. If reasons weren't rational to some sufficient degree, we would not even recognize them as reasons.

What is much more problematic is the idea that a general human tendency to think rationally (together with some specific information about each case) should be enough to guide us in identifying reasons. For any rational belief or intention, there is an indefinite variety of quite different possible reasons, all compatible with the kind of limited evidence that might be available, which would rationally justify it.

Perhaps Lin's assertion that it had been raining wasn't based at all on the observation of the puddles. For all you know, it might have been based on his superior competence at recognizing and interpreting changes in the air's temperature and humidity; or perhaps he had noticed, while you were both in the conference room, someone entering with a wet umbrella; or he may have had quite different reasons that you have no clue about. What this means is that just looking for possible objective reasons that might have been accessible to a person doesn't come near being an effective heuristic to discover that person's actual reasons.

When we talked about mindreading in Chapter 6, we encountered a similar problem. Rationality doesn't by itself provide an adequate basis to attribute beliefs and intentions. The attribution of such mental states, we argued, takes advantage of the modular organization of the mind. What it exploits is not rationality as a general feature but the specific features of various cognitive competencies that humans share (and that jointly make them rational beings). This, we now suggest, is also true of the attribution of reasons. To attribute reasons to others or to themselves, humans rely less on their overall rationality than on the effectiveness of some specific competencies.

Back to Molly at the party. She challenges you to explain why you said she was upset. To answer, you have to draw a backward inference from your initial intuition about her mood to reasons that would justify your intuition. You do not know for sure what might have triggered this intuition, but you are well equipped to make plausible assumptions.

You are not just a rational being; you have the typical human ability to recognize emotions. You have a well-developed expertise to use facial expressions, tone of voice, bodily movements, and so on as evidence of mood, allowing you to "read" the mood of others. While you cannot remember the cognitive process that led to your intuition—processes of intuitive inference are essentially opaque—you remember things you noticed when you met Molly this evening and that are standard indicators of moods: she was not smiling, her tone of voice was strained, and so on. So, you pick from what you remember (or from what you are presently noticing) pieces of evidence that might best justify your intuition that Molly is upset; you then infer that these were the reasons for your intuition.

Typically, your intuition about a person's mood is triggered by a combination of factors, many of which you are not even aware of, and that each makes some contribution to your intuition in the context of all the others. When you single out some pieces of evidence as being the reasons for your intuition you are typically exaggerating their weight as evidence, but this may be the condition for producing a relevant narrative.

Your memory, as we saw in Chapter 3, is not a mere recall of past registrations. It is constructive, and it often "remembers" features that help make better sense of what happened, even when, in fact, you hadn't observed these features at the time. It may be, for instance, that you hadn't noticed that Molly's tone of voice was strained until she asked you why you thought she was upset. Still, because it fits so well with your intuition that Molly was upset, this feature is injected into your memory of what you think caused your intuition in the first place. The strength of the reasons you now invoke is itself inferred. It is inferred from the confidence you have in your own intuition: if your intuition feels right, then your reasons for this intuition must be strong. You look for plausible strong reasons and assume that they are the reasons that motivated you.

We intuitively infer our reasons for some specific intuition not on the general presumption of our own rationality, but on a much narrower confidence in the specific kind of competence that produced this intuition. Our feeling of rightness when we intuit the mood of a friend is based on our sense not that we are rational beings but that we are competent at judging people's mood and particularly the mood of people we know well.

When you have to infer the reasons that led *another* person to a given conclusion, your task is, again, to draw a backward inference from this conclusion to the kind of reasons that could explain and, at least to some extent, justify it.[8] When you walked out of the building with Lin and he said, "It has been raining," you looked around and, seeing puddles, you assumed that their presence provided a good reason for Lin's assertion. Why do you intuit that the puddles provide a reason to infer that it has been raining? Because, just like Lin, you have the competence to recognize evidence of the weather. When Lin's statement causes you to pay attention to the telltale puddles, you yourself intuit that it must have been raining. Since you trust your own inference from the puddles to the rain, you make the higher-level inference that the presence of such puddles is a good reason, for others or for yourself, to conclude that it just rained, and you further infer that this may well have been Lin's reason.

What about cases where we don't share others' intuitions? Suppose that when Lin said it had been raining, you saw the puddles but didn't intuit that it *must,* only that it *might* have been raining. In fact, paying more attention, you notice that the puddles—and there aren't that many—are all in the same limited area in front of you. You are now more disposed to infer that just the grounds in front of you happen to have been watered and that, no, it didn't rain. Still, puddles did evoke in you the possibility of rain: they were a prima facie reason to infer that it might have been raining, but it turns out not a good enough reason. Hence your intuition would still be that Lin's reason for stating that it had been raining probably was that he had noticed the puddles, but you would now judge this to be a poor reason. If, on the other hand, no possible evidence of rain had come to your mind, then you might have no intuition about Lin's reasons. You might just be and remain puzzled by his statement.

The attribution of reasons, to others or to oneself, needn't be more than a rather superficial affair. One searches the environment or memory for some actual or plausible piece of information (Molly's strained voice, puddles) that could be invoked both to explain and to justify an intuition. If such a piece of information is found, it is assumed to be the actual reason for one's own intuition, a probable reason for someone else's.

Contrary to the commonsense view, what happens is not that we derive intuitive conclusions from reasons that we would somehow possess. What we do, rather, is derive reasons for our intuitions from these intuitions themselves by a further process of intuitive backward inference. We infer what our reasons must have been from the conclusions we intuitively arrived at. We typically construct our reasons as an after-the-fact justification.

We attribute reasons to others in the same way: to the extent that we trust their competence, we tend to trust their intuitions and to infer their reasons through the same process of backward inference. When we don't trust their competence, a similar process of backward inference will settle for apparent reasons that we ourselves find too weak or flawed to justify their intuitive conclusions but that they may have found good enough. When we believe others to be mistaken, we are typically content to attribute to them blatantly poor reasons.

We infer our reasons so that they should support our intuitive conclusions. We assess the strength of other people's reasons on the basis of our degree of agreement with their conclusions. Does this mean that this search for reasons is a purely cosmetic affair, a way of dressing up our naked biases just to look good in our own eyes and to have others look good or bad depending on whether we agree or disagree with them? Is there no cognitive benefit to be expected from the process? No, this wouldn't make much sense. If everybody just stood by their initial intuitions, come what may, reasons would be altogether irrelevant.

Reasons, we have argued, are for social consumption. People think of reasons to explain and justify themselves. In so doing, they accept responsibility for their opinions and actions as justified by them; they implicitly commit themselves to norms that determine what is reasonable and that they expect others to observe. In giving reasons, people take the risk of seeing their reasons challenged. They also claim the right to challenge the reasons of others. Someone's reputation is, to a large extent, the ongoing effect of a conversation spread out in time and social space about that person's reasons. In giving our reasons, we try to take part in the conversation about us and to defend our reputation. We influence the reputation of others by the way we evaluate and discuss their reasons.[9]

So, no, we don't invoke reasons for some inane ego-boost, but, yes, the very way we infer our reasons is biased in our favor. We want our reasons to justify us in the eyes of others. Because they are going to be submitted to others' judgment, reasons may be rethought and revised to be better accepted. Sometimes this means revising, moreover, the conclusions that our reasons support: changing opinion or course of action so as to better be able to justify ourselves. Reasons and conclusions may, in the end, have to be mutually readjusted.

There is, we are assuming, a dedicated metarepresentational module, the job of which is to infer reasons, ours and those of others. Its job is not to provide a psychologically accurate account of the reasons that motivate people. In fact, the implicit psychology—the presumption that people's beliefs and actions are motivated by reasons—is empirically wrong. Giving reasons to justify oneself and reacting to the reasons given by others are, first and foremost, a way to establish reputations and coordinate expectations.

Does it follow that the reasons we give and expect others to give are merely adjusted to some local consensus, some culturally constructed notion of rationality, and that one shouldn't expect people's reasons to be rational in an objective sense of the term? No, this doesn't follow at all. The reasons people give play a much more important role than just signaling that they are norm-abiding members of their social group. People get the good reputation they care about when they are seen as reliable sources of information and as effective partners in cooperation. There is no way they could maintain over time such a reputation without the basic kind of objective rationality that makes them draw cognitively sound inferences and act effectively. To serve their reputational purpose, the personal reasons people invoke should be recognized by others as representing objective reasons, and the best way to secure this recognition is, at least, to invoke reasons that, objectively, are good or at least not too bad.

A cultural community may favor certain types of reasons such as reliance on specific authorities. It may unequally recognize the competence of women and men, young and old, socially inferior and superior, in invoking reasons. It may condone some irrational reasons, such as premonitory dreams. What a community cannot do is build a battery of reasons all of its own. Everywhere, people's intuitions about reasons are anchored in cognitive competencies that, to a large extent, they share as members of the human species, competencies that contribute to humans' cognitive efficiency, that is, rationality in a basic sense of the term. Without such cognitive anchoring, we doubt that any norms of rationality could ever emerge and be maintained in a social group.

Our reasons tend to be rational because, in the first place, our intuitions tend to be rational. What humans do and, presumably, other animals don't do is add to their spontaneous inferences higher-level representations of reasons for these inferences. These reasons are not what makes human inferences rational in the biologically relevant sense of cognitively efficient. What these reasons help do, rather, is *represent our inferences as rational* in a different, socially relevant sense of the term where being rational means, precisely, being based on personal reasons that can be articulated and assessed.[10] The public representations of beliefs and intentions as guided by personal reasons are a fundamental aspect of human social interaction. These representations, we suggest, are produced by a dedicated metarepresentational module. All our reasons are, directly or indirectly, outputs of this module.

Could, then, human Reason (with the capital *R* used in classical philosophy) be a module? Have we found it? Should the next step be the localization of the Reason module in the brain? No, no, and no. Classical ideas about Reason are not about a psychological mechanism but about an essential and transcendental feature of the human mind as a whole. In any case, we have not *found* any module; we are merely speculating, with, we hope, sensible arguments, that the identification of reasons might well be the job of a dedicated module. If we are right and there is such a module, it would have indeed to be realized in some neural structure, but that structure needn't occupy—and occupy alone—a single locus in the brain. In any case, such a module for inferring reasons wouldn't correspond to Reason as classically understood.

Descartes and many other philosophers have sung praises of Reason (while other thinkers have been less lyrical). The module we are talking about, if it exists, would not be something humans would want to brag about as they have bragged about Reason. Still, the closest thing to classical Reason to be found in the human mind/brain may well be this module. We will therefore call it the *reason module* with a modest, lowercase *r*.

Can the Reason Module Reason?

The reason module is, at least in part, aimed at producing justifications and is very much biased in our favor. How could the reasons it produces ever improve on our intuitive inferences if they are inferred from them through backward inference? How could it help in reasoning? How could our evaluation of other people's reasons ever be more than a projection of our self-serving prejudices? How could we ever be convinced by the reasons of others to change our own views?

Part of the answer is that our first-order intuitions (about Molly's mood, the rain, and the vast variety of things about which we have such intuitions) are delivered by a great many modules, while our metarepresentational intuitions about reasons for our first-order intuitions are delivered by one metarepresentational module that just works on reasons. First-order modules draw all kinds of inferences about objects in their domain of competency by exploiting regularities in their domain. The metarepresentational module involved draws inferences in its domain of competence; it draws, that is, inferences about the relationship between reasons and conclusions. To do so, it attends to relevant properties of this relationship. Some intuitions may come easily to us and feel quite strong, but finding reasons that feel as strong as these intuitions themselves may not come so easily. Low confidence in reasons for an intuition may undermine initially high confidence in that intuition.

We may, for instance, have an evolved disposition to accept important risks for exceptional opportunities. Such a disposition, possibly advantageous on average in a distant ancestral past, may, in the modern world, easily be exploited by swindlers.

Jeb, for instance, has an immediate intuition that if he responds favorably to a message he just received from the widow of a rich banker asking him to help her transfer millions of dollars to his country, he will become extremely rich, but then he may have trouble finding credible reasons—reasons that he could share with his family and friends—in support of this intuition. He may initially dismiss his friend Nina's warning that this is a scam, but then he might soon enough find good reasons for her warning. His skeptical intuitions about reasons are different from and better than his enthusiastic intuition about his good fortune.

Others often express intuitions that differ from ours. When trying to explain these intuitions that we do not share, we may nevertheless intuitively attribute to them reasons that we find good or even compelling, leading us to revise our initial intuition. A friend of Nikos and Sofia asks them to solve the following problem (actually, a problem much studied in recent psychology of reasoning).[11] A bat and a ball together cost 1.10 euros. The bat costs one euro more than the ball. How much does the ball cost? For Nikos, the intuitive answer is that the bat costs ten cents. To his surprise, Sofia answers, "The ball costs five cents." Seeing him puzzled, she continues, "If the ball costs five cents, then the bat must cost 1.05 euros . . ." Before she finishes, he sees it: the difference is one euro, as required! This, he intuits, must be Sofia's reason for her answer, and it is a good reason. He therefore rejects his own initial intuition that the ball must cost ten cents and accepts Sofia's answer.

The strength of our first-order intuitions and that of our corresponding metarepresentational intuitions need not match. Reasons may strengthen or undermine our first-order intuitions, and sometimes lead to revisions. Reasons, then, need not be mere stamps of approval on our first-order intuitions.

Don't, however, take Jeb's or Nikos's example as typical. Jeb's initial intuition went against common wisdom. It was clear enough that he would have to justify himself if he acted on it; plausible justifications were hard to come by. And then he had Nina's help in finding reasons to reconsider. Similarly, Nikos might not have revised his solution to the bat-and-ball problem if Sofia had not come up with a different solution and started explaining it. Few of our intuitions are as blatantly stupid as Jeb's or as demonstrably false as Nikos's. Most of our first-order intuitions are at least plausible, and backward inference usually yields plausible second-order reasons to justify them.

Unless our reputation is at stake, we are unlikely to seriously examine our own first-order intuitions in the light of our metarepresentational intuitions about reasons. Even if we do, the reasons that come easily to our mind are likely to confirm or even strengthen our initial intuitions. And even if there is a mismatch between our first-order intuitions and our second-order intuitions about reasons, we may not automatically trust the latter more than the former. Victims of scams like Jeb may have some higher-order doubts but fall into the trap all the same: first-order intuitions are too strong. So the fact that our first- and second-order intuitions might not match doesn't, by itself, make us wise.

We began this book with a double enigma, the second part of which was: How come humans are not better at reasoning, not able to come, through reasoning, to nearly universal agreement among themselves? It looks like now we might have overexplained why different people's reasons should fail to converge on the same conclusion and ended up with the opposite problem: If the reason module is geared to the retrospective use of reasons for justification, how can it be used prospectively to reason? How come humans are capable of reasoning at all, and, at times, quite well?

9

Reasoning: Intuition and Reflection

Reasons can justify an opinion already formed or a decision already made—this is the retrospective use of reasons. But what if you have a question in mind that you don't know how to answer, a decision to make but you are not sure which? Isn't this where a prospective use of reasons—reasoning proper—should help? In principle, yes. If your intuitions on the issue are not clear or strong enough to sway you one way or another, nothing could be better, it seems, than to think in an impartial way of reasons showing which answer is right or which decision is best. Is this, however, what the reason module helps you do?

Intuitive Arguments, Reflective Conclusions

The intuitions the reason module provides are not, we have stressed, about facts that could be a reason for some unspecified conclusion; their form is not "*P* is a reason" (for example, "That Amy has a fever is a reason"). These intuitions are about facts taken together with the conclusion that they support; their form is "*P* is a reason for *Q*" (for example, "That Amy has a fever is a reason to call the doctor").

In both justification and reasoning, the output of the reason module is a higher-order conclusion that there are reasons for a lower-order conclusion. In the case of justification, the lower-order conclusion has already been produced and possibly been acted upon; say, Amy's parents have already called the doctor. What is now added is the higher-order conclusion that Amy having a fever justified this decision. This retrospective justification may work even if Amy's parents called the doctor just because she was feeling dizzy, before they actually took her temperature.

In reasoning, the reason module produces not one but two new conclusions, the second conclusion embedded in the first. The first conclusion is the higher-order argument itself, that is, the metarepresentational intuition that such-and-such reasons support a particular conclusion: Amy's fever is a good reason to call the doctor now. The second new conclusion is the conclusion—Let's call the doctor!—embedded in the overall argument and supported by it.

If you catch a fish that has just swallowed another fish whole, you catch two fish at once. Similarly, if you infer that some new conclusion is supported by good reasons, you infer two new conclusions at once: the whole argument and the conclusion the argument supports. When the embedded conclusion is relevant on its own, as is generally the case in reasoning, you may disembed it and assert it, store it in memory, or use it as a premise in further reasoning.

Two illustrations:

You are quite confident that you have left a book you need in your study, but however hard you look, you fail to find it. You try to think where else it might be, and you think, yes, it might also be in the living room. At this point, you reach a conclusion in the form of an argument:

> *Argument:* Since I'm pretty sure the book is either in the study or in the living room and I cannot find it in the study, I should look for it in the living room.

This is the kind of argument one typically accepts without considering higher reasons for it. Given that an intuition, simply defined, is a conclusion accepted without attention to, or even awareness of, reasons that support it, your argument as a whole is definitely an intuitive conclusion, an intuition. This intuitive conclusion, however, is *about* reasons and about the support these reasons give to a second conclusion, which is embedded in the argument:

> *Embedded conclusion:* I should look for the book in the living room.

Is the conclusion that you should look for the book in the living room also an intuition? No, because it doesn't fit the definition of an intuition. It is, after all, supported by reasons, reasons that you very much have in mind. In fact, it occurs embedded in a representation of these very reasons.

If the conclusion embedded in an argument is not an intuitive conclusion, what kind of conclusion is it? It is a reasoned conclusion, or, to use a term common in the literature on reasoning and that we have often used ourselves,[1] a *reflective conclusion,* a conclusion accepted because of higher-order thinking (or "reflection") about it. As this example illustrates, the amount of reflection that goes into reaching a reflective conclusion may be minimal. Still, there is a difference between intuitively believing, without any reasoning involved, that you should look for the book in the living room and having intuitive reasons for this conclusion.[2]

Here is a second example of a reflective conclusion, this time involving a little more reflection. You are sitting with a friend in a café. She challenges you: "I offer you two bets. Accept one of the two bets, and if you win it, I'll pay for the drinks." The two bets she offers are the following:

> *Bet A:* You win if at least three of the next five people to enter the café are males.
> *Bet B:* You win if at least three of the next six people to enter the café are males.

Without having to consider higher-order reasons, you conceive of an intuitive argument that helps you choose one bet.

> *Argument:* The chances of winning bet B are greater than the chances of winning bet A, making bet B the better choice.
> *Embedded conclusion:* Bet B is the better choice.

Your conclusion that bet B is preferable is supported by your intuition about your relative chances of winning the two bets. This intuition provides a reason for what is, therefore, a reflective conclusion.

What these two examples show is that a reflective conclusion need not be the output of a *mechanism* of reflective inference that could be contrasted to

Table 1 Outputs of the reason module

	Types of conclusion	Form of conclusion
Direct output	Intuitive argument	Reasons $R_1, R_2, \ldots, R_n$ support conclusion C
Indirect output	Reflective conclusion	C

mechanisms of intuitive inference. A reflective conclusion may be an indirect output of a process of intuitive inference. The *direct* output of this process is an intuitive conclusion about reasons $R_1, R_2, \ldots, R_n$ for some conclusion C—in other terms, an argument; the *indirect* output is the reflective conclusion C. This is represented in Table 1.

We want to make an even stronger claim: not just some but *all* reflective conclusions in human thinking are indirect outputs of a mechanism of intuitive inference about reasons.

In the basic and most common cases (for instance, in the misplaced book or the choice of bet illustrations we just gave), a reflective conclusion is embedded in an intuitive argument. In slightly more complex cases, there may be not one but two levels of embedding, with an intuitive argument supporting a reflective argument that itself supports a final reflective conclusion.

Take the bet example. We described you as having the intuition that the chances of winning bet *B* are greater than the chances of winning bet A. This intuition would give you a strong argument for choosing bet B. You might, however, be more reflective and accept this argument in favor of bet B only on the basis of a higher-level intuitive argument:

> *Intuitive argument:* If only two of the next five people to enter the café are males, you would lose bet A, but you might still win bet B. All it would take is that the sixth person to enter the café be male. Hence, the chances of winning bet B are greater than the chances of winning bet A, making bet B the better choice.
> *Reflective argument:* The chances of winning bet B are greater than the chances of winning bet A, making bet B the better choice.
> *Reflective conclusion:* Bet B is the better choice.

In this example, the higher-level argument (in which the reflective argument and the final conclusion are embedded) is, for most people, likely to be intuitive. For someone who doesn't find this more general argument intuitive or is unwilling to accept it without reflection, it could itself be embedded in an even more general argument about probabilities that would support it. There may be more than two levels of embedding. Still, ultimately, any reflective conclusion is the indirect output of a process of intuitive inference.

It would be a mistake to assume that higher-order explicit arguments make conclusions easier to understand and accept. To take an extreme example, for most people, it is a plain, incontrovertible fact that $1+1=2$. Alfred North Whitehead and Bertrand Russell famously devoted hundreds of pages of their landmark book, *Principia Mathematica,*[3] to deriving this conclusion through a long chain of complex arguments. Very few people are able to follow these arguments. They were not aimed, anyhow, at strengthening the rock-solid common intuition that $1+1=2$ but at demonstrating that it is provable and, in so doing, at providing logical foundations for mathematics.

What proportion of the many conclusions humans come to in their life are reflective rather than intuitive? What proportion of these reflective conclusions are themselves embedded in reflective rather than intuitive arguments? How common are higher and higher-order levels of reflection? These are empirical questions that have not been properly studied.

Note that the people who would be most likely to express with the confidence of experts their opinion on the importance of reflection in human inference are logicians, philosophers, and psychologists of reasoning. These experts are hardly typical of the population at large; they commonly resort to higher-level arguments as part of their trade. They might be prone, therefore, to mistake their own professional twist of mind for a basic human trait. We surmise that most human reasoning, even excellent reasoning of the kind that is supposed to make us humans so superior, rarely involves more than one or two levels of arguments.

We had suggested in Chapter 8 that the reason module produces, as direct output, intuitions about reasons and, as indirect output, reflective conclusions supported by these reasons. If this is right, then there is no need—and, in fact, no room—for a psychological mechanism ("system 2" or otherwise), the job of which would be to directly produce reflective conclusions. Reasoning can be more or less reflective depending on the degree to which the arguments involved are themselves embedded in higher-order arguments. More reflective, however, does not mean less intuitive. From elementary reflective conclusions directly embedded in an intuitive argument to the many-level reflections characteristic of some of the most impressive scientific achievements, reasoning is always an output of a mechanism of intuitive inference.

In describing reasoning as a use of intuitions about reasons, we adopt a purely psychological approach quite different from more traditional logicist approaches to reasoning. Doesn't logic play a central role in reasoning? Aren't we missing something essential?

To better understand the issue, compare the psychology of reasoning to that of mathematics. The science of mathematics itself is about objective mathematical facts. The psychology of mathematics is about the way in which people learn about these facts and use them to calculate, for instance, accrued interests or the surface of a garden. The psychology of mathematics cannot be approached in a purely psychological way; mathematical facts have to be part of the picture. Similarly, it could be argued, reasoning is the use of objective logical facts. Intuitions about reasons that are not about their logical properties are no more part of reasoning than beliefs about, say, lucky numbers are part of mathematical thinking. If this argument is correct, then the psychology of reasoning should have logic at its core.

The acquisition and use of mathematical competence rely on the existence of mathematical symbols for numbers, operations, and so on. Even before the invention of writing, many words in spoken languages served as mathematical symbols and allowed a modicum of mathematical competence to develop. Logic, likewise, relies on symbols to represent propositions, logical relationships, and so on. The use of a whole range of specialized symbols for logic is relatively recent, but just as in the case of mathematics, many words and expressions of ordinary language can serve as logical symbols. Language, so the argument goes, made reasoning possible well before the development of more formal logic.

So, according to the classical approach to reasoning, language is essential to reasoning. We agree, but for completely different reasons.

Reasoning Relies on Language

Reasons, we have argued, are for social consumption. To be socially shared, reasons have to be verbally expressed and, indeed, reasons appear on the mental or public scene in verbal form. Reasons may serve to justify oneself, to evaluate others, or to convince people who think differently. All this involves verbal communication.[4] Even when you think of reasons on your own, often it is as if you were mentally answering what others have said or might say, as if you were readying yourself to question others' opinions and actions. Even when you think of reasons to answer your own questions, it is as if you were engaging in a dialogue with yourself. For this, you resort to inner speech.

Language is uniquely well adapted to represent reasons. To understand a reason is to mentally represent the relationship between at least two representations: the reason itself and the conclusion it supports; in other words, it is a metarepresentational task. Language and metarepresentation are closely associated (even if metarepresentation, and in particular mindreading, may well be possible without language).[5] Language is a uniquely efficient tool for articulating complex metarepresentations and for communicating them. Language, in particular, may be uniquely well suited to metarepresent relationships between reasons and conclusions.

Linguistic expressions can be embedded within linguistic expressions, and in particular, sentences can be embedded within sentences: "It is nice" is a sentence, and so is "Yasmina said that it is nice." Several sentences can be embedded in a more complex sentence to articulate a reason. For instance,

> Molly isn't smiling.
> Molly is upset.

are sentences that may represent states of affairs. They can be combined, as in

> The fact that Molly isn't smiling is a reason to believe that she is upset.

This complex sentence metarepresents the relationship between a reason and the conclusion it supports.

To express all kinds of metarepresentational relationships and in particular reason-conclusions relationships, language offers a variety of linguistic devices. Nouns such as "argument," "reason," "objection," and "conclusion" can describe statements and their relationships, and so can verbs such as "argue," "object," and "conclude." A number of so-called discourse markers, such as the connectives "so," "therefore," "although," "but," "even," and "however," have an argumentative function: they focus attention on a reason-conclusion relationship and give some indication of its character.[6]

Compare, for instance, these two possible answers to the question, "How was the party?"

a. It was a nice party. Pablo had brought his ukulele.
b. It was a nice party but Pablo had brought his ukulele.

Both answers state the same facts, but they put them in a different perspective. In the (a) answer, the information that Pablo had brought his ukulele would be understood as an elaboration and a confirmation of the statement that it was a nice party; the music contributed to the success of the party. In the (b) answer, on the contrary, what "but" does is suggest that some of the consequences you might have inferred from the assertion that it was a nice party, such as that everything went well, do not actually follow in this case: yes, it was a nice party, but this was in spite of rather than because of Pablo's music.

Argumentative devices such as "but" play a heuristic role in argumentation. What they do is facilitate inference and suggest which implications should and shouldn't be derived in the context. According to the classical view, however, the main role in verbal reasoning belongs to verbally expressed logical symbols—in particular, logical terms such as "or," "if . . . then," "only if," "and," and "not" (and other logical devices such as quantifiers and modals). Such logical devices are what make it possible to construct a valid deductive argument. They are the linguistic tools that make reasoning (or at least so-called deductive reasoning) possible. Really? Are the logical devices the protagonists of reasoning, and other linguistic devices mere supporting characters? We want to tell a different story.

The Adventure of the Stolen Diamond and the Missing Premise

Contrary to received wisdom, sound syllogisms (and other sound formal deductions) do not provide compelling arguments in ordinary reasoning. We illustrate why not with a new Sherlock Holmes adventure.

A telegram is brought to Sherlock Holmes. He reads it, turns, and addresses his friend Dr. Watson:

> *Holmes:* The butler didn't steal the diamond. So, it is the gardener who did.

Watson, trusting Holmes's powers of deduction, might be convinced.

According to the standard logicist view, however, Holmes failed to express a proper deductive argument. Note that he used an argumentative device, "so," but no logical terms. At best, his statement is an enthymeme, that is, a truncated version of a true logical argument. If so, Holmes could have been more explicit:

> *Holmes:* Either the butler stole the diamond or the gardener did *[first premise]*. The butler didn't *[second premise]*. So the gardener stole the diamond *[conclusion]*.

This logical version with the logical terms "or" and "not" corresponds to a disjunctive syllogism:

Premises:	1. P or Q
	2. not P
Conclusion:	Q

Such a syllogism is said to be *valid.* When the premises of a valid syllogism are true, the syllogism is said to be not only valid but *sound:* the conclusion of a sound syllogism is necessarily true. Perhaps Watson recognized the full syllogism that Holmes had expressed in a truncated form and was convinced not just by his trust in Holmes's logical acumen but by the actual logic of the argument.

Does, however, a syllogism that you know to be sound provide you, by itself, with a sufficient argument in favor of its conclusion? It is a common mistake to think so.

Suppose the telegram Holmes has just received is from the chief of police and reads: "The diamond was stolen by the gardener." Holmes, who until then had no idea who might have stolen the diamond, rightly accepts this as a fact, but rather than just report it to Watson, he decides to convey the information in syllogistic form. He might say the following, speaking truly:

Holmes: Either the pope stole the diamond or the gardener did. The pope didn't. So the gardener stole the diamond.

Holmes's pseudo-argument is blatantly circular, and Watson should easily recognize this: since the pope was never considered a possible culprit, Holmes's only plausible ground for asserting the first premise of the argument ("Either the pope stole the diamond or the gardener did") is that he already knew that the gardener stole the diamond and could have said so directly. Sure, the syllogism is sound, but Watson would recognize it not as a genuine argument but as a somewhat quirky way for Holmes to assert what he knew independently. Watson's reason to accept that the gardener stole the diamond would, in that case, have nothing to do with the logical soundness of Holmes's pseudoargument and everything to do with his trust in Holmes.

In the same circumstances, a disingenuous Holmes could have been just as logical and strictly truthful but nevertheless misleading. He could have spoken as follows:

Holmes: Either the butler stole the diamond or the gardener did. The butler didn't. So the gardener stole the diamond.

Replacing the pope with the butler doesn't change a bit the logic of the syllogism, nor does it alter its soundness, but it is likely that this time Watson would mistake it for a genuine argument, a reason to accept its conclusion. Why? Because, unlike the pope, the butler might have been suspected, and Holmes might have deduced that the gardener stole the diamond from information that the butler didn't. Watson could think that Holmes had produced not only a sound syllogism but also a genuine argument.

More generally, for any fact you happen to know, you can always construct a sound syllogism that has this fact as its conclusion and that suggests that you came to know this fact thanks to your powers of deduction. Such a syllogism, however, doesn't give any genuine support to its conclusion. For a syllogism to serve as a bona fide argument in support of its conclusion, there must be good, noncircular reasons to accept its premises. This is not a logical requirement. From a logical point of view, circularity isn't a problem. Noncircularity, on the other hand, is a requirement of good reasoning. Hence, when a full syllogism is presented on its own as an argument, the argument itself is incomplete; it carries an implicit premise. It implies that there are good, noncircular reasons to accept the explicit premises and, as a consequence, the conclusion.

The common wisdom is that most arguments ordinarily used in conversation are, in fact, truncated syllogism (which they sometimes are, but much less often than is generally assumed). What we have just shown is that arguments consisting just of a syllogism, even a fully explicit one, are themselves truncated arguments: they implicitly convey that their premises are not just true but provide genuine reasons to accept the conclusion. There goes the alleged systematic superiority of explicitly laid-out syllogisms over informal statements of reasons: both, in fact, provide incomplete arguments.

In Reasoning, Logic Is a Heuristic Tool

Syllogisms are not better arguments for inquisitive or argumentative reasoning. They are an altogether different kind of thing.[7] Syllogisms (and deductions generally) are abstract formal or semiformal structures that make explicit a relationship of logical consequence between premises and conclusion. They may be used for a variety of purposes, but in themselves, they don't *do* anything. The arguments used in reasoning, on the other hand, are not defined by their structure, which is quite variable, but by what they do, namely, provide reasoners with reasons to come to some conclusion.

It is a commonplace that merely recognizing a syllogism as valid is not a reason to accept its conclusion. We have shown, moreover, why recognizing a syllogism as sound isn't a sufficient reason to accept its conclusion either. Only if you have independent reasons to accept its premises may a sound syllogism give you a reason to accept its conclusion. What the syllogism does, in such a case, is help you see how the reasons in favor of the premises are also reasons in favor of the conclusion.

What if you take a given syllogism to be sound and you do have good noncircular reasons to accept the premises? Shouldn't you, in this case, accept the conclusion? The answer is again no. What you have now is a reason either to accept the conclusion or to change your mind about at least one of the premises. This is far from being a mere theoretical possibility or a rarity. Actually, a common use of syllogisms in reasoning is to bring people to revise their beliefs by showing them that these beliefs entail consequences that they have strong independent reasons to reject.

A reductio ad absurdum, as this form of argumentation is called, consists in arguing against a claim by showing that it leads to an absurdity, or at least to a manifest falsehood.[8] Suppose that Watson answered Holmes's apparent demonstration that the gardener was the thief by informing him, "My dear Holmes, your syllogism is, as usual, impeccable but your conclusion cannot be true: the gardener had a heart attack and died the day before the diamond was stolen; I happen to be the doctor who signed the death certificate!" Watson's testimony would override the words of the chief of police. The very logicality of the syllogism would then force Holmes to reject at least one of its two premises: perhaps the thief was neither the butler nor the gardener but someone else, or else the thief must have been the butler after all.

Most ordinary inquiries and arguments are about empirical facts. They involve premises that, even when we accept them as true, are less than certain and admit of exceptions. A conclusion derived from such premises inherits their precariousness. A syllogism with premises that are not strictly true is not a strict proof. As a result, having a good or even a compelling reason to reject the conclusion of a syllogism should cause one to reconsider the premises—to reconsider them, yes, but not necessarily to reject them completely. What may happen rather is that a premise, rather than being simply rejected, is recognized to have exceptions.

To illustrate, we adapt the case of Mary having an essay to write (which Ruth Byrne used to argue that *modus ponens* deductions can be "suppressed," as we saw in Chapter 1).

Tang and Julia are Mary's flat mates. They wonder whether she will be back in time for dinner:

Tang: Well, if Mary has an essay to write, she stays late in the library.
Julia: Actually, she does have an essay to write. Let's not wait for her!

So they prepare the dinner, set the table, and open the wine, and then, just as they are about to sit down, Mary arrives.

Julia: We thought you had an essay to write and you would stay late in the library.
Mary: I do have an essay to write, and if the library had remained open, I would have stayed there.
Tang: A glass of wine?

Tang's and Julia's initial statements provided the two premises of a conditional syllogism, the conclusion of which was that Mary would stay late in the library (and hence not be back in time for dinner). They both initially accepted this conclusion but then, obviously, had to reject it when Mary appeared. Mary's explanation, however, showed that their premises had been credible, and their reasoning sensible; the problem was just that the circumstances were not quite normal.

A statement like Tang's, "If Mary has an essay to write, she stays late in the library," was not intended or understood to express a necessary truth or an undisputable empirical fact. Like most such statements in ordinary life, it expressed high probability in normal conditions—but then, conditions are not always normal. Tang had no need to qualify his statement by saying something like "probably" or "in normal conditions"; this is how Julia would have understood it anyhow. Mary's unexpected return shows that the two premises of the syllogism were not a sufficient condition to accept its conclusion—not because one of them was false, but because one of them admitted of exceptions, as do most of our ordinary life generalizations.

So, it turns out, reasons to reject the conclusion of a syllogism do not even force you to reject any of its premise; they might just make you more aware of the in-normal-conditions (also called "ceteris paribus") character of at least one of the premises.

How to account for all this? One might deny that "if" and other connectives such as "and" and "or" have the logical sense classically attributed to them. Perhaps their sense is based on a nonmonotonic logic; perhaps it is probabilistic. These are semantic solutions: they consist in revising our understanding of the sense of words. There is, however, another way to go. The philosopher Paul Grice argued for an alternative, pragmatic solution that would explain what is happening with logical connectives (and, actually, with words and utterances in general) by focusing less on what words mean and more on what speakers mean when they use these words. He thought that the classical semantics of logical connectives could be preserved and that apparent counterexamples could be explained in pragmatic terms. (Of course, the semantic and the pragmatic approaches are not incompatible; on the contrary, in a full account, they should be integrated).[9]

Here we make a brief remark on the pragmatics of connectives and of syllogisms generally. Grice's insights have been developed in modern pragmatics and in particular in relevance theory. A basic idea of relevance theory is that the linguistic sense of words and sentences is used not to *encode* what the speaker means but merely to *indicate* it—indicate it in a precise way but with room for interpretation.

A tourist asks a Parisian a question:

Tourist: How far is the Eiffel Tower?
Parisian: It is near.

In the Parisian's answer, the word "near" may convey "within a short walking distance" if the tourist who is asking is on foot, and "within short driving distance" if the tourist is in a car. The word "near," of course, has a linguistic sense, but this sense is not identical to the meaning the speaker intends to convey; rather, this sense makes it possible to infer what the speaker means, given the context.

A word such as "near" has a vague sense that must be made more precise in context. But what about words, such as "straight," that have a precise sense? These too serve as a starting point for inferring the speaker's meaning:

Tourist (standing in front of the American Church on the Quai d'Orsay): What is the way to the Eiffel Tower?
Parisian: Keep walking straight ahead. You can't miss it.

Actually the Quai d'Orsay is not a straight street. It follows the curves of the river Seine. Were the tourist to follow the advice literally and to keep walking straight ahead, she would end up in the river. In the context, however, "straight" conveys that she should stay on the same street, even if this makes for a curved rather than a straight trajectory. Here, the word "straight" is used loosely. So words can be used to indicate a meaning that is narrower or looser than the sense linguistically encoded.[10]

Logical connectives (as well as quantifiers and modals) behave like ordinary words. They don't encode the communicator's intended meaning but merely indicate it. Take the case of "or." The sense of "or" is such that a statement of the form "*P* or *Q*" is true if one of the two disjuncts *(P, Q)* is true. So, for instance, if the gardener stole the diamond, then Holmes speaks truly when he says, "Either the butler stole the diamond or the gardener did." But, as we pointed out, Holmes would be typically understood to mean not only what his utterance literally and explicitly means but also implicitly that he has some reasons to assert the disjunction other than just knowing that one of the disjuncts (that the gardener stole the diamond, in this case) is true. Typically, a "*P* or *Q*" statement conveys a greater confidence in the disjunction itself than in each of the disjuncts. Thus, in most ordinary contexts, the word "or" conveys more than the logical sense it encodes.

"Or" can also be used to convey less than its logical sense. Imagine the following dialogue:

Police chief: Either the butler stole the diamond or the gardener did.
Holmes: Are we sure? Either the butler stole the diamond or the gardener did, or some other member of the household did, or another inhabitant of the village, or of the county, or of the country.

If it turned out that the thief wasn't even an inhabitant of the country but just a visitor from abroad, it wouldn't make much sense to say that Holmes had been wrong. He is clearly using a multiple disjunction (". . . or . . . or . . . or . . .") to express a kind of implicit meaning that utterances with "or" typically convey: that each of the individual propositions connected by "or" is in doubt, and he is doing so without committing to the truth of the whole disjunction itself. He is, in other terms, expressing both more and less than the literal sense of his utterance.

Other logical connectives such as "and" and "if" (as well as quantifiers such as "some" or "all") can be used to convey not (or not just) their literal meaning but some meaning inferred in context. For instance, it is well known that an "If *P*, then *Q*" statement can be intended and understood not only to convey that *P* is a sufficient condition for *Q* (which corresponds to the literal sense of "if") but also to convey that *P* is a necessary and sufficient condition for *Q* (corresponding to "if and only if") or that *P* is a necessary condition for *Q* (corresponding to "only if"). When interlocutors use "if" in such nonliteral ways, they are not violating any rule of logic or of semantics. They are just making a normal use of language. Logical connectives, it turns out, can be used in the same way as discourse markers such as "therefore" and "but" to suggest implications that should be derived in the context even though they are not entailed by the literal meaning of the utterance.

It might have been tempting to view verbal expressions of logical relationships as similar to verbal expression of arithmetic relationships (as in "A hundred and thirty-five euros divided by five equals twenty-seven euros"). Whether they are done with written digits and special symbols or with words, arithmetic operations obey strict rules of construction and interpretation. Even when they are performed verbally, arithmetic operations are not interpreted on the basis of pragmatic consideration; number words and words such as "divided" and "equal" are used literally.

Unlike verbal arithmetic, which uses words to pursue its own business according to its own rules, argumentation is not logical business borrowing verbal tools; it fits seamlessly in the fabric of ordinary verbal exchanges. In no way does it depart from usual expressive and interpretive linguistic practices.

Statements with logical connectives (or other logical devices), and even sequences of such statements that more or less correspond to syllogisms, are just part of normal language use. They are used by speakers to convey a meaning that cannot be just decoded but that is intended to be pragmatically interpreted. Not only the words used but also the force with which premises and conclusions are being put forward are open to interpretation. They may be intended as categorical or as tentative assertions, hedged by an implicit "in normal conditions."

When you argue, you do not stop using language in the normal way, nor does your audience refrain from interpreting your statements using the same pragmatic capacities they use all the time. In argumentation, ordinary forms of expression and interpretation are not overridden by alleged "rules of reasoning" that might be compared to rules of arithmetic. Rules of arithmetic are taught and are not contested. There is no agreement, on the other hand, on the content and very existence of rules of reasoning. What is sometimes taught as rules of reasoning is either elementary logic or questionable advice for would-be good thinking or good argumentation (such as lists of fallacies to avoid, which are themselves fallacious).[11]

We have been focusing on argumentation, but what about reasoning done "in one's head"? Such individual reasoning, we would argue, is a normal use of silent inner speech.[12] The pragmatics of inner speech have not been properly studied. Given, however, that much inner speech rehearses or anticipates conversations with others, the pragmatics involved are probably not that different from the pragmatics of public speech. When we speak to ourselves, we use words loosely or metaphorically as often as we do in public speech. Our assertions are just as likely to be hedged or qualified with an implicit "probably" or "in normal conditions." There is no reason to assume that when we reason in our head, we follow logical rules that we typically ignore in public argumentation.

Couldn't all the evidence showing that ordinary reasoning isn't governed by the rules of classical logic suggest that it is governed by another kind of logic or by a probabilistic system of inference? Classical deductive logic is "monotonic." This means that if some conclusion logically follows from some initial set of premises, it also follows from any larger set of premises that includes the initial set. As the example of Mary having an essay to write well illustrates, ordinary human reasoning is not monotonic. Not only in daily life but also in scientific, technical, medical, or legal reasoning, conclusions are typically tentative. They may be revised or retracted in the light of new considerations (such as Mary showing up in time for dinner after all). In fact, monotonic inference is, at best, of very limited use for a real-life cognitive system.

In early experimental psychology of reasoning, the nonmonotonic character of ordinary reasoning had been largely ignored or idealized away. In many recent approaches to reasoning, on the contrary, it has been given a central role. Some scholars, such as the cognitive scientist Keith Stenning and the logician Michiel van Lambalgen, aim to replace classical logic with a "nonmonotonic logic" that would provide better insight into the way people actually reason. Others, such as psychologists Mike Oaksford and Nick Chater, argue that reasoning is best viewed not as a logical but as a probabilistic—and more specifically Bayesian—form of thinking.[13]

The project of replacing standard logic with nonmonotonic logic or of replacing logic altogether with probabilities shares a basic presupposition with the traditional approach: that the study of inference must be based on a general and formal understanding of norms of good inference. We are not convinced. We have argued for an evolutionary and modularist view of inferential processes. Every inferential module aims at providing a specific kind of cognitive benefit, and at doing so in a cost-effective way. In this perspective, investigating a given module is a matter of relating its particular procedures to its particular function. The function of the reason module, in particular, is much more specific than that of optimizing knowledge and decision making in general. Investigating general norms of inference, while interesting from a philosophical or from a machine intelligence point of view, may not tell us that much about any specific inference module. It may tell us little, in particular, about the reason module (just as a general theory of locomotion would be of limited use in understanding how bats fly or snakes crawl).

While we doubt that recent nonmonotonic or probabilistic approaches provide the key to understanding reasoning proper, we of course agree with the critics of classical logic that it fails to provide a plausible norm for human inference. Shouldn't we then also agree that classical logic is altogether irrelevant to the study of reasoning? Well, no. Let us venture a limited defense of classical logic, one that defenders of classical logic may not be too happy about. Logic can be used not just as a norm or as a procedure but also as a heuristic tool that clarifies questions and suggests answers. This, we claim, is a main role that logic plays in reasoning. Of course, this goes against the standard view that the function of logic is precisely to overcome the limitations of heuristic thinking.

If syllogisms can be interpreted with some freedom and according to the context, if, even when they are logically sound, they don't necessarily compel rational reasoners to accept their conclusion, what good do they do? What is the point of using them at all? The answer we want to suggest is that they often highlight reasons to accept a conclusion that is not immediately intuitive or to reject a conclusion that is. The very schematism of syllogisms (and deductive relationships generally) tends to exaggerate the degree of logical dependence among our assumptions. In particular, it dramatizes mere incoherencies (where two or more ideas might each give reasons to reject the others) as straight inconsistencies or logical contradictions. Just as exaggerating contours in a picture helps recognition and just as leaving out details in a narrative helps one follow the story, leaving out hedges and ignoring exceptions help focus on reasons that may lead to adopting or rejecting a conclusion. Incoherencies are often hard to detect and to reflect upon. Dressing them up as logical inconsistencies makes them salient targets for the reason module.[14]

Reasoning itself, as we have described it, involves higher-order intuitions about how lower-order intuitions may support some conclusion. True, in principle, the higher-order intuitions might be just about logically relevant properties of reasons and conclusions and might ignore other aspects of their content, but why should they be? This, anyhow, is not what happens. When reasoning about a given issue, higher-order intuitions are about various properties of lower-order intuitions, whether "logical" or not, that are relevant to their value as reasons. Higher-order intuitions in reasoning are metacognitive rather than just "metalogical." Reasoning is based on rich and varied intuitions about intuitions.

Take the kind of reasoning that would classically be represented by means of a disjunctive syllogism with a main premise of the form "*P* or *Q*." Depending, so to speak, on the *P* and the *Q* involved—depending, that is, on content properties that from a standard logicist point of view shouldn't be taken into account—higher-order intuitions may be developed in different directions.

You are reasoning about the location of a book you need and have the intuition that the book must be either in the study or in the living room. You have been searching for it in the study, but so far without success. You decide to look for it in the living room. Of course, if asked to explain why you are now moving to the living room, you might articulate your answer in syllogistic form: "I thought that the book is either in the study or in the living room. It is not in the study; hence it must be in the living room." This would make it sound as if all that was involved was a bit of logic. In fact, your reasoning is more attuned to the particulars of the case. In your mind, you are not ruling out the possibility that you might have missed the book in the study; it is just that the more you looked and failed to find it, the more this became an intuitive reason to conclude that it must be in the living room (or else that you were mistaken in your initial intuition that it had to be in one of these two rooms). The way your reasoning proceeds is sensitive to the fact that a book cannot be in the two rooms at the same time, to the fact that in both rooms there is so much clutter that you might miss it, and to the fact that your initial intuition about the possible location of the book is based on less-than-certain memory. Even when you express it as a syllogism, you expect your audience to take this as just a schematic rendering of your thinking and to use their own richer intuitions to understand and evaluate your reasons.

Take now a different example, where the schematism is the same but the relevant intuitions quite different. Seeing Molly frown, you intuit that she must be upset about something that occurred yesterday or worried about something that might occur. Which is it? She tells you that what occurred yesterday didn't upset her and you assume that she is sincere. You conclude that she is probably worried. Your initial intuition that Molly might be upset or worried didn't quite exclude that she might be both. Her sincerely saying that she was not upset didn't rule out that she might be upset without realizing it. Your intuitive reasons favor, then, a somewhat tentative conclusion: you tend to believe that she is worried rather than upset, or at least more worried than upset. Suppose now that you tell Ramon, a common friend, "Molly is worried!" and he answers, "I think she is just upset about what happened yesterday." You might then argue in syllogistic form: "Looking at Molly's face, you cannot tell whether she is upset or worried; she says, however, that she is not upset, not about yesterday, not about anything, and she is manifestly sincere; hence she must be worried." Again, this would schematize your reasoning, but in so doing, it would present Ramon with a clear challenge and with reasons to revise his beliefs and either to conclude just that Molly is worried or to view things in a more nuanced way, as you do yourself.

In argumentative reasoning in particular, the use of logical relationship plays a heuristic role for one's audience. It helps challenge them to examine and enrich or revise their beliefs or else to defend them with arguments in their turn. Thanks in part to its logical garb, argumentation, if not always convincing, is at least quite generally challenging.

More generally, however, reasoning is not the use of logic (or of any similar formal system) to derive conclusions. But what, then, is the method of reasoning (if there is one)?

Is There a Method for Reasoning?

Reasoning, as we have described it so far, is rather limited. Humans reason when they are trying to convince others or when others are trying to convince them. Solitary reasoning occurs, it seems, in anticipation or rehashing of discussions with others and perhaps also when one finds oneself holding incompatible ideas and engages in a kind of discussion with oneself. Just as with justifications, the production of arguments proceeds by means of backward inference, from a favored conclusion to reasons that would support it.

Surely, something must be missing in this picture: one is often prompted to reason not by a clash of ideas (with others or within oneself) but by self-addressed questions. Moreover, isn't the point of such individual, inquisitive reasoning to discover the right answer to a question of one's own rather than to confirm an already favored conclusion?

There are indeed questions that we can and do approach on our own, without any bias, any hunch in favor of this or that answer, and that we are able to answer by reasoning forward, from premises to conclusion. These, however, tend to be special kinds of questions that are approached with ad hoc reasoning methods rather than with everybody's everyday reasoning dis-

5	3			7				
6			1	9	5			
	9	8					6	
8				6				3
4			8		3			1
7				2				6
	6						8	
			4	1	9			5
				8			7	9

Figure 15. A Sudoku grid.

positions. The simplest examples of such questions are found in games or puzzles devised to entertain, teach, or test people.

Take the game of Sudoku (see the example in Figure 15). It is played on a square grid of eighty-one cells, some already containing a digit at the beginning of the game, the others blank.

The task is to find the digit that each of the blank cells should contain, knowing that every digit between 1 and 9 must occur once and only once in every vertical column, every horizontal row, and every three-by-three box indicated by thick lines.

Sudoku players approach the task in an impartial manner. They have no a priori hunch and no stake whatsoever in any particular solution. They know that in every grid there is one and only one right solution for each and every cell, and the players' goal is to find them all. That much is clear. What is less clear is how the players proceed.

In the simplest cases, it is possible to find the digit that goes in a given cell by means of a simple elimination method until only one possibility is left. Which digit, for instance, should go in the central (grayed in the figure) cell?

Looking at the grid, you find reasons to eliminate digit 1 (it already occurs in both the horizontal and the vertical rows that contain the grayed square), digits 2 and 3 (they already occur in the three-by-three box), and so on with all nine digits except 5, which therefore must be the digit that goes in this central cell. Sudoku is a perfect illustration of Sherlock Holmes's famous maxim, "When you have eliminated the impossible, whatever remains, however improbable, must be the truth" (and of the very limited usefulness of "thinking like Sherlock Holmes").

The elimination method we just illustrated may help players find a few missing digits, but it will not help them fill the whole grid. To fully solve a Sudoku puzzle, even a simple one, players must use more complex methods. The real challenge for psychologists, however, is not so much to understand how players apply methods they have been taught but how some of them at least discover useful methods on their own.[15] After all, the reasoning involved in discovering these methods is much more impressive than that involved in solving a puzzle once you know what method to apply.

How do ordinary Sudoku players discover new methods? We argued that in general, when people reason, they start from an intuitive bias or a hunch in favor of a given conclusion and look for reasons in its favor through backward inference. While this is not the way players *use* methods to solve a Sudoku puzzle, it is, we suggest, the way players *discover* these very methods in the first place (when they don't learn them from others). Merely understanding the rules of Sudoku makes the use of the simple elimination method we have described fairly intuitive. Practice familiarizes players with various subtle regularities of Sudoku grids and provides them with intuitive insight into more and more elaborate methods for which successful applications provide confirming reasons.

The discovery of effective Sudoku methods may well be, then, yet another instance of basic biased reasoning: a search for reasons in favor of an initial intuition—not, in this case, an intuition about the digit that goes in a given cell, but a higher-order intuition about the kind of considerations that may allow the player to narrow down the range of possible digits for every cell. If we are right, this means that people do not use a general higher-order explicit method to discover the various, more specific, explicit methods needed to solve a Sudoku puzzle.

Sudoku puzzles are problems that, with practice and explicit methods, people can solve. Does this generalize to reasoning problems across domains? Can one, with the right method, effectively answer questions of any kind? The short answer is no.

Humans are, it is true, capable of applying methods that they have been taught or that they have discovered on their own to do a great variety of things: solve a Sudoku puzzle, construct a Lego castle, find a word in an alphabetic dictionary, buy from a vending machine, bake a cake, convert Roman into Arabic numerals, learn the basic moves of tap dancing, find solutions to arithmetic or geometric problems, or use databases to answer queries about, say, legal precedents, life expectancy, or the yield of equities. This ability to understand and apply step-by-step methods is a hugely important aspect of human psychology. It plays a major role in the development and transmission of cultural skills, including specialized problem-solving skills. The psychologist Keith Stanovich has argued that the ease with which people acquire and apply such methods (or "mindware") correlates with individual differences in rationality.[16] Still, reasoning doesn't consist in applying such methods, and in general, it doesn't need them. Applying these methods is no more quintessential reasoning than military marching is quintessential walking.

Most of the questions that people encounter in their daily life or in the pursuit of their longer-term interests cannot, in any case, be answered by following instructions. Every year, new psychological counseling books offer to instruct you on how to reason better in business, love, friendship, and games, promising to do much better than last year's books that made the same promises. While they may at times include some sensible advice, all these books come short of delivering on their promises. Achieving such desirable results involves understanding a great many aspects of the world around us, and knowing which to take into account and which to ignore in any given situation; it involves identifying and tackling an endless variety of issues along the way; there are no adequate instructions for reasoning effectively about most real life problems.

What differentiates Sudoku puzzles from most issues on which we might reason is that they are perfectly well-defined. The wider context is not relevant, the evidence is all here, there is one and only one correct solution, and any incorrect solution leads to clear inconsistencies. In psychology, such problems with a definite and recognizable solution and some effective methods to discover it have often been studied under the label of "problem solving" rather than "reasoning." Solving such problems when the right method is not known typically involves some reasoning but also trial-and-error tinkering and task-specific forms of insight.

Many games and many serious problems in mathematics, logic, programming, or technology involve problem solving in this sense. Scientist, technicians, and laypersons have discovered and developed methods to address problems that can be solved in such a way. Some of these methods have become cultural success stories because of their practical applications, scientific relevance, intellectual elegance, or appeal as leisure activities.

To psychologists of reasoning, problems and puzzles that can be solved in clear and effective ways, especially problems devised on purpose by experimentalists such as the Wason selection task, may seem to provide optimal material for experimental investigation. The temptation has been to assume not only that some of these problems are easy and fun to experiment with but also that they provide crucial evidence on the basic mechanisms of reasoning. Such narrowly circumscribed tasks are, however, special cases among the quite diverse challenges that reasoning helps us confront. Problem solving involves some reasoning, but most reasoning doesn't even resemble problem solving in the narrow sense in which psychologists use the phrase.

When people reason on moral, social, political, or philosophical questions, they rarely if ever come to universal agreement. They may each think that there is a true answer to these general questions, an answer that every competent reasoner should recognize. They may think that people who disagree with them are, if not in bad faith, then irrational. But how rational is it to think that only you and the people who agree with you are rational? Much more plausible is the conclusion that reasoning, however good, however rational, does not reliably secure convergence of ideas.

Scientists, it is true, do achieve some serious and, over time, increasing degree of convergence. This may be due in part to the many carefully developed methods that play a major role in conducting and evaluating scientific research. There are no instructions, however, for making discoveries or for achieving breakthroughs. Judging from scientists' own accounts, major advances result from hunches that are then developed and fine-tuned in the process of searching, through backward inference, for confirming arguments and evidence while fending off and undermining counterarguments from competitors.

Arguably, science is the area of human endeavor where rationality and good reasoning are most valued (as we will discuss in detail in Chapter 18). Reasoning, however, does not cause scientists to spontaneously converge on the best theories. It causes them, rather, to elaborate and vigorously defend mutually incompatible competing theories. It also helps them—with a higher, even if still imperfect, degree of convergence this time—to evaluate competing theories and reach some degree of tentative agreement on the current winners of the competition. Winning in this competition may well be, to a large extent, a matter of epistemic luck—having invested in a better initial hunch for which there are stronger evidence and better arguments to be found—more than of better reasoning in producing new ideas. What scientific practice suggests is, on the one hand, that various specific methods can be relevant to specific reasoning tasks and, on the other hand, that there is no such thing as a general method for reasoning.

The study of reasoning has been dominated by a normative goal and a descriptive expectation: the goal of discovering and making explicit a general method that could produce good reasoning and the expectation that actual human reasoning would be guided by an approximation of such a method. Both the goal and the expectation have, so far, been disappointed. This failure is not an accident. There is a principled explanation for it. The procedures of intuitive inference, we have argued, are unconscious, opportunistic, and diverse. The idea that intuition might consist in following a general method makes little sense.

Our higher-order intuitions about reasons for some intuitive conclusion take into account many of the properties of the lower-order intuitions they are about, properties that may vary greatly from one domain to another. This is not bad, let alone fallacious, reasoning. In putting our reasons in argumentative form to convince others, sometimes we appeal to their own nuanced higher-order intuitions, sometimes we choose to be schematic, and often we do a bit of both. The schematism we employ, rooted in ordinary language, has inspired one of the great intellectual achievements of human history: the development of the science of logic. Logic, however, tells us neither how we reason nor how we should reason. There is no general method that we could and should follow when reasoning, either on our own or in dialogue with others.

10

Reason: What Is It For?

Look at the piece of furniture that appears in Figure 16 on the following page. Hardly a well-designed chair, is it? The seat is so low that you'd practically have to squat on it. The back is too high; the upper bar is a horizontal slab on which you could not lean comfortably. How come? Actually, this is not a chair. It is a church kneeler. The height is right for kneeling. The "back" is in fact the front, and the upper slab is there to rest not one's back but one's praying hands. Once this artifact's real function is recognized, what looked like flaws turn out to be well-designed features.

Like the object in Figure 16, reason seems to have an obvious function: to help individuals achieve greater knowledge and make better decisions on their own. After all, if using reason doesn't help one reach better beliefs and choices, what is it good for? However, like a kneeler used as a chair, reasoning serves this function very poorly.

We've already examined quite a few failures of reasoning, and we will look at many more in Part IV, but the upshot of experimental research on how we reason is that we do so in a way that is biased and lazy and that often fails to solve even simple but unfamiliar problems. It would be easy to stop here and simply conclude, as many psychologists have done, that human reason is just poorly designed.

Another conclusion is possible. The true function of reasoning may have been misunderstood. Reasoning might be the kneeler-mistaken-for-a-chair of modern psychology. In Chapters 7, 8, and 9, we have argued that reasons are commonly used in the pursuit of social interaction goals, in particular to justify oneself and to convince others. Here, we adopt an evolutionary approach

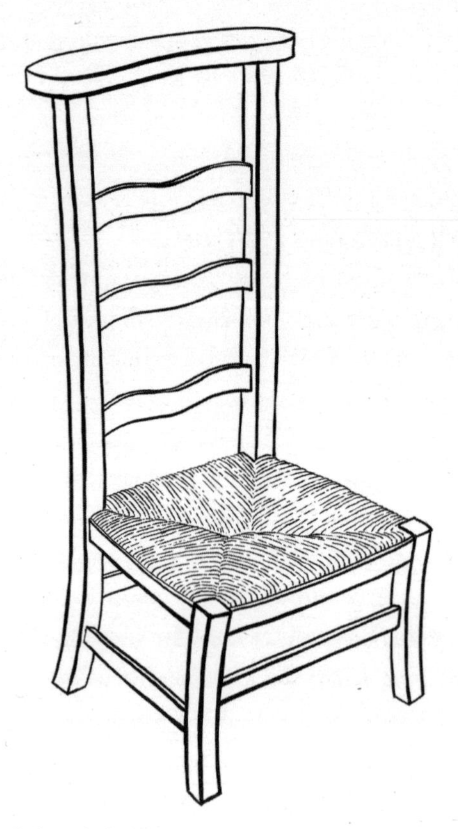

Figure 16. A poorly designed chair?

to argue that these social uses of reason are not side effects or minor functions of a cognitive mechanism having as its main function the enhancement of individual cognition and that, on the contrary, the main function of reason is social. Why resort to an evolutionary approach? Because this is the only approach that explains the fact that complex inheritable traits of living things tend to produce beneficial effects. Outside of an evolutionary perspective, it is quite unclear why human reason, or anything else for that matter, should have any function at all.

Functions before and after Darwin

That human reason has a function is an idea commonly taken for granted and one that we want not to challenge but to develop on a new basis. To begin with, let us be clear about the very notion of a function: discussions about the evolution of reason are often marred by a superficial understanding of the notion.

What is a function? How can one decide what is the function of an artifact such as a chair, a natural quality such as a flower's scent, a body part such as a wing, or a mental capacity such as human reason? The two questions—what is a function, and how do we identify a particular function—are closely related. For artifacts, a rough answer is generally easy to produce. The function of an artifact is what it is made for. Who decides what an artifact is for? Whoever designed it. A chair is made for sitting, a kneeler for kneeling, and so on.

Attributing a function to a biological trait is much less straightforward. No one made wings for flying or the heart for pumping blood. And yet, who would doubt that wings are for flying? Who would contest William Harvey's discovery of the function of the heart?

Before Darwin, attributing a function to some trait of an animal or a plant involved little more than answering the question: What is it good for? Find a good effect of a biological trait and just assume that what it is good for is what it is for. What could be more useful than flying? So, the function of wings is to enable flying. Such a commonsense notion of function is not useless; it has played an important role in science before Darwin, such as in Harvey's discovery of the function of the heart, and it still plays at least a heuristic role in modern science. What question such an ahistorical notion of function raises and fails to solve is: Why, in the first place, should we expect biological traits to have useful effects? We don't expect anything of the sort in the case of other natural objects such as stones, planets, and subatomic particles. These don't have functions. For a long time, the only answer was that God, the maker of all things, has made wings for flying, hearts for pumping blood, and so forth—an answer that replaced a series of interesting problems with an unfathomable mystery.

It was Darwin's theory of natural selection that provided the basis for a coherent and useful notion of a function based on a genuine explanation of why organs and other traits of living things have the functions they have.

Any inheritable trait of an organism has many effects. Most of these effects are without consequences for reproductive success. The scent of a flower, for instance, might make some local animals dizzy without this harming or benefiting the plant in any way. Some effects of a trait may happen to enhance reproductive success: the scent of a flower may attract insects and recruit them as pollinators. Still other effects of a trait may actually compromise reproductive success: a scent might attract florivores and entice them to eat the flower, thereby diminishing the amount of pollen available to pollinators.

When overall the effects of a heritable trait are more beneficial than harmful to the reproductive success of its carriers, then this trait is likely to be selected and to propagate over generations. Selected traits have at least one of their effects that has contributed to the reproductive success of their carriers and thereby to their own propagation. The scent of wildflowers, for instance, has been selected because by attracting pollinators, it contributed to the reproductive success of the plants. When biologists talk of the function of a selected trait, they refer to such a beneficial effect.

Selected traits may have more than one function, and these functions may change in the course of evolution. Penguins, for instance, are descendants of birds that had used their wings to fly, but the function of penguins' wings is to help them swim. At some point in their evolution, these wings must have served both functions, as they do in some seabirds, such as in razorbills.

Traits that owe their propagation to their function(s) are *adaptations*. Not all features of an organism are adaptations. Some features result from physical or chemical constraints on biological development. The fact that all organisms contain carbon, for instance, is a basic chemical property of life itself (as it is found on earth, at least). Not all features of an adaptation are adaptive, either. Many are side effects. The regular contractions of the heart are an adaptation. They have the function of causing blood flow throughout the body. The noise produced by these contractions, on the other hand, isn't adaptive. It is a side effect.

The Function of Reason from Aristotle to Twentieth-Century Psychologists

Just like ideas about the function of wings, ideas about the function of reason emerged well before Darwin and the development of a scientific notion of function. Even after Darwin, much evolutionary thinking on reason shares with pre-Darwinian views a sense that the function of reason, just as that of wings, is obvious enough. Reason is a means for individuals to acquire superior knowledge and to make better decisions. Reason, by performing this intellectual function, elevates humans above all other animals.

Darwin himself expressed this same general idea: "Of all the faculties of the human mind," he wrote in *The Descent of Man,* "it will, I presume, be admitted that *Reason* stands at the summit."[1] He put, however, an evolutionary twist on this classical view. Through his intellectual faculties, he argued, man "has great power of adapting his habits to new conditions of life. He invents weapons, tools, and various stratagems to procure food and to defend himself. When he migrates into a colder climate he uses clothes, builds sheds, and makes fires; and by the aid of fire cooks food otherwise indigestible." Because these intellectual faculties are variable and tend to be inherited, "they would have been perfected or advanced through natural selection. . . . In the rudest state of society, the individuals who were the most sagacious . . . would rear the greatest number of offspring."[2]

The functional effects of reason are roughly the same for Darwin as they were for Aristotle. What is new with Darwin, however, is the use of these effects to explain why reason should have evolved. And because Darwin is making a more precise claim about the function of reason, his claim is more open to a scientific challenge. If reason evolved to help individuals think on their own, then it should really provide for truly better thinking in terms of both cognitive benefits and mental costs, and if it does not, as much of modern psychology of reasoning suggests is the case, then we are indeed faced with a serious problem.

Psychologists who discovered what looked to them like major flaws of human reason must have, you would think, understood and discussed the challenge this presented for a post-Darwinian understanding of the function of reason. Actually, they did not. Until the 1990s, evolution, natural selection, and biological function were hardly ever mentioned in the psychological literature on reasoning. Psychologists just took for granted that the function of reason—with "function" understood as a commonsense rather than biological notion—was to enhance individual cognition. They then concluded that reason was not performing its function as well as had been assumed.

If anything, psychologists saw flaws in reasoning as evidence against an evolutionary approach. Evolved mechanisms should perform their function well enough to have been selected. Reasoning, with all its flaws, is not properly geared to the pursuit of knowledge and good decision. Hence, many concluded, the case of reasoning shows the irrelevance of the evolutionary approach to human psychology.[3] Of course, psychologists might have rethought the function of reason instead of uncritically accepting common wisdom in the matter, but they did not.

While the foundations of an evolutionary approach to psychology were laid by Darwin himself and some relevant work had been done over the years, especially in the study of animal behavior, a general and systematic discussion of the evolution of human cognitive mechanisms and of reasoning in particular started only when psychologist Leda Cosmides and anthropologist John Tooby outlined an ambitious program of "evolutionary psychology" in the 1980s.[4]

"Has Natural Selection Shaped How Humans Reason?"

From the start, human reasoning has been a major topic of research in evolutionary psychology and a main focus of polemics for scholars opposed to this approach. The polemic erupted when, in 1989, Leda Cosmides published what became the most controversial article in the history of psychology of reasoning. It was entitled "The Logic of Social Exchange: Has Natural Selection Shaped How Humans Reason?"[5] (hence the title of this section).

There were in Cosmides's article two major theoretical claims. From an evolutionary point of view, she argued, one should expect reasoning mechanisms to have evolved as responses to specific problems that had been recurrent in the ancestral environment. Specialized adaptations do a much better job of addressing the problems they evolved to handle than would any general problem-solving ability. In fact, it is not even clear what an ability to solve problems in general might consist of (if not, precisely, of a complex articulation of a great many more specialized mechanisms). A general reasoning ability, if such a thing could evolve at all, might, at best, help in dealing with the residue of problems not properly handled by specialized mechanisms. On tasks of no evolutionary significance (such as those typically given in reasoning experiments), Cosmides argued, one should expect reasoning to be much less effective than on problems for which dedicated mechanisms have evolved.

Cosmides's second and central claim was that among the many cognitive adaptations most likely to have evolved, there had to be an inferential mechanism aimed at solving a major problem raised by cooperation. Cooperation is an interaction between two or more individuals where each incurs a cost and receives a benefit. Each member of a party of big game hunters, for instance, spends time and takes risks to help the group catch a prey. Provided that the benefits are greater than the costs and are shared fairly, cooperation is advantageous to all cooperators. However, in many cases, each cooperator stands to gain even more by sharing in the benefits of cooperation without paying the full cost, in other terms by cheating or free riding. Hunters who avoid taking risks but take the same share of the meat as others are at an advantage. Of course, if none of the hunters take risks, the prey will escape. Widespread free riding is likely to result in failure of cooperation. For cooperation to be profitable and hence to endure, cheaters must be identified and either controlled or excluded.

Cosmides and Tooby argued that the identification of cheaters should be seen as a major problem in human evolution, favoring the emergence of a specialized module capable of computing the rights and duties of cooperators as a basis for detecting cheaters. To test this hypothesis, they relied massively on the Wason selection task, which we talked about in Chapter 2 and which we don't see as adequate for the job.[6] Our goal here, however, is not to evaluate Cosmides's arguments for the existence of an evolved ability to detect cheaters (which we see as strong) or her experimental evidence (which we see as weak). It is to use her approach as a source of inspiration and of contrast to propose another, no less radical and even more ambitious evolutionary hypothesis about the function of human reason in general, and reasoning in particular.

Like Cosmides and Tooby, we expect an evolved mind to consist principally in an articulation of modular mechanisms. Modules are specialized—each helps solve some specific type of problem or take advantage of some specific type of opportunity in a way that contributes to biological fitness. Reasoning as classically understood, on the other hand, is allegedly in the business of addressing any kind of problem and finding ways of taking advantage of any kind of opportunity. Classical reasoning is not a specialist but a generalist, not a narrow module but a broad faculty.

If the job of reasoning is much too broad to be that of a module, there seem to be only two possible conclusions: either the central and most distinctive component of the human mind is not modular, or else reasoning as classically understood doesn't even exist. Cosmides and Tooby favor the second conclusion: contrary to a long-held dogma of philosophy and psychology, there is no such thing as reasoning in general.[7] Other researchers working on the evolution of the mind, such as the psychologist Cecilia Heyes and the philosopher Kim Sterelny, favor the first possible conclusion: the existence of domain-general reasoning is evidence that the human mind is much less modular than Cosmides and Tooby and others have argued.[8]

Are these the only two possibilities? No. As we have shown (in Chapter 6), a metarepresentational module processes a very special type of object, representations, and attends to properties specific to these representations. Such a module may nevertheless indirectly provide information about the very topics of the representations it processes. Mindreading, for instance, informs us about not only the thoughts of others but also all the things and events these thoughts are about. Reasoning, we have argued, is produced by a metarepresentational module, the specific domain of which is the relationship between reasons and the conclusions they support. These reasons and conclusions, however, can themselves be about any topics. As a result, inferences about reasons-conclusions relationships indirectly yield conclusions in all domains, indirectly providing a kind of virtual domain-generality.

So, to answer the question posed in the subtitle of Cosmides's article and in the title of this section, we agree that natural selection has shaped how humans draw all kinds of inferences and has produced a wide variety of specialized inferential modules. One of these, we add, is a reason module. The justifications and the arguments that the reason module produces may have, embedded in them, conclusions relevant to all domains of knowledge and action. This virtual domain-generality does not, however, make reason, and the organization of human mind generally, any less modular.

What functions does the reason module fulfill? We have rejected the intellectualist view that reason evolved to help individuals draw better inferences, acquire greater knowledge, and make better decisions. We favor an interactionist approach to reason. Reason, we will argue, evolved as a response to problems encountered in social interaction rather than in solitary thinking. Reason fulfills two main functions. One function helps solve a major problem of coordination by producing justifications. The other function helps solve a major problem of communication by producing arguments. (Our earlier work has been focused on reasoning and on developing, within the interactionist approach, an "argumentative theory of reasoning.")[9]

The Challenge of Coordination and the Justificatory Function of Reason

Human cooperation is exceptional not only by its scale but also by the open-ended variety of the forms it takes. Other animals may have a few types of cooperative interactions in their behavioral repertoire with little or no place for creative improvisation. What each cooperator may expect of the others is largely predetermined. When these expectations are not met because of the incompetence or the defection of one of the cooperators, cooperation is likely to fail. Humans, on the other hand, rely less on predefined expectations. Existing forms of cooperation are deployed with great flexibility and often readjusted on the fly. New forms are often tested. This flexibility and creativity of human cooperation can be highly advantageous, but only if massive cognitive resources are invested to secure effective coordination.

To achieve the degree of fine-grained coordination that their multiple forms of cooperation require, humans need mutual expectations that have to be constantly updated to remain reliable. Members of a party of warriors, or of a sports team, for instance, readjust or even redefine their tactics and their mutual roles on each occasion. They must, moreover, be able to rely on each other not only when things go as anticipated but also when they don't. So must, in different ways, friends, spouses, coworkers, and business partners.

Even among competent cooperators, there may well be differences in the understanding of the common goals and of the part each individual must play. Differences of interests, instead of leading to defection or cheating, are often recognized and handled on the basis of mutual commitment to what is seen as fair; here too, however, there may be differences of interpretation leading to failures of mutual expectations.

For humans, knowing what to expect of each other is a crucial cognitive challenge. How is this challenge met? How do humans succeed in forming, if not perfect, at least adequate mutual expectations? The most common answer consists in invoking two mechanisms: norms at the sociological level, and understanding of the mental states of others at the psychological level.

At the level of the social group, there are shared norms of various kinds: moral, legal, religious, prudential, technical, and so on. They may be explicitly codified or not and enforced with sanctions or not. These norms regulate a great variety of social interactions.[10]

Some norms aim directly at securing coordination in a way that is beneficial to all the people involved. Without traffic rules, for example, driving a car would be absurdly risky. Very precise rules of coordination, however, tend to be highly specialized. The kind of interaction traffic rules regulate, for instance, is unlike any other; at every moment, drivers have a relatively narrow, well-defined range of options. If they don't coordinate their decisions, their lives are at risk. Effective coordination is in the interest of all of them.

Legal, moral, or religious norms, on the other hand, contribute to all kinds of social coordination in a great many ways, but securing effective coordination is not their sole purpose and needn't be their main one. These norms leave much room for interpretation and many relevant issues untouched. In some milieus, for instance, not arriving late at a dinner party may be a faux pas. People, moreover, don't at all times have the same interest in these various norms' being obeyed. Elected officials coming up for reelection, for instance, overtly encourage all their constituents to follow the norm and vote while making it harder to do so for those less likely to reelect them.

Do the many and diverse norms found in one's society constrain people's options to the point of indicating what people can expect from each other in the fine-grained manner needed for effective coordination? Of course not! There are, for instance, many norms—legal, moral, religious—telling spouses what they can expect of each other. Still, to achieve the level of mutual expectations that daily life coordination requires, spouses need to understand much more about each other than the fact that they are supposed to abide by a socially sanctioned set of norms.

In many forms of social interaction, some degree of creativity and improvisation is socially condoned or even required. You throw a party, for example. Whom should you invite? Conventional or moral norms might suggest that you invite people who have invited you in the past, so perhaps you should invite Olga, who invited you to her party last month. For other possible guests, norms of reciprocity won't help at all, but there are still issues of coordination to be addressed. If you invite Diego, for instance, you had better not invite Ruth. If you invite Ruth, she will want you to invite Chao.

Your future interactions with others will depend on many small, interconnected decisions you make. So what should guide you in solving coordination problems that norms leave unanswered? Here the standard answer is: mindreading. You have to understand the states of mind of Olga, Diego, Ruth, and other possible guests in order to anticipate their reactions and secure the conditions for a successful party that will enhance rather than compromise your future relationships with others.

Just like norms, mindreading plays an essential role in coordination. Still, the picture is far from complete. What is blatantly missing is the fact that individuals don't just infer what they can expect from each other on the basis of what they know of other people's minds and of the norms they share. People's mutual expectations are reviewed, discussed, negotiated in detail. Many decisions on how to interact are themselves taken interactively.

Gossip provides rich evidence of what can be expected of third parties. Individuals, however, need not be passive objects of gossip. They can participate in the ongoing conversation about themselves, explain and justify their views and their decisions, and, in so doing, to some extent, safeguard their own reputation. Officials trying to suppress the vote of constituents unlikely to reelect them justify their actions by saying they are just trying to prevent voting fraud. Guests arriving too late at a party justify themselves by explaining how they were delayed. You can tell Diego that, alas, you had to invite Ruth because she had invited you before; you can tell Ruth that, of course, she can come with Chao because her friends are your friends; and so on.

By giving reasons to explain and justify yourself, you do several things. You influence the way people read your mind, judge your behavior, and speak of you. You commit yourself by implicitly acknowledging the normative force of the reasons you invoke: you encourage others to expect your future behavior to be guided by similar reasons (and to hold you accountable if it is not). You also indicate that you are likely to evaluate the behavior of others by reasons similar to those you invoke to justify yourself. Finally, you engage in a conversation where others may accept your justifications, question them, and invoke reasons of their own, a conversation that should help you coordinate with them and from which shared norms actually may progressively emerge.

Reducing the mechanisms of social coordination to norm abiding, mindreading, or a combination of these two mechanisms misses how much of human interaction aims at justifying oneself, evaluating the reasons of others (either those they give or those we attribute to them), criticizing past or current interactions, and anticipating future ones. In these interactions about interactions, reasons are central.[11]

Justificatory reasons, in fact, bridge the gap between norms and mindreading. When we justify ourselves, we present our motivations as normatively apt, and we present norms as having motivating force. In other terms, we psychologize norms and "normalize" mental states. In doing so, our goal is not to give an objective sociological or psychological account of our actions and interactions; it is to achieve beneficial coordination by protecting and enhancing our reputation and influencing the reputation of others.

The role of reasons in social coordination has often been highlighted in philosophy, psychology, and the social sciences. The dominant view, however, is that attributing reasons is the most elaborate form of mindreading. We would argue that as far as mindreading goes, the attribution of reasons is typically misleading. The causal role it gives to reasons is largely fictitious; the reasons people attribute to themselves or to others are chosen less for their psychological accuracy than for their high or low value as justifications. The explanatory use of reasons, we suggest, is in the service of its justificatory use: it links reasons to persons so that good reasons are seen as justifying not just a thought or an action but also the thinker of that thought, the agent of that action.

The ability to produce and evaluate reasons has not evolved in order to improve psychological insight but as a tool for defending or criticizing thoughts and actions, for expressing commitments, and for creating mutual expectations. The main function of attributing reasons is to justify oneself and to evaluate the justifications of others.

The Challenge of Communication

Even more than human cooperation, human communication stands apart by its scale, its diversity, and the complexity of its mechanisms. Other animals use a small repertoire of signals to convey a few basic messages about the here-and-now: warnings about the presence of a predator, threats to competitors, and mating calls, for example. Humans have languages with huge vocabularies and powerful syntax with which they can produce an unbounded variety of linguistic utterances. With the help of language, they are able to communicate simple or complex ideas about any conceivable topic, about events distant in space and time, about general facts, or about abstractions, topics that are absent from animal communication and that are at the center of human cultural knowledge.

Humans, who have by far the richest codes, use them to communicate even more than what they manage to encode. As we have already noted, when speaking, people do not fully encode what they mean. What they do is give partly encoded evidence of their meaning. Hearers infer the speaker's intended meaning from this linguistic evidence taken together with contextual information. When, for instance, Azra tells Marco, "This case is too hard," he understands her to mean that the legal case on which he had chosen to write an essay for the tort law class is too difficult and that he should select an easier case to write about. In so understanding Azra, Marco goes well beyond decoding the sentence she uttered—a sentence that in other contexts could be used to convey utterly different meanings—but he does not go beyond the ordinary pragmatic procedures of comprehension we all use all the time.

Securing comprehension is, however, only half of the goal of communication. A speaker typically wants not only to be understood but also to be believed (or obeyed), to have, in other terms, some influence on her audience. A hearer typically wants not just to understand what the speaker means but, in so doing, to learn something about the world. This occurs when the hearer not only understands what he is told but also accepts it. Azra's intention in telling Marco "This case is too hard" was to have her implicit advice accepted. His goal in paying attention to what she was saying was to gain some guidance from it. Still, Marco may have understood what Azra was telling him and disagreed. In human communication, comprehension does not automatically secure acceptance. (Do you, for instance automatically believe everything you understand when reading a book like this one? Of course not!)

Without acceptance of the information communicated, communication wouldn't be beneficial to either communicators or addressees. If it were not beneficial to both, communication could not evolve. Would, then, a disposition in addressees to automatically accept communicated information ensure optimal communication? Far from it. Imagine what communication would be like if people automatically believed or did what they were told. Communicators might be quite satisfied; their ability to influence gullible and docile listeners would be without limit. Addressees, on the other hand, accepting everything they were told, would be prey to all kinds of misinformation and manipulation, not a satisfactory condition at all.

Automatic trust would make more sense only if there were a corresponding disposition in communicators to be automatically trustworthy. The eighteenth-century Scottish philosopher Thomas Reid affirmed that God had equipped humans with such a pair of dispositions:

> The wise and beneficent Author of nature, who intended that we should be social creatures, and that we should receive the greatest and most important part of our knowledge by the information of others, hath, for these purposes implanted in our natures two principles that tally with each other. The first of these principles is a propensity to speak truth. . . . [The second principle] is a disposition to confide in the veracity of others, and to believe what they tell us.[12]

An omnipotent creator might have decided to make humans both trusting and truthful. Natural selection is much less likely to produce and maintain such perfect harmony. It will, if possible, favor communicators who do not hesitate to deceive trusting receivers when it is in their interest to do so. Likewise, it will favor receivers who are not easily deceived.

Communication can still evolve under natural selection when the biological interests of emitters and receivers are, if not identical, at least sufficiently aligned, such as in the case of honeybees (which in the case of sterile workers contribute to the propagation of their genes by all helping the queen reproduce). Communication can also evolve when it is limited to topics of common interest where the communicator wouldn't gain by deceiving its audience or the audience by distrusting the signal, such as when a female baboon signals to potential mates, by means of a large swelling of her posterior, that she is ready to mate.

Human communication, however, is definitely not limited to topics of common interest where truthfulness and trust are mutually advantageous to the interlocutors. Linguistic signals can be produced at will to inform or to mislead. Unlike the honeybee's waggle dance or the female baboon's sexual swelling, linguistic signals are not intrinsically reliable. Human communication takes place not only among close kin or cooperators but also with competitors and strangers. Lying and deception are in everybody's repertoire. Even children start expecting and practicing some degree of deception around the age of four.[13]

Humans stand to gain immensely from the communication of others. Without communicating a lot, they couldn't even become competent adults or lead a normal human life. As a result of their dependence on communication, humans incur a commensurate risk of being deceived and manipulated. How, notwithstanding this risk, could communication have evolved into such a central, indispensable aspect of human life?

Communication is a special form of cooperation. The evolution of cooperation in general poses, as we saw in Chapter 9, well-known problems. It might seem reasonable to expect that theories that explain the evolution of human cooperation might also explain the evolution of communication.[14] Do they really?

Actually, communication is a very special case. In most standard forms of cooperation, cheating may be advantageous provided one can get away with it. For any given individual, doing fewer house chores, loafing at work, or cheating on taxes, for example, may, if undetected, lower the costs of cooperation without compromising its benefits. This being so, cooperation can evolve and endure only in certain conditions—in particular, when organized surveillance and sanctions make cheating, on average, costly rather than profitable, or when the flow of information in society is such that cheaters put at risk their reputation and future opportunities of cooperation. For communication to evolve, however, the conditions that must be fulfilled differ considerably from this.

Although this is hard to measure and to test, there are reasons to think that communicators are spontaneously honest much of the time, even when they could easily get away with some degree of dishonesty. Why? Because when humans communicate, doing so honestly is, quite commonly, useful or even necessary to achieve their goal. People communicate to coordinate with others, to organize joint action, to ask help from friends who are happy to give it. In such situations, deceiving one's interlocutors, even if undetected, wouldn't be beneficial; it would just be dumb.

Still, in other situations that are also rather common, a degree of deception may be advantageous. People generally try to give a good image of themselves. For this, even people who consider themselves quite honest routinely exaggerate their virtues and achievements, use disingenuous excuses for their failings, make promises they are not sure to keep, and so on. Couples who most of the time may be cooperative and honest with one another lie about their infidelities. Then there are cases where candid communication is generally seen as just incompetent—these include politics, advertisement, propaganda, and some businesses where honesty results in too many missed opportunities. Being lied to may be less frequent than being told the truth, but when it happens, it can be quite costly.

On the one hand, even the most sincere people cannot be trusted to always be truthful. On the other hand, people who have no qualms lying still end up telling the truth rather than lying on most occasions, not out of honesty but because it is in their interest to do so. Hence, standard ways of controlling or excluding cheaters might not work so well in the special case of liars. An ad hoc version of a tit-for-tat strategy[15]—you lie to me, I lie to you; or you lie to me, I stop listening to you—might harm us as much or more than it would punish liars. An ad hoc version of a "partner choice" strategy[16]—you lie to me, I cease listening to you and listen to others—would, if applied systematically, deprive of us of much useful information.

More generally, to benefit as much as possible from communication while minimizing the risk of being deceived requires filtering communicated information in sophisticated ways. Systematically disbelieving people who have been deceitful about some topic, for instance, ignores the fact that they may nevertheless be uniquely well informed and often reliable about other topics. Automatically believing people who have been reliable in the past ignores the fact that, on some new topic where it is in their interest to deceive us, they might well do so. Well-adjusted trust in communicated information must take into account in a fine-grained way not just the past record of the communicator but also the circumstances and contents of communication.

Epistemic Vigilance

Communication is, on the one hand, so advantageous to human beings and, on the other hand, makes them so vulnerable to misinformation that there must have been, we have suggested in earlier work,[17] strong and ongoing pressure for developing a suite of mechanisms for "epistemic vigilance" geared at constantly adjusting trust. There is no failsafe formula to calibrate trust exactly, but there are many relevant factors that may be exploited to try to do so. Reaping the benefits of communication without becoming its victim is so important that any mechanism capable of making a cost-effective contribution to the way we allocate trust is likely to have been harnessed to the task.

The mechanisms involved in epistemic vigilance are of two kinds. Some focus on the source of information and help answer the question: Whom to believe? Other mechanisms focus on content and help answer the question: What to believe?

Whom should we believe, when, and on what topic and issue? We accept different advice from a doctor or a plumber. We believe more easily witnesses who have no personal interests at stake. We are on alert when people hesitate or, on the contrary, insist too much. We take into account people's reputation for honesty and competence. And so on. Much recent experimental work shows, moreover, that children develop, from an early age, the ability to take into account evidence of the competence and benevolence of communicators in deciding whom to trust.[18]

Trust in the source is not everything. Not all contents are equally believable. No one could convince you that 2 + 2 = 10, that the moon is a piece of cheese, or that you are not yet born. If you were told such things by the person you trust most in the world, rather than taking them literally and at face value, you would wait for an explanation (for example, that 2 + 2 = 10 in base four, that the moon looks like a piece of cheese, and that you are not yet "born again" in a religious sense). If, on the other hand, the person you trust least in the world told you that a square has more sides than a triangle, that Paris is the capital of France, or that to reach your current age you must have eaten a lot of food, you would agree.

Whatever their source, absurdities are unlikely to be literally believed and truisms unlikely to be disbelieved. Most information, however, falls between these two extremes; it is neither obviously true nor obviously false. What makes most claims more or less believable is how well they fit with what we already believe. If we realize that a claim is incoherent with what we already believe, we are likely to reject it. Still, rejecting a claim that challenges our beliefs may be missing an opportunity to appropriately revise beliefs of ours that may have been wrong from the start or that now need updating.

Vigilance toward the source and vigilance toward the content may point in different directions. If we trust a person who makes a claim that contradicts our beliefs, then some belief revision is unavoidable: If we accept the claim, we must revise the beliefs it contradicts. If we reject the claim, we must revise our belief that the source is trustworthy.

While it is far from failsafe, epistemic vigilance allows us, as audience, to sort communicated information according to its reliability and, on average, to stand to benefit from it. For communicators, on the other hand, the vigilance of their audience diminishes the benefits they might expect from deceiving others—it lowers the chances of success—and if they were caught, it increases the costs they might have to pay in the form of loss of credibility and reputation.

Shouldn't we be vigilant not just toward information communicated by others but also toward the output of our own belief-formation mechanisms? When it comes to beliefs that result from our own cognitive processes, we are the source. Our cognitive mechanisms have evolved to serve us. Unlike other people, these mechanisms do not really have interests of their own. Moreover, there already are, we saw, metacognitive devices that modulate the confidence we have in our own memories, perceptions, and inferences. So, there is little reason to assume that it would be advantageous to apply further vigilance toward the source when we ourselves are the source.

So be it for vigilance toward the source, but what about vigilance toward the content? Wouldn't it be useful to check the coherence of new perceptions and inferences with old beliefs so as better to decide what new beliefs to accept or reject and, possibly, what old beliefs to revise? Actually, coherence checking is not a simple and cheap procedure, nor is it error-proof. Moreover, if it were worth doing, shouldn't its output itself be double- and triple-checked, and so on? If you are willing to distrust yourself, shouldn't you distrust your own distrust? We know of no evidence showing that humans exercise vigilance toward the content of their own perceptions and inferences (except in special cases where they have positive grounds to think that their senses or intuitions aren't functioning well). We know of no good argument, either, to the effect that doing so would be truly beneficial. More generally, people seem to be unconcerned by their own incoherencies unless something—generally someone—comes up to make them salient.

The Argumentative Function of Reason

Precautions have a price. Even when they are on the whole beneficial, they result in missed opportunities. You were wise, for instance, not to pick these mushrooms in the woods since you were not sure they were edible. You were wise even though it so happens that they were delicious chanterelles. Epistemic vigilance is a form of precaution, and it has a price in the same way. A valuable message may be rejected because we do not sufficiently trust the messenger, a clear case of missed opportunity.

Epistemic vigilance, however useful or even necessary it may be, creates a bottleneck in the flow of information. Still, to receivers of information, the benefits of well-calibrated epistemic vigilance are greater than the costs. To communicators, on the other hand, the vigilance of their audience seems just costly. This vigilance stands in the way not only of dishonest communicators but also of honest ones. An honest communicator may be eager to communicate true and relevant information, but she may not have sufficient authority in the eyes of her interlocutor for him to accept it, in which case they both lose—she in influence, he in relevant knowledge.

The more potentially relevant the message you receive from a given source, the more vigilant you should be, but the more costly this vigilance may turn out to be. Here is a dramatic historical illustration. Richard Sorge was a Soviet spy working undercover in Japan during World War II and providing the Russians with much valued intelligence. When, however, he informed them of the imminent invasion of Russia by Nazi Germany in June 1941, Stalin just dismissed his report: "There's this bastard who . . . deigned to report the date of the German attack as 22 June. Are you suggesting I should believe him?"[19] With hindsight, what a mistake! But precisely because the information was both so important if true and so damaging if false but taken to be true, it made sense to be particularly vigilant and indeed skeptical. There are comparable situations in our personal lives where, for instance, we remain skeptical of a warning or a promise precisely because it is too relevant to be accepted on trust.

This is where reasoning has a role to play. The argumentative use of reasons helps genuine information cross the bottleneck that epistemic vigilance creates in the social flow of information. It is beneficial to addressees by allowing them to better evaluate possibly valuable information that they would not accept on trust. It is beneficial to communicators by allowing them to convince a cautious audience.

To understand how argumentation helps overcome the limits of trust, consider what happens when communicators make claims without ever arguing for them. People in a position of power, for instance, typically expect what they say to be accepted just because they say it. They are often wrong in this expectation: they may have the power to force compliance but not to force conviction. Statements of opinion unsupported by any argument are also common in some cultural milieus where arguing, and generally being "too clever," is discouraged. Intelligence, in all the senses of the term, doesn't flourish in such conditions. In both situations, the flow of information is hampered. What argumentation can do, when it is allowed, is ease this flow.

As communicators, we are addressing people who, if they don't just believe us on trust, check the degree to which what we tell them coheres with what they already believe on the issue. Since we all are at times addressees, we are in a position to understand how, when we communicate, our audience evaluates what we tell them. We benefit from this understanding and adjust our messages accordingly. Unless we are trying to deceive our audience, we needn't see their vigilance as just an obstacle to our communicative goals. On the contrary, we may be in a position to use their vigilance in a way that will be beneficial both to them and to us.

To begin with, even if our audience is reluctant to accept our main point, there may be relevant background information that they will accept from us on trust. By providing such information, we may extend the common ground on the basis of which our less evident claims will be assessed. For instance,

(The doorbell is ringing)
Enrico: It must be Alicia or Sylvain.
Michelle: Actually, Sylvain is out of town. I am sure it is Alicia.
Enrico: Right!

Here Michelle produces a piece of information that Enrico believes on trust: Michelle wouldn't say that Sylvain is out of town if she didn't know it for a fact. This piece of information, however, is a reason for Enrico to revise his conjecture about who might be ringing the doorbell.

A good way to convince addressees is to actively help them check the coherence of your claims with what they already believe (including what they have just unproblematically accepted from you) or, even better if possible, to help them realize that given their beliefs, it would be less coherent for them to reject your claims than to accept them. In other words, as a communicator addressing a vigilant audience, your chances of being believed may be increased by making an honest display of the very coherence your audience will anyhow be checking. A good argument consists precisely in displaying coherence relationships that the audience can evaluate on their own.

As an addressee, when you are provided not just with a claim but also with an argument in its favor, you may (intuitively or reflectively) evaluate the argument, and if you judge it to be good, you may end up accepting both the argument and the claim. Your interlocutor's arguments may be advantageous to you in two ways: by displaying the very coherence that you might have to assess on your own, it makes it easier to evaluate the claim, and if this assessment results in your accepting relevant information, it makes communication more beneficial.

How does a communicator display coherence? She searches among the very beliefs her addressee already holds (or will accept from her on trust) for reasons that support her claim. In simple cases, this may involve a single argumentative step, as in this dialogue between Ben and Krisha, two neighbors of Tang, Julia, and Mary:

Krisha: Tang just called. He is inviting us to come over to their place for dinner with him, Julia, and Mary.
Ben: Mary is there? Then she must have finished the essay she had to write.
Krisha: I would be surprised if she had.
Ben: Surely, if she hadn't finished her essay, she would be working late in the library.
Krisha: But you told me yourself this morning that the library would close early today.
Ben: Ah, yes, I had forgotten. So, you're right—Mary might still not have finished her essay after all but be at home all the same.

Here Krisha uses a piece of information that Ben himself had provided (that the library is closed that evening) as a reason to cast doubt on his claim that Mary must have finished her essay. Krisha's argument shows to Ben that he holds mutually incoherent views, one of which at least he should revise.

Making the addressee see that it would be more coherent for him to agree than to disagree with the communicator's claim may involve several argumentative steps, as in this illustration based on the pigeonhole problem we encountered in Chapter 1 (and which, for once, involves genuine deductive reasoning):

Myra: Since you like puzzles, Boris, here is one. Let me read: "In the village of Denton, there are twenty-two farmers. All of the farmers have at least one cow. None of the farmers have more than seventeen cows. How likely is it that at least two farmers in Denton have the exact same number of cows?" So, Boris, what do you say?
Boris: Well, I say it is likely.
Myra: I say it is certain!

Myra and Boris have come to different conclusions, and neither has the authority to persuade the other just by saying, "It is so." Still, where authority fails, argument may succeed.

Myra: Well, imagine you go to the village, gather the twenty-two farmers, and ask them to stand in groups, each group made of farmers who have exactly the same number of cows.

Boris: How does that help? It could be that there are no farmers who have exactly the same number of cows, and then, in each of your "groups," there would be only one farmer.

Boris's reply has an entailment that he is not yet aware of but that Myra highlights:

Myra: This is impossible. Since none of the farmers have more than seventeen cows, there couldn't be more than seventeen groups: a group for farmers with one cow, a group for farmers with two cows, and so on up to a group for farmers with seventeen cows. But then, given that there are twenty-two farmers and at most seventeen groups, there would be at least one group with several farmers, right?

Boris: Yes, and?

Myra: And since farmers in the same group would have the same number of cows, then it must be the case that at least two farmers have exactly the same number of cows. This is certain, as I said, not merely likely.

Boris: You are right. I see it now.

What Myra does is present Boris with intuitive arguments that spell out some clear implications of the situation described in the puzzle. The conclusion of the last intuitive argument, however, directly contradict Boris's initial conclusion. When he realizes this, it is more coherent for him to agree with Myra and change his mind.

The dialogue between Myra and Boris is a trivialized version of the Socratic method: help your interlocutors see in their own beliefs reasons to change their views. When reflection on reasoning began in the Western tradition in the work of Plato and Aristotle, the argumentative use of reasons to try to convince an interlocutor, a court, or an assembly was seen as quite central. The social aspect of reasoning was well in evidence. Socratic reasoning could be seen as reasoning par excellence. Then, however, starting already in the work of Aristotle, the study of reasoning took a different turn.

Important normative questions about what makes an argument valid led to a more abstract approach to reasoning and to the development of logic proper. The use of reasons in argumentation could now be seen as just one practical application among others of more general reasoning principles. A new image of the typical reasoner emerged. Rather than Socrates trying to convince his interlocutor and the interlocutor understanding the force of Socrates's argument, the paradigm of a reasoner became the scientist reasoning on his own (more rarely on *her* own) to arrive at a better understanding of the world. From the point of view of the psychology of reasoning, this has been unfortunate: it has obscured the degree to which reasoning (including scientific reasoning) is a social activity, and the degree to which it is based on intuitions.[20]

Reasons are not arbitrary rhetorical devices. If they were, how would they have any force? Reasons are supported by intuitions that are themselves based on genuine cognitive competencies. Intuitively good reasons are more likely to support true conclusions. Imagine for instance that instead of Myra trying to convince Boris, Boris would have tried to convince Myra that his answer was the right one. How likely would he have been to succeed? Not very. His reasons would have been intuitively wanting. In fact, in trying to formulate them, he might himself have become aware of their inadequacy. In this case, what Myra and Boris are disagreeing about is a logical problem with a demonstrable solution. Even when the question on which two people disagree does not have a demonstrably true answer, there may be intuitively much better argument for one answer than for another. Then, of course, there are cases where two mutually incompatible conclusions can each be supported by plausible but inconclusive reasons. In such cases, argumentation by itself typically fails to change people's minds.

We construct arguments when we are trying to convince others or, proactively, when we think we might have to. We evaluate the arguments given by others as a means—imperfect but uniquely useful all the same—of recognizing good ideas and rejecting bad ones. Being sometimes communicators, sometimes audience, we benefit both from producing arguments to present to others and from evaluating the arguments others present to us. Reasoning involves two capacities, that of producing arguments and that of evaluating them. These two capacities are mutually adapted and must have evolved together. Jointly they constitute, we claim, one of the two main functions of reason and the main function of reasoning: the argumentative function.[21]

Main Functions, Secondary Functions, and Sea Turtles

The central thesis of the book is that human reason has two main functions corresponding to two main challenges of human interaction: the attribution of reasons serves primarily a justificatory function, and reasoning serves primarily an argumentative function.

Why do we say the two *main* functions rather than just the two functions? Frankly, out of prudence. We are not convinced that the reason module has any other function than these two, but we want to leave the possibility open.

We have not denied, for instance, that attributing reasons to an individual may be done for explanatory rather than justificatory purpose. One may attribute reasons without assessing their justificatory value. This is, after all, what historians do when they seek to explain the actions of people from the past in a nonjudgmental way. Couldn't such explanatory use be beneficial? Couldn't it be at least a secondary function of the attribution of reasons?

We have not denied that reasoning can be pursued individually, either to produce arguments aimed at convincing others or to evaluate others' arguments. Couldn't such individual inquisitive reasoning be beneficial? Couldn't producing such benefits be, if not the main function, at least a secondary function of reasoning?

Well, possibly, but to better understand what it would take to establish such claims, consider sea turtles.

Although sea turtles are descendants of land reptiles, they spend all their time in the water, or almost all their time. Females lay eggs out of the water, in nests they dig on the beach, where therefore all sea turtles are born (and then try to get into the water as fast as possible).

The limbs of sea turtles have become exquisitely adapted to life in the sea: their forelimbs have evolved into flippers and their hind limbs into paddles or rudders. What could be better for swimming, which is the ordinary mode of locomotion of sea turtles? On repeated occasions, however, females must use these same limbs to crawl out of the water and to dig nests in the sand. The use of limbs in such a manner, even if rare and clumsy, plays a direct role in reproduction and is clearly adaptive. To argue conclusively, however, that digging nests is a secondary function of sea turtle limbs, one would have to find features of these limbs that are best explained by this use. Otherwise, this might be best described as a beneficial side effect.[22]

Female turtles' nest digging is obviously beneficial, even if the limbs they use are not specifically adapted to the task and if their performance is notably clumsy. Similarly, if reason has beneficial effects different from its two main functions, these might correspond to secondary functions. If so, we would expect to find features of reason tailored to the achievement of these benefits. Otherwise, we might suspect that these are mere side effects of reason.

It is plausible that the capacity to explain people's ideas and actions by attributing to them reasons is on the whole beneficial (even if, as we saw, it is at best a distortion of the psychological processes involved). It is less obvious but not inconceivable that genuine individual inquisitive reasoning (as opposed to mentally simulated argumentation) does more good than harm in guiding beliefs and decisions. Still, even assuming that these two uses of reason are each, on average, beneficial, it does not follow that they are, properly speaking, functions of reason. To argue that they are, one would have to find some specific features of the attribution of reasons on the one hand and of reasoning on the other hand that are geared to the production of these particular benefits. We are not aware of any such evidence. In the case of reasoning, what is more, it is not just that features specifically tailored to the fulfillment of its alleged inquisitive function are lacking; it is that several well-evidenced features of reasoning would actually undermine this function.

We should keep an open mind regarding possible secondary functions of reason, of course, but at present, the challenge is to establish what beneficial effects would explain why reason evolved at all—in other terms, it is to identify reason's primary function or functions. In trying to answer this challenge, we denied that anything like classical Reason, with the capacity to procure better knowledge and decisions in all domains, has ever evolved. What has evolved rather is a more modest reason module—one intuitive inference module among many—specialized in producing intuitions about reasons in the service of two functions, justificatory and argumentative.

In Parts IV and V of this book, we will demonstrate, with evidence rich and varied, that reason is precisely adapted for fulfilling these two functions. We will show, moreover, how this new interactionist approach illuminates the role reason plays in human affairs.

IV

What Reason Can and Cannot Do

In Chapters 7 through 10, we have developed a novel interactionist approach to reason. According to this approach, the function of reason is to produce and evaluate justifications and arguments in dialogue with others. For the standard intellectualist approach, the function of reason is to reach better beliefs and make better decisions on one's own. As we'll see, the two approaches yield sharply contrasting predictions that we are now in a position to test. Is reason objective or biased? Is it demanding or lazy? Does it help the lone reasoner? Does it yield better or worse results in interactive contexts? Looking at the evidence with these questions in mind throws new light on the promises and dangers of reason.

11

Why Is Reasoning Biased?

There is much misunderstanding about the way to test adaptive hypotheses. It might seem that in order to understand what reason is for, why it evolved, we must be able to find out how our ancestors reasoned, using archeological or genetic data. Fortunately, million-year-old skulls and gene sequencing are not indispensable to test hypotheses about the function of an evolved mechanism. Think about it: we can tell what human eyes are for without knowing anything about when and how they evolved. What matters is the existence of a match between the function of an organ, or a cognitive mechanism, and its structure and effects. Do the features of the eye serve its function well? By and large, yes. Do eyes achieve their function well? By and large, yes.

We can use the same logic to guide our examination of the data on human reason. Do the features of human reason serve best the functions posited by the intellectualist or the interactionist approach? Which functions does it achieve best? In many cases reason couldn't serve both functions well at the same time, so there will be plenty of evidence to help us decide between the two approaches. We'll start our tour of what reason can and cannot do with a historical case. How does a certified scientific genius reason? Does reason help him discard misguided beliefs and reach sounder conclusions?

When Linus Pauling received the American Chemical Society's Award in Pure Chemistry—he was only thirty years old—a senior colleague predicted that he would win a Nobel Prize.[1] Actually, Pauling won *two* Nobel Prizes—Chemistry and Peace—joining Marie Curie in this exclusive club. As a serious contender in the race to discover the structure of DNA, he narrowly missed a third Nobel Prize. Indeed, James Watson was long afraid to be beaten by Pauling's "prodigious mind."[2] A former student who would also become a Nobel laureate described him as a "god-like, superhuman, great figure."[3] Undoubtedly, Linus Pauling had a great mind, and he was steeped in the most rigorous scientific tradition.

When he was not busy winning Nobel Prizes, Pauling sometimes pondered the powers of vitamin C. At first, he advocated heavy daily doses as a prophylactic against colds and other minor ailments, stopping short of recommending vitamin C for the treatment of serious diseases. This changed when he met Ewan Cameron, a surgeon who had conducted a small study demonstrating the positive effects of vitamin C on cancer patients—or so it seemed. Pauling and Cameron joined forces to defend the potential of vitamin C as a treatment for cancer. Thanks to Pauling's clout and to his continuous efforts, the prestigious Mayo Clinic agreed to conduct a large-scale, tightly controlled trial.[4] Unfortunately for Pauling and Cameron—and for the cancer patients—the results were negative. Vitamin C had no effect whatsoever.[5]

At this point, Pauling could have objectively reviewed the available evidence: on the one hand a fringe theory and a small, poorly controlled study; on the other hand, the medical consensus and a large, well-controlled clinical trial. On the basis of this evidence, the vast majority of researchers and doctors concluded that vitamin C had no proven effects on cancer. But Pauling did not reason objectively. He built a partisan case. The first Mayo Clinic study was dismissed because its participants had, according to Pauling, not been selected properly. When a second study was performed that addressed this issue,[6] Pauling made up another problem: the new patients had not received vitamin C for an "indefinite time."[7] Pauling's requirements did not match any standard cancer research, only fitting the small trial Cameron had performed.

The most egregious example of biased reasoning on Pauling's part is found in an article, published a few years later, in which he advanced three criteria to evaluate clinical trials of cancer treatments.[8] While "most of the reported results of clinical trials of cohorts of cancer patients satisfy these criteria of validity,"[9] there was one black sheep, "a reported clinical trial that fails on each of the three criteria for validity."[10] Can you guess what this outlier was? A study "*described* as a randomized double-blind comparison of vitamin C (10 g per day) and a lactose placebo" (emphasis added).[11] The study singled out as the only flawed cancer study in a sample of several hundred is the

Mayo Clinic study that embarrassed Pauling in front of the whole scientific community.

There is no reason to accuse Pauling of conscious intellectual dishonesty. He took high doses of vitamin C daily, and his wife used the same regimen to fight—unsuccessfully—her stomach cancer. Still, for most observers, Pauling's evaluation of the therapeutic efficacy of vitamin C is a display of selective picking of evidence and partisan arguments. Even when he was diagnosed with cancer despite having taken high doses of vitamin C for many years, he did not admit defeat, claiming that the disease would have struck earlier without it.[12]

Pauling may have erred further than most respected scientists in his unorthodox beliefs, but his way of reasoning is hardly exceptional—as anyone who knows scientists can testify, they are not paragons of objectivity (more on this in Chapter 18). Undoubtedly, even the greatest minds can reason in the most biased way.

Is Bias Always Bad?

How should cognitive mechanisms in general go about producing sound beliefs? Part of the answer, it seems, is that they should be free of bias. Biases have a bad press, in part because a common definition is "inclination or prejudice for or against one person or group, especially in a way considered to be unfair."[13] However, psychologists often use the term in a different manner, to mean a systematic tendency to commit some specific kind of mistakes. These mistakes need not have any moral, social, or political overtones.

A first kind of bias simply stems from the processing costs of cognition. Costs can be reduced by using heuristics, that is, cognitive shortcuts that are generally reliable but that in some cases lead to error. A good example of this is the "availability heuristic" studied by Tversky and Kahneman.[14] It consists in using the ease with which an event comes to mind to guess its actual frequency. For instance, in one experiment participants were asked whether the letter *R* occurs more frequently in first or third position in English words. Most people answered that *R* occurs more often in first position, when actually it occurs more often in third position. The heuristic the participants used was to try to recall words beginning with *R* (like "river") and words with *R* in the third place (like "bored") and assume that the ease with which the two kinds of words came to mind reflected their actual frequency. In the particular case of the letter *R* (and of seven other consonants of the English alphabet), the availability heuristic happens to be misleading. In the case of the other thirteen consonants of the English alphabet, the same heuristic would give the right answer.

Although the availability heuristic can be described as biased—it does lead to systematic errors—its usefulness is clear when one considers the alternative: trying to count *all* the words one knows that have *R* as the first or third letter. While the heuristic can be made to look bad, we would be much more worried about a participant who would engage in this painstaking process just to answer a psychology quiz. Moreover, the psychologist Gerd Gigerenzer and his colleagues have shown that in many cases using such heuristics not only is less effortful but also gives better results than using more complex strategies.[15] Heuristics, Gigerenzer argues, are not "quick-and-dirty"; they are "fast-and-frugal" ways of thinking that are remarkably reliable.

The second type of bias arises because not all errors are created equal.[16] More effort should be made to avoid severely detrimental errors—and less effort to avoid relatively innocuous mistakes.

Here is a simple example illustrating how an imbalance in the cost of mistakes can give rise to adaptive biases. Bumblebees have cognitive mechanisms aimed at avoiding predators. Among their predators are crab spiders, small arachnids that catch the bees when they forage for nectar. Some crab spiders camouflage themselves by adopting the color of the flowers they rest on: they are cryptic. To learn more about the way bumblebees avoid cryptic predators, Thomas Ings and Lars Chittka created little robot spiders.[17] All the robots rested on yellow flowers, but some of them were white (noncryptic) while others were yellow (cryptic). To simulate the predation risk, Ings and Chittka built little pincers that held the bees captive for two seconds when they landed on a flower with a "spider."

In the first phase of the experiments, two groups of bumblebees, one facing cryptic spiders and the other facing noncryptic spiders, had multiple opportunities to visit the flowers and to learn which kind of predators they were dealing with. Surprisingly, both groups of bumblebees very quickly learned to avoid the flowers with the spiders—even when the spiders were cryptic.

Yet the camouflage wasn't ineffective: to achieve the same ability to detect spiders, the bumblebees facing camouflaged spiders spent nearly twice as long inspecting each flower. This illustrates the cost of forming accurate representations of one's environment: the time spent inspecting could not be spent harvesting.

But there is also an asymmetry in the costs of mistakenly landing on a flower with a spider (high cost) versus needlessly avoiding a spider-free flower (low cost). This second asymmetry also affected the bumblebees' behavior. On the second day of the experiment, the bees had learned about the predation risks in their environment. Instead of spending forever (well, 0.85 seconds) inspecting every flower to make sure it carried no spider, the bumblebees facing the cryptic spiders settled for a higher rate of "false alarms": they were more likely than the other bees to avoid flowers on which, in fact, there were no spiders.

This experiment illustrates the exquisite ways in which even relatively simple cognitive systems adjust the time and energy they spend on a cognitive task (ascertaining the presence of a spider on a flower) to the difficulty of the task on the one hand, and to the relative cost of a false negative (assuming there is no spider when there is) and of a false positive (assuming there is a spider when there is not) on the other hand. This difference in cost results in a bias: making more false positive than false negative errors. This bias, however, is beneficial.

From Surprise to Falsification

While the example of the bumblebees illustrates that bias may be beneficial, it speaks of very specific costs: the costs of predation versus the costs of missing a feeding opportunity. Are there cognitive biases that would be useful more generally?

A main goal of cognitive mechanisms is to maintain an accurate representation of the organism's environment, or at least of relevant aspects of it. It could be argued that only the future, more particularly the immediate future, of the environment is directly relevant: it determines both what may happen to the organism and what the organism can do to modify the environment in its favor. The past and present of the environment may be relevant too, but only indirectly, by providing the only evidence available regarding the environment's future. Another way of making the same point is to say that a main goal of cognition is to provide the organism with accurate expectations regarding what may happen next. Cognitive mechanisms should pay extra attention to any information that goes against their current expectations and use this information to revise them appropriately. Information that violates expectations causes, at least in humans, a typical reaction: surprise. The experience of surprise corresponds to the sudden mobilization of cognitive resources to readjust expectations that have just been challenged.

Indeed, paying due attention to the unexpected is so ingrained that we are surprised by the lack of surprise in others. If your friend Joe, upon encountering two cats that are similar except that one meows and the other talks, fails to pay more attention to the talking cat, you'll suspect there's something wrong with him. Even one-year-old babies expect others to share their surprise. When they see something surprising, they point toward it to share their surprise with nearby adults. And they keep pointing until they obtain the proper reaction or are discouraged by the adults' lack of reactivity.[18]

This doesn't mean that people seek out surprises. If you are lucky enough to find yourself in a nice environment where you know what to expect, deviations from these expectations—surprises—are likely to be bad news (hence the fake Chinese curse, "May you live in interesting times"). However, even if you don't like surprises, when the environment has something surprising in stock for you, you're better off learning about it as early as possible. Being fired is a very bad surprise; better be surprised by indications that you might be fired before it actually happens. As a result, people favor sources of surprising information over sources of repetitive information (provided they are equally reliable): they buy the newspaper with the proverbial "Man Bites Dog" headline, not "Dog Bites Man."

If reason is an instrument of individual cognition aimed at better beliefs and decisions, it should be biased toward information that violates our expectations. Reason should look for counterarguments to our generalizations, reasons to question our decisions, memories that clash with our current beliefs. Failing to pay attention to negative evidence and arguments is generally more costly than failing to pay attention to their positive counterparts.

This insight is so essential that it is one of the few themes that we find again and again in scholars' admonitions about the best way to reach sound beliefs. In the twelfth century, Robert Grosseteste suggested that to eliminate faulty hypotheses, one should take a hypothesis and see if anything follows from it that is illogical or contrary to fact.[19] Four centuries later, Francis Bacon overturned the scholastic tradition to which Grosseteste belonged by putting observation at the center of his philosophy. Still, we find again the idea of looking for counterexamples. Observation should not be the "childish" enumeration of cases that fit with the researcher's hypothesis. Scholars are to look specifically for instances that can prove their hypothesis wrong: one must "analyse nature by proper rejection and exclusion."[20]

Another four centuries later, it is Karl Popper's turn to challenge dominant views of the day, distant descendants of Bacon's philosophy. Popper does so by stressing more than ever before the importance of falsification. For Popper, what demarcates science from other forms of knowledge is not that scientific theories are verifiable but that they are falsifiable: in principle, a single observation of an event that, according to a given theory, could never occur would falsify the theory. In practice, of course, as Popper well recognized, a single observation is not enough to cause scientists to abandon a theory—after all, the observation itself might be mistaken. Still, when falsifying observations are numerous and reliable enough, then the theory must at least be revised if not abandoned. Scientists produce theories that are at risk of being falsified. They improve their theories by looking for falsifying evidence, by rejecting falsified theories, and by holding on only to theories that have withstood repeated attempts at falsifying them.

For all their differences, these scholars agree that counterexamples and other violations of expectations, by allowing us to discard misguided beliefs, play a crucial role in the accumulation of knowledge. As Popper put it, "In searching for the truth, it may be our best plan to start by criticizing our most cherished beliefs."[21]

The Confirmation Bias

Peter Wason devised the clever experimental tasks that were to have such an impact on the psychology of reasoning at University College in London, within

In front of you are four cards. Each card has a letter on one side and a number on the other. Two cards have the letter side up; the two others have the number side up:

Which of these four cards *must* be turned over to find out whether the following rule is true or false of these four cards: *'if there is a vowel on one side of a card, then there is an even number on the other side'*?"

Figure 17. The standard Wason selection task.

walking distance of the London School of Economics where Popper was teaching. The proximity was more than geographic. Wason extended Popper's insight about scientific theories to everyday reasoning, asking: Do people rely on falsification to arrive at better beliefs?

The selection task, which was discussed in Chapter 2 (see Figure 17 for a reminder), was precisely meant to simulate aspects of scientific thinking. Participants, Wason thought, must select cards that provide the right kind of evidence to test the rule, just as scientist must identify the right kind of evidence to test their hypotheses. As we saw, the right selection of cards in this example is that of the card with the vowel E and that of the card with an odd number 7—these are the only two cards that might falsify the rule (if there is an odd number on the other side of the E, or a vowel on the other side of the 7). Most participants, however, fail to select the 7 card. Many select the card with the even number 2 presumably to see whether it has a vowel on the other side. But since the rule does not say that it should, this card is irrelevant to deciding whether the rule is true or false.

On the basis of this and other experiments, Wason suggested that participants suffered from a bias that would later be called "confirmation bias":[22] "The results do suggest . . . that even intelligent adults do not readily adopt a scientific attitude to a novel problem. They adhere to their own explanation with remarkable tenacity when they can produce confirming evidence for them."[23]

Actually, the experiment that convinced so many people that there is a confirmation bias is not such a straightforward example of this bias. Wason and many of his followers made a logical mistake in claiming that participants try to "verify" or "confirm" the rule instead of falsifying it: it is, in fact, exactly the same cards that can falsify the rule if it is false or verify it if it is true. In the preceding example, for instance, the E and the 7 cards could each, when turned over, reveal that the rule is false. If neither falsifies the rule, then the two same cards jointly prove that the rule is true. The quite common selection of the card with the even number 2 not only fails to falsify the rule, it also fails to confirm it. Selecting it is not a confirmation strategy at all. So people who succeed on the task do not show a greater disposition to falsify than to confirm, and people who fail do not show a greater disposition to confirm than to falsify.

If the choice of any given card cannot reveal whether people have a confirmation bias, the *way* they choose their cards can. People's selection of cards, we saw, is rapidly determined not by reason but by intuitions about the relevance of the different cards. After this initial, intuitive reaction, participants tend to reason long and hard about their choice. However, they do not reason as much about each of the four cards. By tracking participants' gaze, researchers established that participants spend all their time thinking just about the cards made relevant by the rule, the cards that they have already selected.[24] Moreover, when participants are asked to think aloud, it becomes clear that they mostly think of reasons supporting their intuitive choice.[25] Here's the real confirmation bias: instead of finding reasons for and against each card, participants find plenty of reasons supporting their initial card choice, neglecting reasons to pick other cards, or reasons to not pick the cards initially chosen.

We will soon argue that this confirmation bias is in fact best understood as a "myside bias," but let's first look at a small sample of the rich evidence demonstrating its existence.[26]

Deanna Kuhn, a pioneering scholar of argumentation and cognition, asked participants to take a stand on various social issues—unemployment, school failure, and recidivism. Once the participants had given their opinion, they were asked to justify it. Nearly all participants obliged, readily producing reasons to support their point of view. But when they were asked to produce counterarguments to their own view, only 14 percent were consistently able to do so, most drawing a blank instead.[27]

Ziva Kunda led participants to believe that extroverts are more likely to be successful than introverts. On a later memory task, these participants found it much easier to recall memories of their own extroverted rather than introverted behavior. Another group of participants was led to believe that it is introverts who are more likely to be successful than extroverts. They found it easier to recall memories of introverted rather than extroverted behavior. Both groups were simply seeking reasons to believe that they had the qualities that would make them successful.[28]

Charles Taber and Milton Lodge gave participants a variety of arguments on controversial issues, such as gun control or affirmative action, and asked them to list their thoughts relative to the arguments.[29] They divided the participants into two groups: those with low and those with high knowledge of political issues. The low-knowledge group exhibited a solid confirmation bias: they listed twice as many thoughts supporting their side of the issue than thoughts going the other way. But knowledge did not protect the participants from bias. The participants in the high-knowledge group found so many thoughts supporting their favorite position that they gave *none* going the other way. Greater political knowledge only amplified their confirmation bias.

The list could go on for pages (indeed for chapters or books, even). Moreover, as the example of Pauling suggests, it is not only ordinary participants who fall prey to the confirmation bias. Being gifted, focused, motivated, or open minded is no protection against the confirmation bias.[30] A small industry of experiments has busily demonstrated the prevalence and robustness of what is "perhaps the best known and most widely accepted notion of inferential error to come out of the literature on human reasoning."[31] As the journalist Jon Ronson quipped, "Ever since I learnt about confirmation bias I've started seeing it everywhere."[32]

A Challenge for the Intellectualist Approach

Psychologists agree that the confirmation bias is prevalent. They also agree that it is a bad thing. The confirmation bias is "irrational,"[33] and it "thwart[s] the ability of the individual to maximize utility."[34] It is the "bias most pivotal to ideological extremism and inter- and intragroup conflict."[35] Raymond Nickerson aptly summarizes the common view:

> Most commentators, by far, have seen the confirmation bias as a human failing, a tendency that is at once pervasive and irrational. It is not difficult to make a case for this position. The bias can contribute to delusions of many sorts, to the development and survival of superstitions, and to a variety of undesirable states of mind, including paranoia and depression. It can be exploited to great advantage by seers, soothsayers, fortune tellers, and indeed anyone with an inclination to press unsubstantiated claims. One can also imagine it playing a significant role in the perpetuation of animosities and strife between people with conflicting views of the world.[36]

A damning assessment indeed. Moreover, this bad thing, far from being hidden in the recesses of human psychology, is well in view. It doesn't take a very shrewd or cynical observer of human nature to realize that humans have a confirmation bias. Why did Bacon, for instance, take such care to warn against the dangers of the "childish" enumeration of instances? Because he was well aware of people's tendency to confirm their beliefs:

> The human understanding when it has once adopted an opinion . . . draws all things else to support and agree with it. And though there be a greater number and weight of instances to be found on the other side, yet these it either neglects and despises, or else by some distinction sets aside and rejects; in order that by this great and pernicious predetermination the authority of its former conclusions may remain inviolate.[37]

Bacon, writing more than two centuries before Darwin's discovery of natural selection, could merely observe that people have a confirmation bias and that this bias hinders the acquisition of sound beliefs. For contemporary defenders of the intellectualist approach to reason who take it for granted that reason is an outcome of Darwinian natural selection, the very existence of the confirmation bias presents a radical challenge.

In particular, a defender of the intellectualist approach should accept the following three claims: (1) reason has a confirmation bias, (2) the confirmation bias makes it harder for reason to help the lone reasoner arrive at better beliefs and better decisions, and (3) the main function of reason is to arrive at better beliefs and better decisions. This makes as much sense as accepting the three following claims: (1) the elk's antlers are enormous, (2) the size of these antlers makes it harder for the elks to avoid predators, and (3) the function of the enormous antlers is to avoid predators.

The only ways out of this conundrum that come to mind are nonstarters. A first way out would be to argue that the confirmation bias is an unavoidable feature of reason. For instance, the extra weight that the bones, muscles, and feathers of avian wings add to the body of a bird makes it harder, not easier, for that body to fly. Yet since neither weightless wings nor wingless flight is an option, the weight of wings does not in itself raise a problem for evolutionary theory. But why would having a confirmation bias be necessary for reason to function at all?

A second escape route would be to claim that the confirmation bias serves a secondary function. Some features of an adaptation may hinder its main function but be well explained by the fact that they serve a secondary function. Razorbills, for instance, have a wing area relative to body mass ("wing loading") that is too low for optimal flight. This feature, however, is explained by the fact that their wings are also adapted for underwater propulsion.[38] Unlike razorbills' low wing loading, the confirmation bias doesn't just make reason's alleged main function a bit harder to achieve; it works directly against it. Moreover, there is no particularly plausible secondary function of reason that would explain the confirmation bias the way underwater propulsion explains low wing loading in razorbills. For the intellectualist approach, the confirmation bias should be a deep puzzle.

Can Intuitions Be Blamed for Confirmation Bias?

Supporters of the intellectualist view who are aware of how problematic the confirmation bias is still have one option: deny that the confirmation bias is a feature of reason proper, shifting the blame from reason to intuition. Indeed, when the dual process approach, briefly discussed in Chapter 2, became popular, one of its main attractions was that it seemed to help solve the enigma of reason: weaknesses of reasoning could now be blamed on type 1 intuitive processes, while type 2 reasoning proper could be exonerated. Intuitions, it was claimed, make mistakes that it is one function of reason to correct.[39] Stanovich, for instance, in his 2004 book *The Robot's Rebellion,* listed the confirmation bias among other alleged cognitive weaknesses of intuition.[40] Similarly, Evans suggested that the confirmation bias resulted from a more general "bias to think about positive rather than negative information."[41]

This solution—blaming the intuitions—doesn't make much evolutionary sense. Mechanisms of intuitive inference guide our thoughts and actions; natural selection has honed some of these mechanisms for hundreds of millions of years. Our survival and reproduction very much depend on the quality of the information provided by intuitions. As we saw, some specific biases may on the whole be advantageous when they lower the costs of cognition or make less likely particularly costly kinds of mistake. The confirmation bias carries none of these advantages.

Unsurprisingly, then, no confirmation bias emerges from studies of animal behavior. A mouse bent on confirming its belief that there are no cats around and a rat focusing its attention on breadcrumbs and ignoring other foods to confirm its belief that breadcrumbs are the best food would not pass on their genes to many descendants. Foraging behaviors adapt to changing environments. Animals abandon food patches as soon as they expect to find better elsewhere.[42] Human intuitions are no worse than the inferences of other animals. Our ancestors passed on to us abilities for foraging and avoiding predators, as well as a great variety of other inferential devices that do not suffer from a confirmation bias.[43]

If anything, as we argued earlier, we should expect intuitions to be biased toward disconfirming, surprising information. An experiment conducted by psychologist Thomas Allen and his colleagues offers a nice demonstration of the respective biases of intuitions and reason.[44] The participants were shown a picture of an individual and two statements describing his behavior, and they were then asked to form an impression of this individual. Some participants saw the picture of "a young adult Black male wearing a black headband and dark sunglasses"[45] followed by two statements: "Swore at the salesgirl" and "Gave up his seat on the crowded subway to the elderly man." If you are familiar with the stereotypes of young black males in the United States, you will have figured out that the first statement was designed to comport with most participants' expectations, and the second to be surprising.

In order to test the role of intuition in impression formation, half the participants were stopped from using reason—by having to hold in mind a long string of digits, a task that monopolizes resources necessary for sustained reasoning. These participants, then, were guided by their intuitions. As we would have predicted, these participants paid more attention to the surprising statement. By contrast, participants who could reason paid more attention to the unsurprising statement. Intuitions aimed at gathering the most useful information while reasoning aimed at confirming the participants' stereotypes.

The Myside Bias—and What Is It For?

So far, we have taken for granted that the bias described was a *confirmation* bias, a bias to confirm whatever view one happens to be entertaining. However, some experiments reveal clearly that this is not a good description of what reasoning does. For instance, we saw earlier that participants have trouble finding counterarguments to their favorite theories. But when participants are asked to reason about ideas they disagree with, they easily find counterarguments.[46]

What these results—and many others[47]—show is that people have no general preference for confirmation. What they find difficult is not looking for counterevidence or counterarguments in general, but only when what is being challenged is their own opinion. Reasoning does not blindly confirm any belief it bears on. Instead, reasoning systematically works to find reasons for our ideas and against ideas we oppose. It always takes our side. As a result, it is preferable to speak of a *myside bias* rather than of a confirmation bias.[48]

This being cleared up, let's recap. A lot of evidence shows that reasoning has a myside bias. Reason rarely questions reasoners' intuitions, making it very unlikely that it would correct any misguided intuitions they might have. This is pretty much the *exact opposite* of what you should expect of a mechanism that aims at improving one's beliefs through solitary ratiocination. There is no obvious way to explain the myside bias from within the intellectualist approach to reasoning.

A strong, universal bias is unlikely to be a mere bug, a bad thing. Instead, it is more likely to be a useful feature. What goal could the myside bias serve? The analogy of the lone reasoner as scientist proposed by Wason does not fit, but another analogy does throw some light on this mysterious bias.

Cicero may be Western civilization's most illustrious lawyer, his writings on rhetoric echoing from the Roman senate down through the halls of medieval and modern universities. This is how, in *De Inventione,* he advised orators to conclude a speech:

> It will be serviceable both to run over the arguments which you yourself have employed separately, and also (which is a matter requiring still greater art) to unite the opposite arguments with your own; and to show how completely you have done away with the arguments which were brought against you. And so, by a brief comparison, the recollection of the hearer will be refreshed both as to the confirmation which you adduced, and as to the reprehension which you employed.[49]

In other words, when you want to convince someone, give only arguments that support your position or that counter the position you oppose. Cicero is bluntly advocating for the myside bias. And it makes complete sense. If a lawyer starts arguing against her client or for the other side, she'll soon be out of business.[50]

The lawyer analogy brings to mind a context in which persuasion is paramount and the myside bias makes obvious sense: when defending a point of view, the myside bias is a good thing.[51] *It is a feature, not a bug.* This fits with the prediction of the interactionist approach. If the function of reasoning, when it produces reasons, is to justify one's actions or to convince others, then it *should* have a myside bias.

Being Our Own Lawyers

If the myside bias is so ingrained in reason, why should Cicero have to remind us to have a myside bias? While the "reason is a lawyer" analogy can be illuminating, especially by contrast with the more typical analogy of reason as a scientist, it shouldn't be pushed too far. For instance, unlike typical reasoners, lawyers often argue for positions they personally do not endorse. Their "side" is that of the client who employs them. Finding arguments for a position we do not support, or even one we disagree with, is difficult. It takes skills and training. We may be lawyers, but only when it comes to defending beliefs and decisions we actually endorse.

There are other reasons not to take the lawyer analogy too far. Besides reason, many other cognitive mechanisms are at play when lawyers prepare their plea. When they consciously suppress arguments against their client or use such arguments to anticipate what the other party might say, they rely on strategic planning rather than only on ordinary reasoning. This type of rehearsed-in-advance consistency in the reasons presented goes beyond what can be expected of lay reasoners. Moreover, a lawyer is committed to defend her client come what may. On the other hand, it may be in the best interest of ordinary reasoners who, in spite of the myside bias, find or stumble upon counterarguments against their own views to take these counterarguments seriously and perhaps even to change their minds.

Most importantly, the lawyer analogy applies only to the *production* of arguments. During a court trial, each actor is ascribed a carefully defined role. To simplify things a bit, lawyers produce arguments, and judges and juries evaluate those arguments. To evaluate arguments, judges and juries also rely on reason, but they, unlike the lawyers, are not supposed to be biased. We will see in Chapter 12 how much the evaluation of arguments fits with the idealized picture of the disinterested judge.

In an adversarial trial, the two battling parties are locked in a zero-sum game: one side's win is the other side's loss. While this highlights the utility of the myside bias, it might also unnecessarily tie it to competitive contexts. In fact, even when people have a common stake in finding a good solution and are therefore engaged in a positive-sum game, having a myside bias may still be the best way to proceed.

Imagine two engineers who have to come up with the best design for a bridge. Whichever design is chosen, they will supervise the construction together—all they want is to build a good bridge. Ella favors a suspension bridge, Dick a cantilever bridge. One way to proceed would be for each of them to exhaustively look at the pros and cons of both options, weigh them, and rate them. They would then just have to average their ratings—no discussion needed, but a lot of research.

Alternatively, they can each build a case for their favored option. Ella would look for the pros of the suspension bridge and the cons of the cantilever; Dick would do the opposite. They would then debate which option is best, listening to and evaluating each other's arguments. To the extent that it is easier to evaluate arguments presented to you than to find them yourself, this option means less work for the same result: Ella and Dick each have to find only half as many arguments to thoroughly review the pros and cons of each option.

The myside bias doesn't turn argumentation into a purely competitive endeavor. Argumentation is a form of communication and is typically pursued cooperatively. At its best, the myside bias becomes a way of dividing cognitive labor. In Chapter 12, we will see a similar dynamic at play in the way reason evaluates, rather than looks for, arguments.

12

Quality Control: How We Evaluate Arguments

The Cicero we met in Chapter 11 recommending complete one-sidedness was a mere "boy," whose advice was "rough and incomplete." Many years later the Roman orator, rich of the "experience which [he] gained from so many and such important causes as [he has] pleaded,"[1] tells a different tale:

> If ever a person shall arise who shall have abilities to deliver opinions on both sides of a question on all subjects, after the manner of Aristotle, and, from a knowledge of the precepts of that philosopher, to deliver two contradictory orations on every conceivable topic, . . . and [who] shall unite with those powers rhetorical skill, and practice and exercise in speaking, he will be the true, the perfect, the only orator.[2]

Accumulating arguments for one side had become child's play for this veteran of many senate fights. To make a great case, an orator must combine rhetorical skills with the ability to anticipate and take into account potential counterarguments. While a speaker might still only present arguments for her side, her mind must be agile enough not to be blinded by the myside bias.

We claim that the myside bias makes sense when the goal is to justify one's ideas and actions or to convince others. But wouldn't justification or conviction be better served if reason allowed us also to find arguments against our side, even if only to refute them? A skilled orator spends time honing her argument—its content, its form, its delivery. From Cicero and Quintilian to contemporary speechwriters and spin doctors, massive efforts are expended to devise convincing arguments. If reasoning, as we claim, evolved to change others' minds, shouldn't it look for such well-crafted arguments?

Here, then, is a potential problem for our interactionist theory: a substantial amount of evidence—and a passing familiarity with Internet comments—shows that people are far from being natural Ciceros. In Chapter 11 we encountered Deanna Kuhn's study of argumentation demonstrating the difficulty most people have in finding counterarguments. The same study showed that even in support of their own point of view, people often give rather weak arguments. Asked about her opinion on school failure, one of Kuhn's participants identified poor nutrition as the main culprit. Here's an excerpt from the interview:

> *Experimenter:* If you were trying to convince someone else that your view is right, what evidence would you give to try to show this?
> *Participant:* The points that they get in school. The grades that they get in school to show . . .
> *Experimenter:* What would that show?
> *Participant:* That they are lacking something in their body. That the kids who were failing lack something in their body.[3]

As Kuhn points out, the participant "makes it clear that the existence of the phenomenon itself is sufficient evidence that it is produced by the cause the [participant] invokes."[4] Other participants offered explanations that were close to a restatement of their theory. Most could not think of what would constitute evidence supporting their ideas. Kuhn's bleak assessment of untrained participants' argumentative skills is shared by other eminent psychologists, such as Richard Nisbett and Lee Ross, who suggest that people are generally content with the first reason they stumble upon,[5] or David Perkins, who asserts that many arguments make only "superficial sense."[6]

That reasoning shouldn't always be able to home in on excellent, knockdown arguments is obvious enough: looking for good reasons is a costly, time-consuming business. There has to be a trade-off between the quality of the reasons and the effort put in finding them. However, what these psychologists claim is not that we find very good, even if imperfect, reasons but that we find quite superficial, weak reasons.

This is another problem for the intellectualist approach. If reason's function is to improve the lone reasoner's beliefs and decisions, it had better provide good reasons for them. In fact, however, people's criteria for their own reasons are pretty lax. So not only do people mostly find reasons that support their intuitions (the myside bias), they don't even make sure that the reasons are much good. It's hardly surprising that reason should by and large fail at correcting mistaken intuitions.

What about reason aimed at changing others' minds? Wouldn't people be more persuasive if they acted like lawyers, investing time and effort to anticipate potential counterarguments and find better arguments? Reason evolved to be used not in law courts but in informal contexts. Not only are the stakes of an everyday discussion much lower, but its course is also very different. A discussion is interactive: instead of two long, elaborate pleas, people exchange many short arguments. This back-and-forth makes it possible to reach good arguments without having to work so hard.

Interactive Argumentation Is Easier

Sociolinguists stress how much interlocutors help each other communicate effectively, for instance by providing constant feedback that they follow what the speaker is saying: all these "Hm hms," "Yeahs," nods, and so forth. Drawing on this tradition, the anthropologist and linguist Steven Levinson has argued that humans are endowed with an "interaction engine."[7] Our communicative abilities are tailored to the interactive context in which they naturally function. For instance, when we want to refer to someone, we often have a wide array of options: "Ms. Catherine Turk," "Ms. Turk," "Kate," "the head of the accounting department," and so forth. Which option is most appropriate in a given context depends, inter alia, on how well the interlocutor knows the person. Happily, if we use the wrong option, the miscommunication can easily be fixed:[8]

Michael: I had lunch with Kate.
Rob: Who?
Michael: Kate Turk.
Rob: I just arrived here. I don't think I know her.

Michael: Sorry. She's the head of the accounting department.
Rob: Ah, right.

Here Michael starts off with the most conventional way of referring to a familiar person in the United States: the first name. Rob's answers allow him to refine his description until he reaches the appropriate level.

Without using feedback from the interaction partners, deciding the most appropriate form to use would involve much reflection:

Michael, reflecting: Everybody here knows Kate. But Rob just arrived. However, Kate always makes an effort to introduce herself to everybody when they arrive. Rob got here two days ago, so he should know her. But I haven't seen Kate yesterday or the day before, so perhaps she was sick and didn't come. I'd better specify who she is just in case.

The solution people adopt—starting out with what seems to be the best option and, if necessary, refining it with the help of feedback—is the most economical. For communicators, being "lazy"—using the shortest form likely to be understood—is being smart.[9]

The feedback from the interlocutor is useful for at least two reasons. One reason is that it is the interlocutor's understanding that matters in the end, so it is easier to let her decide whether she understands or not. Another reason is that the interlocutor can do more than indicate she doesn't understand what the speaker means: she can actively guide the speaker's efforts, as Rob did when he told Michael why he doesn't know who Kate is.

Reasoners face a similar challenge: figuring out what is the best way to get their message across. They, too, could engage in elaborate guesswork to find the best argument in every situation. This strategy is both time and effort intensive, and far from being foolproof, as illustrated by the following exchange:

Sherlock Holmes meets his friend Watson at a coffee shop.

Holmes: My dear Watson, I didn't dare interrupt—you were in such charming company! I've been observing the two of you for a few minutes, and I must absolutely advise you to see this woman again!

She's a perfect match for you. You may say that she's younger than you, but the age difference is not that large. I know you have a thing for brunettes, and she's blond, but that should easily be overlooked. I also observed that while you were talking of personal matters, there were no signs of physical intimacy, but I'm persuaded that would come quickly enough. Watson, I'm sure this woman is perfect for you!

Watson: She's my sister.

Of course, Conan Doyle always rigged the situation so that his Holmes would not commit this kind of faux pas. Our Holmes, who is also fictional but a tad more realistic, was unable to guess that Watson's charming companion was his sister. Had he paused after exclaiming, "You were in such charming company!," Holmes would have been told that the lady was Watson's sister and would have avoided further embarrassment.

Feedback plays an important role even in simple forms of argumentation. Take this banal exchange:

Hélène: We should go to Isami; it's a good restaurant.
Marjorie: I don't know. I had Japanese last week already.
Hélène: But this one is very original.

Isami might be a great place for many reasons—its originality, but also the prices, the freshness of the fish, the crowd—but Hélène doesn't list them all at the outset. Instead, she offers a generic summary assessment: "it's a good restaurant." This first argument doesn't sway Marjorie, but she doesn't simply say "no." Instead, she provides a reason for her dissent: "I had Japanese last week already." Thanks to Marjorie's feedback, Hélène can tailor her next argument, pointing out the quality of Isami that is most likely to change Marjorie's mind. In another context, the exchange might have gone as follows:

Hélène: We should go to Isami; it's a good restaurant.
Marjorie: I don't know. I don't have much money at the moment, and Japanese restaurants can be quite pricy.
Hélène: But this one is quite cheap.

Among all the arguments Hélène could have put forward at the outset, some are much more likely to convince Marjorie than others. Hélène could have tried to anticipate which arguments would be most convincing, but that would have taken some effort—even in the unlikely event that she had access to all the pertinent information, from where Marjorie ate last week to the state of her bank account.

Admittedly, sometimes achieving prompt conviction is crucial. When Voltaire, the high priest of French Enlightenment, was about to be lynched by an English mob, he had to convince them quickly that he was a genuine anglophile (which he apparently did). Fortunately, very few discussions have such urgency. Most aim at deciding who should do the dishes, whether the Joneses should be invited for dinner, or what is the best movie in the theaters at the moment. Failing to immediately prevail in mundane discussions is nearly costless. Even when the stakes are higher—does the customer buy the car, does the policeman accept your account, should your family move to Singapore or stay in Hong Kong—failing with the first reason is rarely critical; more reasons can be tried out.

Providing a stream of poor reasons does carry a cost: making the speaker look daft. As a result, even casual arguers should exhibit a moderate degree of quality control regarding the reasons they provide. What is clear, however, is that our interactionist theory does not predict that humans should be born Ciceros, weaving complex arguments and spontaneously anticipating rebuttals. Reason should make the best of the interactive nature of dialogue, refining justifications and arguments with the help of the interlocutors' feedback.

Refining Reasons

The experiments presented earlier, which prompted psychologists to deplore the poor quality of the reasons put forward by participants, did not take place in a typical dialogic context. When a normal interlocutor is not swayed by a reason, she offers counterarguments, pushing the speaker to provide better reasons. An experimenter, by contrast, remains neutral. She may prompt the participant for more arguments, but she doesn't argue back. If reason evolved to function in an interactive back-and-forth, strong arguments should be expected only when they are called for by an equally strong pushback.

To evaluate people's ability to construct arguments in a genuine argumentative setting, Lauren Resnick and her colleagues asked small groups of students to discuss a controversial topic. Participants were able to exchange several arguments every minute, bringing new ideas to the table and criticizing each other's suggestions and arguments. Thanks to this back-and-forth, the participants performed well and the researchers were "impressed by the coherence of the reasoning displayed. Participants . . . appear to build complex argument and attack structure. People appear to be capable of recognizing these structures and of effectively attacking their individual components as well as the argument as a whole."[10] Deanna Kuhn and her colleagues reached a similar conclusion in a more quantitative study:[11] they found that the students were able to produce much better arguments—about, say, the death penalty—after arguing about that topic with their peers. They clarified the links between premises and conclusions, added evidence to support their opinions, and relied on a wider range of argument types.

At the end of Chapter 11, we saw how the myside bias could turn into an efficient way of dividing cognitive labor, with each individual finding arguments for her side and evaluating arguments for the other side. The process we have just described is another elegant division of cognitive labor. Instead of engaging in a costly and potentially fruitless search for a knockdown argument, reasoners rely on the interlocutors' feedback, tailoring their arguments to the specific objections raised.

Fallacy? What Fallacy?

People exercise low quality control on their own reasons and are easily satisfied with relatively weak justifications and arguments. From the interactionist point of view, this is readily explained by the fact that reason evolved to work in an interactive setting. As predicted, people end up formulating better, more pointed arguments in the back-and-forth of a dialogue than when reasoning on their own.

What about quality control on other people's reasons, though? Here the interactionist approach makes very different predictions. If one can afford to be lax when evaluating one's own reasons, one ought to be more demanding when evaluating others' reasons. Otherwise we would accept the silliest excuses as good justifications and the most blatant fallacies as good arguments. We would be all too easily manipulated. This prediction goes against a common idea that people are easily gulled by sophistry. This common idea, however, is wrong, in part because it relies on misguided criteria for evaluating arguments. When more sensible criteria are used, experiments reveal that people tend to accept reasons when they should, and only when they should.

A common criterion used to distinguish good from bad arguments is whether they can be categorized as an informal fallacy, from the *ad populum* to the *ad hominem*. Lists of such fallacies were already produced in classical antiquity and today can be easily found online. The issue is that for almost each and every type of fallacy in such lists, there are counterexamples in the form of arguments that meet the criteria to be declared fallacious but that in real life are quite acceptable or even good arguments, arguments that might convince a rational audience.[12]

Here is a *tu quoque* ("you too") fallacy:

Yoshi: You shouldn't drink since you'll be driving!
Makiko: Weren't you yourself caught driving under the influence a month ago?

Makiko's objection is supposed to be a fallacious argument against the advice given by Yoshi. After all, the fact that he does not follow his own advice does not make it wrong. On the other hand, if one suspects the speaker has a good reason not follow his own advice, then the *tu quoque* argument would be quite reasonable:

Yoshi: You shouldn't eat these chocolates that Aunt Hélène brought us; they are not very good!
Makiko: Didn't you almost finish the box?

In this case, Makiko's objection does cast a reasonable doubt on the reliability of Yoshi's advice.

Here is an *ad ignorantiam* fallacy (arguing that a claim is true because it is not known to be false):

The policeman: I am convinced that Ishii is a spy. I could find no evidence that he is not.

The policeman's argument is supposed to be fallacious because not knowing that a proposition is false is generally not a good argument that it is true. On the other hand, there are cases where such an argument is indeed quite good:

The policeman: I am convinced that Ishii is a law-abiding citizen. I could find no evidence that he is not.

We could multiply the examples but our point would each time be the same: *tu quoque* is fallacious except when it is not; *ad ignorantiam* is fallacious except when it is not; in fact, most if not all fallacies on the list are fallacious except when they are not. This is often implicitly acknowledged when the fallacies are given a more careful definition. The *tu quoque* fallacy is an *inappropriate* use of the fact that the interlocutor does not abide by her own judgment, the *ad ignorantiam* fallacy is an *inappropriate* use of the fact that the claim examined is not known to be false, and so on.

This way of defining fallacies gives us license to invent indefinitely many new types: the Mount Everest fallacy (inappropriate comparison to the Mount Everest in an argument), the chicken soup fallacy (inappropriate use of facts about chicken soup in an argument), and so on. After all, the political philosopher Leo Strauss invented the *reductio ad Hitlerum* fallacy, which consists in arguing against an opinion by inappropriately comparing it to a view that Adolf Hitler might have held. Fallacious references to Hitler are much more frequent than fallacious references to Mount Everest, so why not enjoy the fun of labeling such references with a special name?

The standard view of informal fallacies has been attacked by psychologists Ulrike Hahn, Mike Oaksford, and their colleagues.[13] They have detailed the variables that should make some arguments more convincing than others. For instance, in the appeal to ignorance, one of the main variables is: How likely are we to find positive evidence if we look for it? The spy argument is weak because it is difficult, even for a policeman, to find evidence that someone is *not* a spy—if spies were easy to recognize, they would not remain spies for long. By contrast, if a policeman looks for evidence that an individual is a delinquent when such is indeed the case, he is quite likely to find the evidence. If he does not, this is in fact evidence that the individual is not a delinquent. Moreover, spies are, fortunately, much rarer than good citizens, further undermining the spy argument.

Hahn and Oaksford's case is not only theoretical. They have tested their hypotheses by manipulating relevant variables and asking people to evaluate the resulting arguments. What they found, with a variety of examples, is that people rate arguments in a rational way and don't easily fall prey to genuinely fallacious arguments.

For instance, here's one of the arguments—an *ad ignorantiam*—that they asked participants to rate:

> This drug is likely to have no side effects because five meticulously controlled, large-scale clinical trials have failed to find any side effect.

When given this argument, people rate it as being quite strong. Participants also react appropriately when the argument changes. For instance, they are less convinced if only two trials failed to reveal any side effect. Similar results were obtained for different types of arguments. On the whole, the evidence shows that when people are presented with truly fallacious arguments, they are reasonably good at rejecting them.[14]

Evaluating One's Own Reasons as if They Were Someone Else's

The experiments we reviewed suggest that there is an asymmetry between how people produce reasons—they are relatively lax about quality control—and how they evaluate others' reasons—they are much more demanding. With Emmanuel Trouche, Petter Johansson, and Lars Hall, Hugo conducted a tricky experiment that aimed at making this asymmetry as plain as possible. It involved getting people to evaluate their own arguments as if they were someone else's.[15]

In the first phase of the experiment, participants tackled five simple reasoning problems regarding the products sold in a fruit and vegetable shop. For instance, they might be told that in the shop, "none of the apples are organic." From this, they had to draw a conclusion as quickly and intuitively as possible, choosing among several options such as "Some fruits are not organic" and "We cannot tell anything for sure about whether fruits are organic in this shop." One participant whom we will call Rawan, for instance, selected as the correct conclusion, "Some fruits are not organic."

In the second phase, participants were asked to give reasons for their intuitive answers to each of the five problems they had answered in the first phase of the experiment. As they did so, they could, if they wanted, modify their initial answer. Given what we know about the production of reasons, we shouldn't expect much to happen at this juncture. Most participants should produce reasons that support their intuition without carefully checking how good their reasons are and without revising their initial selection. Indeed, only 14 percent of participants changed their minds, and the change was as likely to be for the better as for the worse. Rawan was among those who didn't change their minds. For her answer to the organic fruit problem, she offered the following justification: "Because none of the apples are organic, and an apple is one type of fruit, we can say that some of the fruits in the store are not organic."

In the third phase of the experiment, participants were given one by one the same five problems, and reminded of the answers they had given. They were then told about another participant who, on an earlier day, had answered differently, and they were given this participant's answer and argument. On the basis of this argument, they could change their mind and accept the other participant's answer or they could stick to their original answer.

With one of the five problems, we played a trick on the participants and told them that their answer had been different from what it had actually been. We told Rawan, for example, that she had answered, "We cannot tell anything for sure about whether fruits are organic in this shop." Moreover, we told her that someone else had selected the conclusion, "Some fruits are not organic" and given as justification "Because none of the apples are organic, and an apple is one type of fruit, we can say that some of the fruits in the store are not organic" (which had been Rawan's own selection and argument).

We hoped to have the perfect setup to test the asymmetry between the production and the evaluation of reasons: participants were led to evaluate an argument they had given a few minutes earlier as if it was someone else's.

And it worked. About half of the participants, Rawan included, did not notice that they had been tricked into thinking their own reason was somebody else's.

Would participants we had successfully misled at least agree with the argument they had themselves produced a moment before? Well, no. Even though they had deemed the argument good enough to be produced, they became much more critical of it when they thought it was someone else's, and more than half of the time, they found it wanting. Reassuringly, there was a tendency for participants to be more likely to reject their own bad reasons than their own good reasons. Rawan, who had initially given a good reason, found the same reason convincing when she thought someone else had given it.

Accepting Good Arguments

So far, we have stressed one important feature of reason evaluation: it should be demanding enough to reject poor reasons. But it is just as important that it should accept good reasons. Reasoning, we argued in Chapter 10, serves a function both for communicators and for their audiences. In a situation where a communicator wants to make a claim that the audience is unlikely to accept just on her authority, reasoning generates arguments that the audience might evaluate and accept, and hence accept the claim. For the audience, reasoning is a tool of epistemic vigilance. It serves to evaluate arguments provided by a communicator so as to reject claims that are poorly supported and to accept claims that are well supported. Indeed, the whole point of epistemic vigilance is not just to reject dubious information but also to accept good information. For this we must be able to change our minds when presented with good enough reasons to do so.

To test this prediction, with Emmanuel Trouche and Jing Shao we conducted a series of experiments using the following problem, which we can call the Paul and Linda problem:[16]

> Paul is looking at Linda and Linda is looking at John. Paul is married but John is not. Is a person who is married looking at a person who is not married?

The three possible answers are "Yes," "No," and "Cannot be determined." Think about it for a minute—it's a fun one.

Most people answer "Cannot be determined," thinking that knowing whether Linda is married or not is necessary to answer the question for sure. But consider this argument:

> Linda is either married or not married. If she is married, then she is looking at John, who is not married, so the answer is "Yes." If she is not married, then Paul, who is married, is looking at her, so the answer is "Yes" again. So the answer is always "Yes."

If you are anything like our participants (Americans and Chinese recruited online), then you are likely to accept the argument. When we gave the participants the argument, more than half changed their minds immediately. By contrast, if you had figured out the problem on your own and we had told you, "The answer is 'Cannot be determined' because we don't know whether Linda is married or not," then you would never have changed your mind. The way people evaluate these arguments is remarkably robust.

When we gave our participants the argument for the correct answer, we didn't tell them it was *our* argument (as experimenters). We told them it had been given by another participant in a previous experiment. So they had no reason to trust her more than they trusted themselves. To make things worse, we told some participants that the individual who gave them the argument was really, really bad at this kind of task. We told others that the individual who gave the argument would make some money if the participants got the problem wrong. So they expected an argument from someone who they thought was either stupid or out to trick them. And yet when they saw the argument, most accepted it. Indeed, although the participants had said that they did not trust the individual giving them the argument one bit, this lack of trust barely affected their likelihood of accepting the argument.

We also asked some participants to think hard about the problem and to justify their answers. A few of them did get it right. But most got it wrong, and because they had thought hard about it, they were really sure that their wrong answer was right. Most of them said that they were "extremely confident" or even "as confident as the things I'm most confident about." But that didn't make them less likely to change their mind when confronted with the argument above than participants who had had their doubts. Even though they could have sworn that the conclusion was wrong, when they read the argument, they were swayed all the same.

The Two Faces of Reason

In this chapter and in Chapter 11, we have looked at how reasoning produces arguments and at how it evaluates the arguments of other people. The results can be summarized in Table 2.

The "production of reasons" row is really bad for the intellectualist approach. When people reason on their own, they mostly produce reasons that support their decisions or their preconceived ideas, and they don't bother to make sure that the reasons are strong. As we'll see in Chapter 13, this is a recipe for disaster: not only is the solitary use of reason unlikely to correct mistaken intuitions, but it might even make things worse.

The fact that people are good at evaluating others' reasons is the nail in the coffin of the intellectualist approach. It means that people have the ability to reason objectively, rejecting weak arguments and accepting strong ones, but that they do not use these skills on the reasons they produce. The apparent weaknesses of reason production are not cognitive failures; they are cognitive features.

This picture of reason fits with the predictions of the interactionist approach. People produce reasons as predicted by the theory: they find reasons for their side—a good thing if their goal is to change others' minds—and

Table 2 The two faces of reason

	Bias	Quality control
Production of reasons	*Biased:* people mostly produce reasons for their side	*Lazy:* people are not very exigent toward their own reasons
Evaluation of others' reasons	*Unbiased:* people accept even challenging reasons, if those reasons are strong enough	*Demanding:* people are convinced only by good enough reasons

they start out not with the strongest reasons but with reasons that are easier to find, thus making the best of the feedback provided by dialogic settings. People also evaluate others' reasons as expected, rejecting weak reasons but accepting strong enough reasons, even if that means revising strong beliefs or paying attention to sources that don't inspire trust.

If we take an interactionist perspective, the traits of argument production typically seen as flaws become elegant ways to divide cognitive labor. The most difficult task, finding good reasons, is made easier by the myside bias and by sensible laziness. The myside bias makes reasoners focus on just one side of the issue rather than having to figure out on their own how to adopt everyone's perspective. Laziness lets reason stop looking for better reasons when it has found an acceptable one. The interlocutor, if not convinced, will look for a counterargument, helping the speaker produce more pointed reasons. By using bias and laziness to its advantage, the exchange of reasons offers an elegant, cost-effective way to solve a disagreement.

13

The Dark Side of Reason

Arthur Conan Doyle's novel *The Hound of the Baskervilles* begins when a Dr. Mortimer tries to hire Sherlock Holmes's services:

> "I came to you, Mr. Holmes, because . . . I am suddenly confronted with a most serious and extraordinary problem. Recognizing, as I do, that you are the second highest expert in Europe ——"
>
> "Indeed, sir! May I inquire who has the honour to be the first?" asked Holmes with some asperity.
>
> "To the man of precisely scientific mind the work of Monsieur Bertillon must always appeal strongly."
>
> "Then had you not better consult him?"[1]

Who was this Monsieur Bertillon, superior even to Holmes as an expert detective? Alphonse Bertillon was, at the end of the nineteenth century, one of the most respected policemen and forensic scientists of the world. He had developed a scientific method for recording the identity of criminals based on anthropometry: a precise measurement of various traits, from the size of the left foot to the length of the right ear.[2] The most famous component of this system, still used today, is the mug shot, a standardized way of photographing people who have been arrested.

It is for his expertise in the new domain of photography that Bertillon is contacted by investigators from the French army in early October 1884. His task is to take a series of pictures of an extremely sensitive document, a letter known as the *bordereau*, written by a French officer spying for the Germans.

Bertillon obliges. A few years later, this tiny piece of paper will turn him into the most ridiculed man in France.

On October 13 Bertillon is asked to perform one more task, one that lies outside his area of expertise. In order to identify the *bordereau*'s writer, he must tell whether the writing on another letter matches that of the *bordereau*. After a few hours, Bertillon concludes that the same person very likely wrote both letters. That, however, is not conclusive enough for the lieutenant-colonels Henry and du Paty de Clam, who lead the inquest. They want a definitive indictment. And so, two days later, they visit Bertillon once again. Now they give him more information: they have evidence that Captain Albert Dreyfus is the spy who leaked French secrets to the Germans. All they need is a final proof: that the *bordereau* bears his handwriting. Actually Henry and du Paty de Clam have no other piece of evidence, but Bertillon does not know that, and with the extra confidence afforded by anti-Semitism, he happily starts working with the premises that the Jew Dreyfus is guilty.

Bertillon's mind works tirelessly with a single purpose: proving that Dreyfus wrote the *bordereau*. Here's what he has to work with: two letters—the *bordereau* and a sample of Dreyfus's writing—that have some similarities but also marked differences. These differences are sufficient for real experts to conclude that the two letters have not been written by the same person. But Bertillon is smarter than that. Only by imagining what clever deceptions Dreyfus has devised will this connoisseur of the criminal mind be able to prove the traitor's guilt.

Bertillon wonders: What kind of spy would write such a compromising message in his own hand? (The real spy, as it turns out, but no one knows this yet.) In Bertillon's mind Dreyfus, a spy, and a Jew to boot, is too shrewd to make such a glaring mistake. He must have disguised his hand. This explains the differences between Dreyfus's normal writing and the *bordereau*.

But now Bertillon has another problem: How to account for the similarities? Why hasn't that shrewd spy simply used a completely different writing? To answer this question Bertillon comes up with his chef-d'œuvre, the keystone of his system: the theory of the auto-forgery.

Imagining what a shrewd spy might do, Bertillon realizes that transforming one's writing would work only if the potentially incriminating document were found in a nonincriminating place. Then Dreyfus could use the disparities to claim that he was not the author of the *bordereau.* However, if the letter were discovered on Dreyfus's person or in his office, he could not simply claim that it wasn't his. Instead, this master of deception would have to say that he was being framed, that someone had planted the *bordereau.* But if someone were to try to frame Dreyfus, surely they would be careful to reproduce his handwriting. And so *Dreyfus set out to imitate his own handwriting*—he engaged in auto-forgery. Frank Blair, a Chicago lawyer writing at the time of the affair, offers a sarcastic summary of Bertillon's reasoning:

> In short the differences between Dreyfus's natural hand and that of the bordereau, admitted by Bertillon, were, according to him, artfully put in by Capt. Dreyfus to throw off suspicion; while the absolute similarities were put in to enable him, in a proper case, to claim they were traced from his own handwriting, and therefore, done by someone else.[3]

If Dreyfus's handwriting looks like that of the *bordereau,* he is guilty. If it does not, he is guilty. The blatant flaw in the reasoning doesn't deter Bertillon, who states in conclusion: "I have arrived at a set of observations and comments that embrace all the facts with a comprehensiveness so perfect that the conclusions impose themselves indisputably."[4] Partly on the strength of Bertillon's "evidence," Dreyfus is arrested, tried, and sentenced to cashiering and life deportation—in a minuscule cell, on a minuscule island, far away from France.

A year later, Lieutenant-Colonel Georges Picquart takes over as head of military intelligence. Reviewing the files associated with the Dreyfus case, Picquart discovers conclusive evidence that the real culprit is the officer Ferdinand Walsin Esterházy. His handwriting matches that of the *bordereau* perfectly. When Bertillon is presented with this evidence, even he has to admit that Esterházy's writing is strikingly similar to the writing on the *bordereau.* But Bertillon's mind never falls short of reasons to sustain his views; Bertillon claims that "the Jews have been training someone for a year to imitate the writing"—to become a scapegoat, presumably.[5] The higher echelons of the military are similarly unmoved. They already have their culprit; another one would just mess up their story. Thanks to Picquart, Esterházy is tried. Thanks to the generals' influence, the guilty man is proven innocent, the innocent man proven guilty once more, and the whistle-blower disgraced. Dreyfus stays on his island.

It will take a major social and political upheaval, brought about by Jean Jaurès, Leon Blum, Georges Clemenceau, and, most famously, the novelist Emile Zola and his *J'accuse,* for the army to reopen the case. In 1899, the original decision is quashed and a new trial is held in Rennes. Bertillon, again, is one of the expert witnesses. His task has become more difficult, as he must now prove that the *bordereau* is not an undisguised note left by the sloppy spy Esterházy but the carefully designed product of Dreyfus's devious mind. Bertillon is up to the task. Apotheosis of Bertillon's system, the deposition runs for more than fifty pages of dense text, plus pictures and exhibits.

Of that *bordereau,* Bertillon peruses every word, measures every letter, photographs every wrinkle. He sees patterns everywhere. When the thirteenth line of the *bordereau* is superimposed on the thirtieth, three letters are aligned.[6] When the word *intérêt* is taken out and repeated and the two copies put end to end, they measure 12.5 millimeters, a unit size on military maps.[7] Even more damning, a standard subdivision of this unit, 1.25 millimeters, is found everywhere in the word: "length of the *t*'s cross: 3 [units of 1.25 millimeters]; length of the acute accent: 1; width of the circumflex, one and a half, and height of the final *t:* 4, etc."[8] Coincidences? Impossible. The *bordereau* must be the work of a master craftsman who used several templates and a military-issue ruler to create one of the most complex forgeries of modern times.

Such considerations can leave no place for doubt and so, after ten hours of deposition(!), Bertillon gives a forceful conclusion: "In the set of observations and concordances that form my demonstration there is no place for doubt, and it is made strong by a certainty both theoretical and material that, with the feeling of responsibility born of such an absolute conviction, I affirm with all my soul, today as in 1894, under oath, that the *bordereau* is the work of the accused. I am done."[9]

It is hard to tell how impressed the court is with Bertillon's arguments. In any case, the judges find Dreyfus guilty of treason once again, although with "mitigating circumstances." This nonsensical verdict reflects more the need to uphold the status quo than the merits (or lack thereof) of the case. Dreyfus's innocence is plain for anyone to see. Refusing to wait for yet a new trial that may never happen, Dreyfus consents to be pardoned by President Loubet on September 19, 1899, at the cost of accepting the guilty verdict. He will have to wait seven more years for his final rehabilitation: being reinstated to his former rank in the army.

The Bertillon in All of Us

Bertillon offers a fascinating study in the use of reasoning to defend preexisting beliefs. He seems to have been truly convinced by his own arguments. When three experts were tasked with evaluating his system, they found an "incomprehensible jumble" "totally devoid of scientific value" whose absurdity was "so blatant that one is hard put to explain the length of its exposition."[10] Yet they also concluded that the very "naiveté with which [Bertillon] unveiled the secrets [of his system] would lead one to believe in his good faith."[11]

It would be easy to regard Bertillon as a madman—many of his contemporaries did. But that would overlook his otherwise successful professional career, how he rose through the ranks and devised new ways to catch criminals—hardly what you would expect of a lunatic. And lest we feel too smug, every aspect of Bertillon's thinking has been reproduced in the laboratory, showing how reasoning can lead everyone on the wrong track. These experiments have replicated—on a smaller scale, fortunately—the mental processes occurring in Bertillon's mind. Unambiguously, they point to reasoning as the culprit.

When Bertillon mentions the perfect comprehensiveness with which he embraces all the facts, he exhibits clear symptoms of overconfidence. According to the intellectualist approach, reasoning is supposed to make us doubt our own beliefs, especially when they rest on foundations as shaky as Bertillon's. How can reasoning lead to overconfidence instead?

Asher Koriat suggested an answer more than thirty years ago.[12] In one of his experiments, participants had to answer general knowledge questions, such as, "Does Corsica belong to France or Italy?" and to specify how confident they were. Participants were overconfident. If they thought, say, that they would be right in 80 percent of the cases on average, they might have the correct answer only six times out of ten.

While some participants were not given any special instructions, others were told to give reasons supporting their answer. This had no effect on confidence because participants had been doing that all along: piling up reasons supporting their initial hunch. This is why they were overconfident. And it's not as if they were completely unable to think of reasons why they might be wrong—when asked to provide such reasons, they obliged, and became less overconfident. But the myside bias stopped them from spontaneously engaging in this more objective reasoning.

Between his first, not fully conclusive report and the deposition in Rennes fifteen years later, Bertillon has not only become even more confident, his beliefs have also undergone severe polarization. Dreyfus comes out more cunning, the conspiracy wider, Bertillon's testimony more critical than ever. Reason had had years to push Bertillon toward such extremes, but the roots of polarization can be observed in a much briefer time. When participants were made to think about someone for a few minutes, they ended up liking him more if they had liked him at first, and liking him less if they hadn't liked him at first.[13] In this brief interval of time, they had piled up reasons supporting and reinforcing their initial impression.

Initially, the only evidence Bertillon had of Dreyfus's guilt was the alleged resemblance between his handwriting and that of the *bordereau.* When Picquart showed Bertillon Esterházy's handwriting, which exactly matched the *bordereau,* Bertillon should have immediately changed his mind. But between his first encounter with the case and Picquart's intervention, Bertillon had built an unyielding scaffold of reasons. This scaffold upheld his original contention in spite of overwhelming evidence against it, a process known as belief perseverance.

The earliest experimental demonstration of belief perseverance was done by a team led by Lee Ross in 1975.[14] Participants were asked to distinguish between real and fake suicide notes, and were told how well they'd done. They were then left to think about their performance for a little while. During this time, they thought of many reasons why the feedback made sense: they had always been very sensitive, they knew someone who was depressed, and so on. Then the participants were told that in fact the feedback had been completely bogus, bearing no relationship whatsoever with their actual performance. But it was too late. Participants had found many reasons to buttress their beliefs, so that even when the initial support was removed, the beliefs stood on their own. Participants who had received a positive feedback thought they had done better than those who had received negative feedback, even after they'd been told the feedback was bogus.

Reasoning Poorly Together

Bertillon offers a sad example of how "starting with a mistake, a remorseless logician can end up in bedlam."[15] He kept accumulating reasons for his initial belief in Dreyfus's culpability, with a blatant lack of self-criticism, until he reached grandiose and absurd conclusions. Bertillon's mind seems boundlessly fertile, but it is only one mind. Imagine what people can do when they gang up to find reasons supporting their beliefs.

Conspiracy theories often start small, questioning accepted facts: Why does the American flag allegedly planted on the moon appear on the photos to be waving in spite of the lack of wind? Could the World Trade Center really have crumbled on 9/11 the way people say it did? Some of the people who raise these questions go online, discover pamphlets, find kindred spirits. Soon enough the doubt escalates, alternative answers are found, and pointed questions turn into full-blown paranoia. Officials must be lying. The government has to be in on it—NASA, the CIA, the FBI, the NSA. The conspiracy has to be global, pushed by the United Nations, the Bilderberg group, and more often than not, some form of "international Jewry," heirs to the anti-Semitic conspiracy theories of Bertillon's days.[16]

While the development of vast conspiracy theories involves hundreds of people, smaller groups can also be led astray by reason. In the 1960s, the Yale psychologist Irving Janis started investigating when and why small groups make poor choices. He examined in detail the process that led to disastrous decisions such as the failed attack on Cuba launched by the American government in 1961—the infamous Bay of Pigs invasion. For Janis, the culprit was groupthink, the failure of group members to criticize each other's suggestions and to consider alternatives.[17]

The problems caused by a lack of dissent have also been captured in the laboratory, where psychologists have accumulated evidence of group polarization. Put a bunch of people together and ask them to talk about something they agree on, and some will come out with stronger beliefs. Racists become more racist, egalitarians more egalitarian.[18] Hawks increase their support for the military; Doves decrease it.[19] When you agree with someone, you don't scrutinize her arguments very carefully—after all, you already accept her conclusion, so why bother? When like-minded people argue, all they do is provide each other with new reasons supporting already held beliefs. Just like solitary reasoners, groups of like-minded people can be victims of belief polarization, overconfidence, and belief perseverance.[20]

Evolution Doesn't Care How Good We Feel

Reasoning can lead to outlandish territories—Bertillon developing his system, a conspiracy theorist thinking that lizard men control the earth. It is only natural to dismiss people who adopt such beliefs as crazy or stupid—or, more politely, to suggest that they suffer from cognitive limitations.

The facts do not support this interpretation. Bertillon was neither a madman nor a dunce. Few conspiracy theorists suffer from psychosis or cognitive impairment.[21] Moreover, people who cannot be suspected of any mental deficiency share the same plight. Linus Pauling went from seeing vitamin C as a remedy for the common cold to hailing it as a universal cure. The escalation of the Vietnam War was decided by "extraordinarily intelligent, well-educated, informed, experienced, patriotic, and capable leaders."[22]

This spells trouble for the intellectualist approach. Not only does reasoning fail to fix mistaken intuitions, as this approach claims it should, but it makes people sure that they are right, whether they are right or wrong, and stick to their beliefs for no good reason. Historical examples attest that these are not minor quirks magnified by clever experiments, but real phenomena with tragic consequences.

Psychologists sometimes use a distinction between cognitive and motivational explanations. People who do or believe something wrong either must be the victims of a cognitive failure or must have been motivated to go astray. Since cognitive failures cannot explain the surprising outcomes of reasoning, a sensible move is to offer a motivational account, as Ziva Kunda did when she defended the prevalence of motivated reasoning. According to her, when reason leads people astray, it is because of a "wish, desire, or preference" to reach a preordained belief, whether it is accurate or not.[23] People may want to believe something for many reasons, but the most common is hedonic: because it makes them feel good. There would be a desire to believe that leads people to "give preferential treatment to pleasant thoughts and memories over unpleasant ones."[24]

At its most extreme, a feel-good account could simply claim that people adopt whatever belief is pleasurable—an extreme form of wishful thinking. "Wishful thinking" is a phrase commonly used to describe someone's belief when it seems to be grounded not on evidence that the belief is true but on the desire that it were. Moreover, in ordinary conversation, "It's wishful thinking" is often given not just as description but as explanation: "Why does John believe he is popular? It's just a case of wishful thinking—he wishes he were!"

From a psychological point of view, and even more from an evolutionary one, wishful thinking is something to be explained rather than an explanation. Our beliefs are supposed to inform us about the world in order to guide our actions. When the world is not how we would want it to be, we had better be aware of the discrepancy so as to be able to do something about it. Thinking that things are the way one wishes they were just because one so wishes goes against the main function of belief.

It should therefore be no surprise that Kunda and other specialists of motivated reasoning have shown that people do not simply adopt beliefs as they see fit. They look for reasons, making sure that they can provide some justification for their opinions—and they drop even cherished beliefs when they fail to find justifications for them. For instance, people have an overall preference for believing they are better than average—smarter, better at socializing, more sensible, and so forth. However, they do not simply believe what would make them happiest: that they are the best at everything. Instead, they selectively self-enhance, only providing inflated assessments when they are somehow defensible.

For instance, people tend to think they are more intelligent than the average—that's an easy enough belief to defend: they can be good at math, or streetwise, or cultured, or socially skilled, and so on. By contrast, there aren't two ways of being, say, punctual. Since people can't think of ways to believe they are more punctual than the average, they just give up on this belief, or on other beliefs similarly hard to justify.[25]

Even if reasoning is not wishful thinking, its function could still be to make us feel good. A first problem with this hypothesis is that reasoning often has the opposite effect. Bertillon might have been engulfed by the pleasure of complete self-confidence, but he may also have been distressed by every new proof of Dreyfus's devilish ingenuity. The journalist Jonathan Kay interviewed many conspiracy theorists in the writing of *Among the Truthers* (people who do not accept the standard account of 9/11). He found people suffering from "debilitating emotional agony" caused by "sudden exposure to the magnitude of evil threatening the world."[26] If reasoning is supposed to make people feel good, it fails abysmally. Other cases are easy to conjure—the jealous husband who persuades himself his wife is cheating on him, the pessimist who keeps finding reasons why humanity is bound to self-destroy, the hypochondriac who looks for the symptoms of yet another disease.[27]

Not only does a feel-good explanation fail to fit the facts, it doesn't make evolutionary sense, either. It confuses the *proximal* and the *ultimate* levels of explanation.[28] A proximal explanation aims at pinpointing the psychological or neurological causes of a behavior. For instance, if Michael gets thirsty and drinks some water, the pleasure he derives from drinking could be a proximal explanation for his behavior: he drank the water because he anticipated that it would make him feel good.

Ultimate explanations, by contrast, answer questions at the evolutionary level. At the ultimate level, feeling good is no more than a means to an end. For evolution, hedonic states—pleasure, pain, happiness, despair—serve the purpose of motivating animals to perform certain actions critical to their survival and reproduction. We feel pleasure while quenching our thirst because drinking is necessary for survival. We experience pain when touching a burning log so that we withdraw our hand and avoid long-term damage. We like spending time with friends because having partners and allies has been crucial to reproductive success in human evolution. For the same reason, we despair if our friends abandon us. An individual who would find drinking painful but would enjoy feeling his hand roast or who would resent the affection of friends and revel in their loathing would not be well equipped to survive and reproduce. So whether or not reasoning helps people feel good, it cannot have evolved to this end.

Adaptive Lags in Reasoning

The interactionist approach can account for the various epistemic distortions introduced by reason—overconfidence, polarization, belief perseverance. Chapters 11 and 12 pointed out two major features of the production of reasons: it is biased—people overwhelmingly find reasons that support their previous beliefs—and it is lazy—people do not carefully scrutinize their own reasons. Combined, these two traits spell disaster for the lone reasoner. As she reasons, she finds more and more arguments for her views, most of them judged to be good enough. These reasons increase her confidence and lead her to extreme positions.

Many psychologists might agree with this diagnostic. However, such an explanation should only be a first step, soon followed by: Why on earth would reasoning behave that way? When an artifact fails to produce the desired results, this might be because it is broken, but it might also be because it is operating in abnormal conditions. If your pen doesn't work upside down, if your car doesn't start with an empty tank, it is not because they are out of order but because they are not designed to function in such conditions. Biological devices also have normal conditions: the conditions to which they are adapted.[29] The normal conditions for human lungs are formed by the earth's atmosphere around ground level. Our lungs work splendidly in these conditions, but less well or not at all in abnormal conditions—high altitudes, under water, in a tank full of helium, and so forth.

In our interactionist approach, the normal conditions for the use of reasoning are social, and more specifically dialogic. Outside of this environment, there is no guarantee that reasoning acts for the benefits of the reasoner. It might lead to epistemic distortions and poor decisions. This does not mean reasoning is broken, simply that it has been taken out of its normal conditions. In the same way, when objects take on new colors under the sodium lighting of an underground parking lot, our color perception is not broken; it is simply working in an abnormal environment. The artificial lights that have replaced the lighting we encountered during our evolution—chiefly, the sun—mislead our color perception.

This explanation—that reasoning now often works in an abnormal environment—is incomplete. If a bomb explodes inside the bomber plane rather than when it hits the intended target, the engineer in charge does not get kudos by pointing out that the explosion was exactly of the predicted force. When the bomb explodes is at least as important as how it explodes. Similarly, when a cognitive mechanism is triggered is at least as important as how it works once triggered.

The basic trigger of reasoning is a clash of ideas with an interlocutor. This clash prompts us to try to construct arguments to convince the other or at least to defend one's own position. This trigger works also in the absence of an actual interlocutor, in anticipation of a possible disagreement. Sometimes this anticipation may be quite concrete: a meeting is already scheduled to try to resolve a disagreement or to debate opposing ideas. At other times, we might just anticipate a chance encounter with, say, a political opponent, and mentally prepare and rehearse arguments we would then be eager to use. There are even times when we replay debates that have already taken place and think, alas, too late of arguments that we should have used.

Sasha, for instance, is about to ask his mother to let him go to the all-night party at Vanessa's. He has been rehearsing his arguments: he has been working very well at school; his homework for the next week is done; the party will be a small affair, nothing wild, nothing his mother should worry about. The more he thinks about it, the more Sasha becomes convinced that his request is perfectly reasonable and that his mother should, of course, say yes.

Several things can happen then. Sasha might convince his mother that there are no serious objections to his going to the party. Or his mother might convince him that it is not such a good idea after all—she has heard from other parents that the party might be crashed by older kids who would bring beer and perhaps even drugs. Also, he seems to be forgetting that there is an exam next week for which he has not yet prepared. Listening to his mother's arguments, Sasha might want to argue back. At the end of the back-and-forth, either one will have convinced the other, or at least both will have given reasons to justify their points of view.

By contrast, if his mother just said no without paying attention to his argument, or if he never mustered the courage to ask, Sasha would probably see the arguments he never gave as compelling; he would see himself as a victim of parental injustice and incomprehension. Reasoning in anticipation of a discussion is fine—as long as the discussion actually takes place.

What is problematic isn't solitary reasoning per se, but solitary reasoning that remains solitary. Reasoning, however, is bound to sometimes remain in one's head, as people cannot fully anticipate when they will be called to defend their opinions. Just as one can be taken aback by an unanticipated quest for justification, one can prepare for a confrontation of points of view that never materializes. The latter case may well be more common because of the difference in costs between the two types of failures. Being caught unprepared to defend an opinion or an action that others might object to is likely to be worse, and hence less common, than rehearsing a defense that in the end will not serve.

Modern environments distort our ability to anticipate disagreements. This is one of many cases in which the environment changed too quickly for natural selection to catch up. For example, our modern environments make some psychoactive substances, from coffee to cigarettes to alcohol, widely available. Some of these substances, such as cigarettes, are clearly bad for their users' fitness (in both meanings of the word). Yet we haven't evolved an innate disposition to avoid these substances in the same way that we have innate dispositions to avoid poisonous foods. Arguably, the explanation is that these substances would have been much rarer during our evolution and that they became common enough too recently for our brains to adapt to the change.

Have environmental changes thrown off-balance our ability to anticipate disagreements in the same way they made our reactions to psychoactive substances dangerous? Life in a modern, affluent society is different in myriad ways from life in the ancestral environment, and some of these differences are bound to affect the way we reason. For instance, before the invention of the printing press and the advent of modern media, people were typically made aware that somebody in their own group had opinions different from theirs thanks to interaction with that person. Finding out about difference of opinion and trying to resolve them commonly occurred through repeated exchanges of arguments that could be anticipated and mentally rehearsed. Nowadays we are inundated with the opinions of people we will never meet: editorialists, anchormen, bloggers. We are also expected to have an opinion on many different topics—from politics to music to food—and to be able to defend this opinion when challenged, giving us reasons to prepare for a variety of debates that might never occur.

And this only scratches the surface of the problem. More dramatic changes affect the workings of reason. Big-city dwellers meet more strangers in any single day than their ancestors did in their lifetime. Many of these strangers have different cultural backgrounds. It is easy to see how this novel mix generates possibilities for disagreement, making it considerably more complex to properly anticipate the need for justifications.

Some cognitive mechanisms have been so fully repurposed by the modern world that they bear only a small resemblance to their ancestral form—witness the transformations brought by literacy to our ability to recognize simple arbitrary shapes.[30] While we do not believe that reason has undergone such dramatic transformations, environmental changes have certainly had an effect on when reason is triggered, on how it functions, and even on what goals it achieves. Reason is used now in a variety of ways that differ from its evolved function—from displaying one's smarts in a formal debate to uncovering the laws of physics. Unfortunately, some of those new uses of reason, such as preparing for debates that never come, turn out to be potentially harmful to the reasoner. As Keynes put it, "It is astonishing what foolish things one can temporarily believe if one thinks too long alone."[31]

14

A Reason for Everything

In Chapter 13, when solitary uses of reason led people astray, it was because they started out with a strong intuition—that Dreyfus was guilty, that this was the right answer to the problem, and so on. The myside bias, coupled with lax evaluation criteria, make us pile up superficial reasons for our initial intuition, whether it is right or wrong. Often enough, however, we don't start with a strong intuition. On some topics, we have only weak intuitions or no intuitions at all—a common feeling at the supermarket, when faced with an aisle full of detergents or toilet papers. Or sometimes we have strong but conflicting intuitions—economics or biology? Allan or Peter? Staying at home with the kids or going back to work?

These should be good conditions for the individualist theory to shine. Reason has a perfect opportunity to act as an impartial arbiter. When the reasoner has no clear preconception, the myside bias is held at bay and reason can then guide the reasoner's choice, presumably for the better. Perhaps it is from such cases that beliefs about the efficiency of reason are born in the mind of philosophers. They make it a duty to examine precisely these cases in which intuitions are weak or conflicting. And if there is not enough conflict, philosophers excel at stirring it up: Are you sure other people exist? Think again! There is a philosophical theory, solipsism, that says other people don't exist or, at least, that you can never be sure that they do. Situations where intuitions are absent, weak, or conflicting might provide perfect examples of reason working in line with the expectations of the classical theory: reach a status quo between different intuitions, and only then let reason do its job. Let's look at what reason does in such cases, starting with a clever experiment conducted by Lars Hall, Petter Johansson, and Thomas Strandberg.

As you walk along the street, a young man approaches you with a clipboard and asks whether you would be willing to take part in a short survey. For once, you accept. He hands you the clipboard with two pages of statements on political, moral, and social issues such as "If an action might harm the innocent, it is morally reprehensible to perform it." You must indicate how you feel about each statement on a scale going from "Completely disagree" to "Completely agree." You fill in the survey and hand the clipboard back. You're not quite done, though: the young man passes the clipboard back and asks you to explain some of your ratings. You happily do so—after all, you take pride in being an informed, thoughtful citizen with sound opinions.

What you haven't realized is that in the few seconds during which he held the clipboard, the young man—who is in fact an experimenter—has, by means of a simple trick, replaced some of the statements on the page with statements having the exactly opposite meaning. For instance, the statement about harming the innocent would now read, "If an action might harm the innocent, it is morally permissible to perform it" (with "permissible" having replaced "reprehensible") If some statements have been flipped, the answers haven't, so that for these statements, the sheet now indicates that you hold the exact opposite of the opinion you asserted one minute earlier. If you had indicated that you strongly agreed with the first statement, the sheet now says that you strongly agree with the second statement, which means the opposite.

Fewer than half of the participants noticed that something was wrong with the new answers. The majority went on justifying positions contrary to those they had professed a few minutes earlier, especially if their opinions weren't too strong to start with.[1]

Our boundless ability to produce reasons for just about anything we believe (or we think we believe, as shown by Hall and colleagues) has become a staple of social psychology since the pioneering experiments of Richard Nisbett and Tim Wilson in the 1970s that we evoked in Chapter 7. In one experiment, Nisbett and Wilson were standing outside malls pretending to sell stockings.[2] Some passersby stopped at their stall, made a choice, and, when asked, happily justified their decision: "This one looks more resistant"; "I prefer the color of that one." But the psychologists knew all these explanations to be bogus: they had mischievously displayed strictly identical pairs of stockings. That all the stockings were the same did not stop people from expressing preferences, which must have been based, then, on the position of the pairs in the display (many participants showed a right-most bias, for instance).

In such experiments participants start out with weak intuitions. In the substitution of statements experiment, it is mostly those participants who had expressed only mild agreement or disagreement and who therefore didn't have strong intuition on the issue who failed to detect the manipulation. In the second experiment, the stockings were all the same, so whatever preference was created by their position would have been very weak. Still, reason doesn't do the job classically assigned to it. It does not objectively assess the situation in order to guide the reasoner toward sounder decisions. Instead, it just finds reasons for whatever intuition happens to be a little bit stronger than the others. Humans are rationalization machines. As Benjamin Franklin put it, "So convenient it is to be a reasonable creature, since it enables one to find or make a reason for everything one has a mind to do."[3]

There are cases, however, where reason has a demonstrable impact on people's decision—but not one that fits with the intellectualist approach.

When Reasoning Makes a Difference

We find Tim Wilson again, except that this time he is dealing in posters, not stockings, and there is no trick: all the posters are different. The experiment is straightforward. Some participants are asked to rate five posters, period. Others have to rate the same five posters, but also to explain their ratings.[4] Being asked to reason affected participants' choices, such as by making them give higher ratings to humorous posters.

Many other studies have demonstrated that reason can make a difference. Some experiments require people to justify their decisions;[5] others give participants some extra time to reflect on their choices;[6] still others pit decisions based on feelings against decisions based on reasoning.[7] Each time, people who reason more act differently from those who reason less or not at all. A mere rationalization machine is not supposed to influence decisions. What is happening? Is reason helping people make better decisions?

Not quite. In every one of these experiments, more reasoning led to worse decisions. For instance, in Tim Wilson's experiment, the participants were given the poster they had ranked higher to take home. Asked a few weeks later about their appreciation of the poster, those who had had to explain their preferences were less satisfied than those who had relied on their unfiltered intuitions.

To understand why reason can mess up people's decisions even when the myside bias is not the main culprit, we must look more precisely at how reason affects decisions.

Itamar Simonson performed an early experiment on this topic.[8] He started by designing two products—for example, beers—that would be equally preferred by most people. Let's call the first brand Beer Deluxe. It's a fancy product, with a quality rating of 75 out of 100, worth $18 for a six-pack. The second is Beeros, a less sophisticated—rating at 65—but cheaper—$12—alternative. When people had to choose between these two brands, they were indifferent, picking either beer about as often. Then the experimenter introduced a third brand, Premium Beer. At $18, Premium Beer is as expensive as Beer Deluxe, but it is also less well rated—a 70 out of 100. Given that Premium Beer is simply inferior to Beer Deluxe, it should not make a difference in people's choices. In fact, it did: once Premium Beer was introduced, people were more likely to pick Beer Deluxe.

Christopher Hsee conducted one of the most original experiments in the area. He asked participants which of two treats they would prefer to receive as a gift for having completed a task. Both gifts were chocolates, but one was a small (0.5 ounce), cheap ($0.50), heart-shaped chocolate while the other was a big (2 ounces), expensive ($2), roach-shaped chocolate. When participants relied more on their feeling, they were about split between the two choices. But when they reasoned to make a decision, most picked the big roach-shaped chocolate.[9]

Debora Thompson explored the phenomenon of feature creep: the multiplication of useless features that burdens so many gadgets and, in the end, reduces their usability. With her colleague Michael Norton, they showed that when people feel they must provide reasons for their decisions, they are more likely to pick a feature-rich item—such as a digital video player with dozens of functions—even though they realize it would be less convenient to use.[10]

Here's the common thread in all these results: in each case, reason drives participants toward the decision *that is easier to justify*. "Beer Deluxe is better but not more expensive than Premium Beer, so I'll pick it." "Given it's a gift, it would be irrational not to pick the bigger and more expensive chocolate just because of its shape—it's not as if it was a real roach anyway." "Why buy a digital video player that does fewer things?"

This common phenomenon is known as reason-based choice: when people have weak or conflicting intuitions, reason drives them toward the decision for which it is easiest to find reasons—the decisions that they can best justify.

Paying for Reasons

While these results are difficult to reconcile with the intellectualist theory—reason should lead people to better decisions, not to worse decisions—they are what the interactionist approach predicts. Reason doesn't stop being a social device in the absence of a point of view to uphold. Instead, it samples potential reasons for the different options available and drives the reasoner toward the decision that is the easiest to justify—whether or not it is otherwise a good decision.

In many cases, it looks as if reasoning is driving people toward worse, less rational decisions. The introduction of an obviously inferior option—Premium Beer—should not influence the decision between two superior options. Psychologists studying disgust can tell you that people will not enjoy eating that roach shaped chocolate, however big.[11] A gadget bloated with useless features will become a source of anxiety, not enjoyment.[12] Refusing to buy a jam simply because there are more jams to pick from doesn't make much sense.

Even more strikingly, people are willing to pay simply to have a reason for their decision. Amos Tversky and Eldar Shafir, who were among the first, with Itamar Simonson, to explore reason-based choice, asked a first group of participants to imagine the following scenario:

> You have just taken a tough qualifying examination. It is the end of the fall quarter, you feel tired and run-down, and you are not sure that you passed the exam. In case you failed, you have to take the exam again in a couple of months—after the Christmas holidays. You now have an

opportunity to buy a very attractive 5-day Christmas vacation package in Hawaii at an exceptionally low price. The special offer expires tomorrow, while the exam grade will not be available until the following day.

Would you

x. Buy the vacation package.
y. Not buy the vacation package.
z. Pay a $5 nonrefundable fee in order to retain the rights to buy the vacation package at the same exceptional price the day after tomorrow—after you find out whether or not you passed the exam.[13]

A second group of participants was asked to imagine that they had passed the exam and a third group that they had failed.

Most of the participants told that they had passed the exam decided to buy the vacation package—they reasoned that it was a well-deserved reward for their success. Most of the participants told that they had failed the exam also decided to buy the vacation package—they reasoned that they direly needed a break to recover from this failure.

Combined, these two results imply that it would be rational for most participants in the first group, who didn't yet know whether they had passed or failed, to buy the package and not to waste five dollars to postpone their decision. Whether they passed or failed, they would buy it. However, participants in this group chose to pay the fee and to wait a couple of days in order to know the exam results. Their problem: the reasons for buying the package were incompatible, one being "I deserve a reward for success" and the other "I need a break after failure." And so they paid to wait, effectively buying a reason to make a decision that they would make either way.

Social Rationality

When assessing decisions, it might seem clear that we should focus on the fit between the content of our decisions and our practical goals. We should buy posters we will enjoy more. When buying electronic devices, we should pick a model that best meets our needs. However, if being rational is striving for the best decision, all things considered (and not just our practical goals), then making a good decision gets more complex.

Humans constantly evaluate one another. Are the people we interact with competent and reliable? Is their judgment sound? As we argued in Chapter 7, much of this evaluation is done in terms of reasons: we understand others' ideas and actions by attributing to them reasons, we evaluate the goodness of these reasons, and we evaluate people's reliability on the basis of their reasons. The way we rely on reasons may distort and exaggerate their role in thought and action, but it is not as if a better, more objective understanding were readily available. After all, psychologists themselves are still striving to develop such an understanding, and they disagree as to what it would look like. Reason-based understanding, for all its shortcomings, has the advantage of addressing two main concerns: providing a basis for an evaluation of people's performance, and providing an idiom to express, share, and discuss these evaluations.

Just as we evaluate others, they evaluate us. It is important to us that they should form a good opinion of us: this will make them more willing to cooperate and less inclined to act against us. Given this, it is desirable to act efficiently not only in order to attain our goals but also in order to secure a good reputation. Our reasons for acting the way we do shouldn't just be good reasons; they should be reasons that are easily recognized as good.

In some situations, our best personal reasons might be too complicated, or they might go against common wisdom, and hence be detrimental to our reputation. In such a case, it may be more advantageous to make a less-than-optimal choice that is easier to justify than an optimal choice that will be seen as incompetent. We might lose in terms of the practical payoff but score social points yielding a higher overall payoff.

Reason influences our decisions in the direction of reputational gains. For instance, those participants who picked Beer Deluxe because it was the easiest decision to justify may not have maximized their product satisfaction, but they scored social points: their decision was the least likely to be criticized by others.[14] Customers who ended up with a device burdened with useless features are (ironically) regarded as technologically savvy.[15] Trying to look rational, even at the price of some practical irrationality, may be the most rational thing to do.

In the type of choices we have examined in this chapter, people's intuitions are generally weak. Having such weak intuitions is often a reliable sign that the decision at issue is not so important. After all, when it comes to dealing with the most pressing aspects of our ancestral environment, specific cognitive mechanisms are likely to have evolved and to provide us with strong intuitions. So, when our intuitions are weak, being guided by how easy it is to justify a particular decision is, in general, a simple and reasonable heuristic.

Our environments, however, have changed so much in the past millennia, centuries, and even decades that having weak intuitions is no longer such a reliable indication of the true importance of a decision. For most people, for instance, buying a car is an important decision. Much of their money goes into buying and maintaining a car; much of their time goes into using it; and their life may depend on its safety features. There are no evolved mechanisms for choosing cars in the way we have dedicated mechanisms aimed at selecting safe foods or reliable friends. As a result, intuitions give only weak and limited guidance. Does this mean that looking for an easily justifiable choice in these evolutionarily novel situations—choosing a car that is popular and well-reviewed, for instance—is unreasonably risky? Not really. In most cases, the decisions that are the easiest to justify in the eyes of others and hence that are the most likely to contribute to our reputation are also the best decisions to achieve our goals.

When the reasons that are recognized as good in a given community are objectively good reasons, people guided by reputational concern may still arrive at true beliefs and effective decisions. But this is not always the case—far from it. Throughout the centuries, smart physicians felt justified in making decisions that cost patients their lives. A misguided understanding of physiology such as Galen's theory of humors created a mismatch between decisions easy to justify—say, bleeding the patient to restore the balance between humors—and the fact that the condition of some patients deteriorated after being bled, a fact that must have given pause to some of these physicians. Still, if they were eager to maintain their reputation, they were better off bleeding their patients, and anyhow, there was no clear alternative. By contrast, today's doctors, relying on vastly improved, evidence-based medical knowledge, may make decisions guided in good part by the sense of what the medical community would approve and, in so doing, preserve both their reputation and the health of their patients.

When Justification and Argumentation Diverge

The message of this chapter might seem bleak. Reason improves our social standing rather than leading us to intrinsically better decisions. And even when it leads us to better decisions, it's mostly because we happen to be in a community that favors the right type of decisions on the issue. This, however, cannot be the whole picture. Justifications in terms of reasons do indeed involve deference to common wisdom or to experts. What implicitly justifies this deference, however, is the presumption that the community or the experts are better at producing good reasons. But there is a potential for tension between the lazy justification provided by socially recognized "good reasons" and an individual effort to better understand and evaluate these reasons, to acquire some expertise oneself.

In the beginning of the nineteenth century, for instance, the doctor Joseph Victor Broussais was the most respected medical authority in Paris. He insisted that all fevers are caused by inflammation and should be treated by bloodletting. The younger doctor Pierre-Charles-Alexandre Louis didn't really doubt the efficacy of bloodletting, but he wanted to evaluate it precisely. For this, he compared two groups of patients that had been bled for pneumonia and discovered that, contrary to his expectations, those who had been bled early in their illness had died in greater numbers than those who had been bled late, showing that not only had bloodletting not cured them, it had worsened their condition. Louis had now compelling evidence and arguments, if not against bloodletting in general, at least against the systematic usage recommended by Broussais. Louis's pioneering work in evidence-based medicine played a crucial role in the progressive abandonment of bloodletting as a major medical procedure in the nineteenth century. In criticizing the overextended practice of bloodletting, Louis was taking immediate reputational risks, but precisely because he had good arguments to do so, in the end his ideas prevailed and his reputation grew.[16]

It would be nice to think that, when there is a conflict between the goal of having good reasons in the eyes of others and that of having demonstrably good reasons, argumentative strength trumps ease of immediate justification and that the best reasons ultimately win. Well, things are somewhat more complicated.

Consider the following scenario:

> You have a ticket to a basketball game in a city sixty miles from your home. The day of the game there is a major snowstorm, and the roads are very bad. Are you more likely to go to the game if:
>
> a. You paid $35 for the ticket.
> b. You got the ticket for free.
> c. Equally likely.[17]

Most people answer that they would be more likely to face the snowstorm if they had bought the ticket than if they had got it for free. According to psychologists Hal Arkes and Peter Ayton, this decision is based on a reason such as "wasting is bad."[18] Most people would understand better why someone would brave a snowstorm for a ticket they bought than for one they got for free; they might even disapprove of somebody who bought the ticket and didn't go.

Economists call the price of the ticket in this situation a "sunk cost." The money has already been spent and cannot be retrieved. It is as good as sunk. Decisions are about the future, which can be altered, not about the past, which cannot. The only question that really matters, then, is: Would you be better off now facing the snowstorm to get to the game, or doing something else? If you would be better off doing something else, then undertaking an unpleasant and potentially dangerous drive simply makes you worse off. People who accept this argument should, it seems, answer that whether they bought the ticket or got it for free would not affect their decision to go or not to go to the game.

The argument against this so-called sunk-cost fallacy, on the other hand, is clear. It even has the backing of many philosophers and economists. If, however, you are convinced by the argument and you decide, say, to stay at home in spite of having paid for the ticket, you might be ill-judged by people who are not aware of this argument, and you might not have the opportunity to explain the reasons for your choice. So you might, ironically, be seen as making an irrational decision when in fact you made a decision based on a sound reason. To the extent that you care what these people think of you, their judgment will pull toward making the socially acceptable decision.[19] (Mind you, if you are a student in economics and make a decision based on a sunk cost not to be ill-judged by your family, you may end up being deemed incompetent by your fellow economics students.)

But why should the sunk-cost fallacy be common and indeed be seen not as fallacious but as the right thing to do? Here is a speculative answer. One of the qualities people look for in friends, partners, or collaborators is that they should be dependable. Some degree of stubbornness in carrying through any decision made, in pursuing any course of action undertaken, even when the expectation of benefits is revised down, gives to others evidence that one can be relied upon. People who persevere in their undertakings even when it might not be optimally rational from their individual point of view may, in doing so, strengthen their reputation of reliability. It may be rational, then, at least in some cases, not just to knowingly commit the sunk-cost fallacy in order to signal to others that one can be counted upon but also to have a better opinion of people who commit the fallacy than of people who don't.

Attending to the interactional functions of reason not only makes better sense of it but also shows its limit. Justificatory and argumentative reasons are fundamental tools in human interaction, but which type of reason trumps the other when they diverge may depend not only on the quality of the reasons involved but also on the social, and in particular reputational, benefits involved. The reason module cannot pretend to the commandeering position that the classical approach assigned to capital *R* Reason.

15

The Bright Side of Reason

At the beginning of the movie *12 Angry Men* (spoilers ahead), a youngster stands accused of stabbing his father to death. His life is in a precarious position: in the jury room, the arguments for conviction are piling up. One witness has seen the boy do it; another heard the fight and saw the accused flee out of the apartment; the boy's alibi doesn't hold water; he has a motive, and a long record of violence. Group polarization is lurking, ready to convince the jurors that the boy should be sent to the electric chair. But one juror is less confident than the others. While this juror is not convinced of the defendant's innocence, he's not quite sure of his guilt, either. He "just wants to talk." When urged to provide arguments, he starts with a weak one: the evidence against the boy is too good; it's suspiciously good. Unsurprisingly, this doesn't sway any of the other jurors. From then on, however, this juror does a better job at poking holes in the prosecution's case.

He unearths inconsistencies in the incriminating arguments. One witness claims to have seen the murder from the other side of the street, through the windows of a passing train of the elevated subway. Another witness claims to have heard the boy threaten his father—"I'll kill him!"—and then a body fall a few seconds later. But how could the second witness have heard anything amid the deafening sound of the train?

More inconsistencies emerge in other jurors' arguments. The boy's fingerprints cannot be found on the knife. Not a problem for juror four: the boy is a cold-blooded murderer who wiped the knife, still tainted with his father's blood. The defendant was caught by the police walking home three hours after the crime. Why come back to the crime scene? Juror four has an answer: "He ran out in a state of panic after he killed his father, and then, when he finally calmed down, he realized that he had left the knife there." But then, points out the more skeptical juror, how do you square that with the fact that "he was calm enough to see to it that there were no fingerprints on the knife?"

Inconsistencies can take the form of double standards. The boy is from the slums, and juror ten knows they're "born liars"; "you can't believe a word they say." Yet he has no problems accepting the testimony of the witness who claims she saw the boy commit the murder, even though "she's one of them too," as our juror points out.

Sometimes the inconsistency surfaces so blatantly that it doesn't even need pointing out. Juror three has vehemently defended the guilty verdict, relying in large part on the testimony of the man who said he heard the fight and saw the kid leave the apartment right after. To see this, however, the witness claims that he got up and crossed the length of the whole building in fifteen seconds—something impossible for this old man with a limp. But that is not an issue for our juror three: the witness must have been mistaken in his estimate—after all, "he was an old man, half the time he was confused. How could he be positive about anything?"

These inconsistencies slowly sway most of the jurors toward reasonable doubt, but not all of them. Juror ten remains impervious to rational considerations. In the end, they must shame him to make him relent. Still, most of the work is done by argumentation. It is argumentation that allows the holes in the prosecution's case to surface. It is argumentation that highlights double standards. It is argumentation that lays bare inconsistencies. It is argumentation that raises doubt in the jurors' minds. In *12 Angry Men*, argumentation saves a boy's life.[1]

Argumentation Is Underrated

In Chapters 11 through 14, we have emphasized the "bad" sides of reason, those that have given rise to an enigma in the first place: reason is biased; reason is lazy; reason makes us believe crazy ideas and do stupid things. If we put reason in an evolutionary and interactivist perspective, these traits make sense: a myside bias is useful to convince others; laziness is cost-effective in a back-and-forth; reason may lead to crazy ideas when it is used outside of a proper argumentative context. We have also repeatedly stressed that all of this is for the best—in the right context, these features of reason should turn into efficient ways to divide cognitive labor.

Crucially, this defense of argumentative reasoning depends on how people evaluate others' reasons: they have to be sensitive to what the philosopher Jürgen Habermas has called the "forceless force of the better argument."[2] They have to be able to reject weak arguments while accepting strong enough ones, even if that means completely changing their minds. In Chapter 12, we presented evidence that people are good at evaluating others' arguments, but these experiments still targeted solitary reasoners provided with a single argument to evaluate—a far cry from the back-and-forth of an argumentative exchange.

It is now time to make good on our promise and to show that in the right interactive context, reason works. It allows people to change each other's minds so they end up endorsing better beliefs and making better decisions.

For more than twenty years, Dave Moshman, psychologist and educational researcher, had asked his students to solve the Wason four-card selection task (the one from Figure 17) individually and then in small groups. While individual performance was its usual low—around 15 percent correct—something extraordinary was happening in the course of group discussions. More than half of the groups were getting it right. When Moshman teamed with Molly Geil to conduct a controlled version of this informal experiment, groups reached 80 percent of correct answers.

It may be difficult for someone who hasn't read article after article showing pitiful performance on the Wason four-card selection task to realize just how staggering this result is. No sample of participants had ever reached anywhere close to 80 percent correct answers on the standard version of the task. Students at the best American universities barely reach 20 or 25 percent of correct answers when solving the task on their own.[3] Participants paid to get it right still fail abysmally.[4]

What Moshman and Geil have achieved is the equivalent of getting sprinters to run a 100-meter race in five seconds by making them run together.

You'd think that such an extraordinary result would get the attention of psychologists. Not at all: it went completely neglected. Perhaps no one really knew what to do with it. The only researchers who paid attention to Moshman and Geil's result were those whose theories were compromised. They asked for replications[5]—not unfairly, given the suspicious tendency of many psychology experiments not to replicate.[6] While not always as dramatic as in the original experiment, the improved performance with group discussion has proven very robust.[7] It also works very well with other tasks, such as the Paul and Linda problem we introduced in Chapter 12.[8] Try the experiment with friends, colleagues, or students—it works unfailingly.

Skeptical researchers also suggested that argumentation had little to do with the improvement in performance. Rather than paying attention to the content of each other's arguments, group members, they suggested, rely on superficial attributes to decide which answer to adopt. Perhaps people simply follow the most confident group member.[9] This alternative explanation makes some sense: confidence can be an important determinant of the outcome of group discussion, for better or worse.[10]

This lower-level interpretation, however, offers a very poor description of what happens when groups discuss a reasoning task. Looking at the transcripts, it is apparent that those whose views prevail are not just saying "I know that for a fact" with a confident tone. They put forward one argument after the other.[11] We also know that a single participant with the correct answer can convince a group that unanimously embraces the wrong answer, even if she is initially less confident than the other group members.[12]

How does the exchange of arguments fare when it's impossible to demonstrate, in the strict logical sense, that a given answer is correct and that the others are mistaken? When argumentation lacks such demonstrative force, other factors become more relevant in convincing people or in evaluating claims, such as who seems more competent, or how many people support a given opinion. Still, even for problems that do not have a single definite solution, group performance is generally above that of the average group member. In some cases it is even superior to that of the best individual in the group.[13] Incidentally, even when groups fail to surpass the answer of the best individual performer, one is still better off going with the answer of the group unless there is a clear way to tell who is the best performer in the first place.

When Groups Work and When They Don't

Skepticism toward group efficiency is not entirely misplaced. When argumentation is not involved, group performance is disappointing. A hundred years ago, the agronomical researcher Maximilien Ringelmann noticed a weird pattern: tractors, horses, and humans seemed to be less efficient when performing a task jointly.[14] For instance, people pushed less hard to move a cart when they were doing it together.

Since the decrease in performance held for machines and animals as well as humans, Ringelmann assigned most of the blame to coordination problems: the strength is not applied simultaneously, which decreases the total strength exerted at any given time. However, observing prisoners powering a flour mill, he also noted that motivation could be an important factor for humans: "the result was mediocre because after only a little while, each man, trusting in his neighbor to furnish the desired effort, contented himself by merely following the movement of the crank, and sometimes even let himself be carried along by it."[15] Several decades later, social psychologists would show that such motivational factors are often the main culprits for group underperformance, labeling this phenomenon "social loafing."[16]

Groups can have disappointing performance not only when pooling physical force but also on a variety of cognitive problems. Brainstorming is a typical example. By and large, group brainstorming doesn't work. In a typical brainstorming session, participants are told not to voice their criticisms, so that they feel free to suggest even wild ideas. This doesn't work: a brainstorming group typically generates fewer and worse ideas than if the ideas of each individual working in isolation had been gathered.[17] By contrast, telling people that "most studies suggest that you should debate and even criticize each other's ideas" allows them to produce more ideas.[18]

That group performance should be disappointing in many domains only makes the successes of argumentation even more remarkable. When people argue, even about seemingly dull mathematical or logical tasks, there is no social loafing or cognitive disruption. Instead, their motivation is increased by the dialogical context. They respond to each other's arguments and build on them. Many great thinkers have noted the importance of a lively debate to fuel their intellect. Here is Montaigne:

> The study of books is a languishing and feeble motion that heats not, whereas conversation teaches and exercises at once. If I converse with a strong mind and a rough disputant, he presses upon my flanks, and pricks me right and left; his imaginations stir up mine; jealousy, glory, and contention, stimulate and raise me up to something above myself; and acquiescence is a quality altogether tedious in discourse.[19]

For a wide variety of tasks, argumentation allows people to reach better answers. The results reviewed so far, however, stem from laboratory experiments conducted in a highly controlled setting with participants who have not met before and will not see each other after the experiment. In the real world, things are different. Problems can be tremendously difficult and may not have a complete and satisfactory solution. Scientists look for the principles that govern the universe. Politicians try to get laws passed in deeply divided and confrontational parliaments. Judges search for a way to give due respect to legitimate but conflicting interests. In these situations, personal biases and affinities interfere or even take precedence. Strongly held convictions and values are attacked and staunchly defended. Does argumentation still have a positive role to play in managing more complex problems and overcoming emotional convictions?

How to Make Better Predictions

Prediction is hard, especially about the future. (Ironically this common aphorism also illustrates the difficulty of learning about the past, since it has been attributed, in one form or another, to everyone and their cousin, from Confucius to Yogi Berra.)[20] Phillip Tetlock, the expert of expert political judgment, wanted to find out just how hard it is to make good predictions in politics.[21] In the late 1980s he recruited 300 political experts, many with PhDs and years of experience, and asked them to do their job: make predictions about political events. Fast-forward fifteen years. The predictions are compared to the actual outcomes. How do the experts perform? Very poorly. They barely beat the proverbial dart-throwing chimp—random answers—and they are easily topped by simple statistical extrapolations from existing data.[22] Prediction *is* hard.[23]

In a way, the main problem faced by the experts isn't so much that they weren't accurate but that they weren't aware that they weren't accurate. The world is a complicated place, and even experts face stringent cognitive limitations on the amount of information they can acquire and process. But that didn't stop them from making extreme forecasts: the experts were often saying that a given event was nearly certain to happen or not to happen.[24] Experts were much too confident in the power of their pet theories to predict the future.

In line with the experiments on polarization and overconfidence described in Chapter 14, Tetlock found that reasoning biases were responsible for these extreme and, more often than not, mistaken predictions. He observed that when making their predictions, experts have "difficulty taking other points of view seriously"[25] and that their "one-sided justifications are pumping up overconfidence."[26]

Reason also creates distortions in the way experts revise their beliefs. When an event happens as predicted by their favored theory, the experts grow more confident. But when things don't happen as they were expected to, our political experts turn into expert excuse finders. The war that failed to erupt is just about to be declared. A small accident of history prevented their predictions from coming true. Politics is too complicated anyway . . . [27] Some experts were so skilled at finding excuses that they became even more convinced that their theories were right after the events had proven them wrong.[28] Would the experts have been better off if they had been able to talk things over?

Predictions may have never mattered more to our species than during the Cold War, when the risk of an all-out atomic war was barely prevented by an "equilibrium of terror." The U.S. Air Force was one of the actors looking for better forecasts about the effects of nuclear war, and for this it turned to the RAND Corporation. Averaging the opinions of several experts offers a simple and efficient solution to improve on their forecasts. Yet two of RAND's researchers, Norman Dalkey and Olaf Helmer, thought they could do better by giving each expert information about what the other experts answered—such as the average answer, for instance. Experts made new predictions based on this information; the predictions were averaged and again provided to all the participants, who got to make another prediction; and so forth for a few rounds.

This reiterated averaging technique, known as Delphi, was first used to figure out how many bombs would the Russians have to drop on U.S. industrial targets to reduce their output by three-quarters.[29] Fortunately, Dalkey and Helmer never found out if their method had yielded accurate forecasts in this specific case, but since the 1950s, many studies have shown that it can improve a variety of predictions, from defense issues to medical diagnoses.[30]

The Delphi method offers several advantages over face-to-face discussions. In face-to-face discussions, providing the best forecast may be less important than pleasing a senior colleague or keeping with the consensual opinion. Face-to-face discussions also require getting a group of busy experts in one room at the same time, which is not always easy to arrange. Delphi's anonymous questionnaires solve both problems.

Yet if the argumentative theory is right, the original form of Delphi is missing out on a major way of improving predictions: the exchange of reasons. If Zoe believes Italy has an 80 percent chance of winning the next soccer world cup, and Michael tells her he thinks it's only 20 percent, what should Zoe do? Balance the two opinions and adjust the odds to 50/50? On average, people put in this situation only go part of the way toward the other opinion. Zoe could settle on 60 percent, for instance. After all, she knows why she thinks Italy should have good odds, but she doesn't know the reasons for Michael's opinion.[31] If she knew why Michael is giving Italy low odds, she might be more inclined to change her mind.

In the Delphi method, instead of receiving only an average of others' forecasts, the experts can also be given the reasons for others' forecasts. Gene Rowe and George Wright looked at the difference this makes to the predictions.[32] As a matter of fact, the reasons did not make the experts change their minds more often. Indeed, they were more likely to cling to their initial opinion than when provided only with averages. But reasoning was not failing; it was merely revealing its discriminating power.

Not all reasons are good reasons. If Michael tells Zoe he thinks Italy has a 20 percent chance to win based on the predictions of an octopus,[33] she would be crazy to update her estimate at all. But if Michael tells Zoe he has insider information about the failing health of Italy's key forward, she might simply adopt Michael's odds. This is exactly what the participants in Rowe and Wright's experiment were doing. They were not changing their minds more, but they were doing it more discriminately. They were changing their minds when they should, in the direction they should. And they were making better predictions.[34]

Providing a one-time summary of reasons is a good step forward, but it falls short of making the best of reasoning. Reasoning thrives in the back-and-forth of conversation, when people can exchange arguments and counterarguments. Online communication enables groups of experts to exchange arguments at a distance, opening up prospects for even better forecasts.

Twenty years after his original study of expert political judgment, Phillip Tetlock, together with Barbara Mellers and other colleagues, launched an even more ambitious experiment.[35] Over 1,300 participants were recruited and asked to make geopolitical predictions. Participants from a first group worked alone, kept in the dark about other participants' forecasts so as to maintain the independence of their judgments. Participants assigned to the second group also worked alone, but as in the early versions of the Delphi method, they were provided with statistical information about others' forecasts.

The third group was divided into teams of about twenty people who were allowed to discuss the forecasts together, online. Nearly all of their predictions proved more accurate than those of the independent forecasters, and they also beat the second group on two-thirds of the forecasts. Exchanging arguments had allowed them to produce significantly better predictions.

In Defense of Juries

So far we have seen that participants provide better answers to a variety of tasks when they are allowed to discuss them in small groups. We have seen that groups of experts can also take advantage of argumentation to improve their forecasts. But this chapter opened in a jury room—albeit a fictitious one—not a well-controlled experiment or an online chat among experts. The jury room is a typical face-to-face situation: tempers may flare, the search for consensus become paramount, errors never be dispelled. Jurors are the archetypal *non*experts, with only a tenuous understanding of the law. In *12 Angry Men,* the discussion is filled with prejudice, and many other biases cloud jurors' decisions.

All too often, jurors form the wrong intuition regarding the verdict after hearing the evidence. With its myside bias, solitary reasoning is unlikely to help correct this initial intuition. Deliberation could turn into group polarization, amplifying rather than correcting shared prejudices. In light of these limitations, *12 Angry Men* seems optimistic indeed, an edifying and wishful ode to the power of argumentation. Perhaps we should follow the advice of Cass Sunstein and his colleagues: take some decisions out of juries' hands and give them to "specialists in the subject matter."[36]

Before damning juries, we should consider the alternative: judges and experts. Sir Edward Coke was undoubtedly such an expert. This English jurist of the late sixteenth and early seventeenth centuries was "possibly the most learned common lawyer of all time."[37] He was able to base his opinions "on innumerable medieval texts, most of them manuscript rolls, which he had perused with indefatigable zeal." All this perusing, however, was not antiquarian fancy. Coke "was clearly hoping to find precedents that would suit his legal and political convictions," and he "sometimes misinterpreted precedents to support his case." It may have been Coke that Sir William Blackstone had in mind when he warned, in his hugely influential *Commentaries on the Laws of England* of 1766, that a judge's knowledge and intelligence are no guarantee of fair opinion: "in settling and adjusting a question of fact, when entrusted to any single magistrate, partiality and injustice have an ample field to range in; either by boldly asserting that to be proved which is not so, or more artfully by suppressing some circumstances, stretching and warping others, and distinguishing away the remainder."[38]

Readers should not be surprised that judges, however competent, have a myside bias, using their erudition to defend preconceived opinions rather than arrive at an impartial verdict. Jurors are obviously not exempt from the myside bias, but, as Blackstone realized, deliberation has the potential to compensate for each juror's biases. Having berated judges, he continued: "Here therefore a competent number of sensible and upright jurymen . . . will be found the best investigators of truth, and the surest guardians of public justice." Two centuries after Blackstone, Justice Harry Blackmun of the U.S. Supreme Court would defend juries in similar terms: "the counterbalancing of various biases is critical to the accurate application of the common sense of the community to the facts of any given case."[39]

Blackstone may have been right in his pessimistic assessment of judges,[40] but was he also right that jury deliberation could balance out jurors' biases? This is not an easy question to answer. We do not have access to the arguments exchanged by jurors. At best, postdeliberation interviews can show that a view represented by a minority at the outset sometimes becomes the final verdict—we can tell at least that deliberation can change jurors' minds.[41] To know more about the effects of deliberation, we must rely on studies of mock juries.

In the early 1980s, Reid Hastie, Steven Penrod, and Nancy Pennington conducted a very important study of mock juries.[42] In order to make their experiment as realistic as possible, they recruited people who had been called for jury duty and showed them a three-hour video reenactment of a real trial before sending them in to deliberate. In this trial, the defendant stood accused of having stabbed a man to death during a fistfight that escalated. While the killing was well established, the verdict could plausibly range from not guilty—if the defendant were found to have acted in self-defense—to first-degree murder—if the defendant were found to have premeditated his crime. While it is impossible to know what the correct answer is for sure, according to the opinion of many legal experts, the appropriate verdict was second-degree murder; it was also the verdict delivered in the real trial that inspired the study.

Right after the jurors in the experiment had seen the three-hour video, they had to say which verdict they favored. The most common answer was manslaughter. Only a quarter of the jurors favored second-degree murder. In other words, most jurors initially got it wrong. Most juries, however, reached the best verdict. Even though the verdict of second-degree murder must have been defended by only a few jurors in some groups, these jurors often managed to convince the whole jury.[43] Deliberation had allowed many juries to reach a better verdict. Just like Blackstone expected, deliberation allowed jurors to counterbalance their respective biases.

Indeed, this is exactly what Phoebe Ellsworth, a scholar of law and psychology, observed while replicating the experiment of Hastie and his colleagues:

> Individual jurors tended to focus on testimony that favored their initial verdict preference: Testimony about the previous confrontation between

> the two men was generally raised by jurors who favored a murder verdict, whereas testimony that the victim punched the defendant immediately before the killing was generally raised by jurors who favored manslaughter or self-defense. This tendency is not a weakness, but rather a benefit of the deliberation process—the opportunity it affords for comparing several different interpretations of the events along with the supporting factual evidence.[44]

12 Angry Men is not that fanciful. As in the movie, individual jurors in real trials may be biased, they can make mistakes, and they certainly defend dubious interpretations. Yet deliberation makes jurors review the evidence more thoroughly and more objectively, compensating for individual biases and allowing juries to reach better verdicts. Ellsworth concluded her review of the movie with these optimistic words: "*12 Angry Men* is an Ideal, but it is an achievable Ideal."[45]

Argumentation Works

The interactionist approach to reason predicts that people should be good at evaluating others' reasons, rejecting weak ones and changing their mind when the reasons are good enough. Although this might seem like a trivial prediction, it runs against a general pessimism regarding the power of argumentation. For instance, people asked to estimate how easily participants would solve a logical task either on their own or in small groups don't think that groups would do much better than individuals.[46] Even psychologists of reasoning—who should know better—underestimate how well groups perform.

The results reviewed here belie this pessimistic but common view of argumentation. Again and again, we see people changing their minds when confronted with good arguments. Whether they solve logical tasks or look for new solutions to open-ended problems, whether they are experts or laypeople, whether they reflect on geopolitics or ponder what verdict to deliver, people reach better conclusions after debating the issue with their peers.

In Chapters 16 through 18, we will discover still more settings in which argumentation allows good ideas to spread and groups to outperform individuals, showing that Thomas Jefferson was not unduly optimistic when he wrote that

> Truth is great and will prevail if left to herself, that she is the proper and sufficient antagonist to error, and has nothing to fear from the conflict, unless by human interposition disarmed of her natural weapons, free argument and debate, errors ceasing to be dangerous when it is permitted freely to contradict them.[47]

V

Reason in the Wild

Chapters 11 through 15 have painted a picture of reason that unambiguously supports the novel interactive approach over the standard intellectualist approach. But haven't we focused on situations in which we were most likely to observe results that fit our pet theory? Much of the evidence we used came from laboratory experiments using psychology students in American universities as subjects. Isn't it a bit risky to draw conclusions about how human reason works and why it evolved from this tiny and arguably rather unrepresentative sample of humanity? In Chapters 16 through 18, we expand the range of our inquiry and look for evidence that the fundamentals of reason can be found in a wide variety of contexts: remote Mayan communities in Guatemala, kindergarten playgrounds, citizens' forums, laboratories meetings, and more.

16

Is Human Reason Universal?

With huge temperature swings—from minus 40° to plus 40° Celsius—and scarcely any rain, Uzbekistan is not a very fertile land. Its unforgiving climate and its remoteness conspired to maintain a feudal system until the early twentieth century.[1] Modernization only started with integration into the Soviet Union in the 1920s, as Moscow decided to open hundreds of schools throughout the country.[2] This offered Alexander Luria, of the Moscow Institute of Experimental Psychology, the perfect opportunity to test his ideas. Following his master Lev Vygotsky, Luria thought that humans acquire most of their cognitive skills through learning, including school learning. Uzbekistan provided him with people who had just begun the schooling process but were otherwise identical to the illiterate peasants from nearby villages. By comparing these two populations, Luria could pinpoint precisely the effect that even a modicum of schooling had on cognition.

One of the objectives of the "psychological expedition to Central Asia"[3] Luria launched in 1931 was to investigate logical reasoning. The Russian psychologist had no doubt that illiterate peasants were capable of reasoning with familiar materials. Indeed, they could probably win an argument with any outsider about cotton growing. What he was looking for was different, an ability to draw the conclusion of an argument for its own sake, irrespective of whether its premises are true or false—a skill he thought lay beyond the abilities of unschooled populations. Luria used problems that were logically trivial but whose content was unfamiliar to the participants, so they would have to evaluate the logic of the argument itself:

In the Far North, where there is snow, all bears are white. Novaya Zemlya is in the Far North. What color are bears there?[4]

After one or two years of formal education, the Uzbeks found this problem trivial. But when unschooled peasants were interviewed, the vast majority seemed at a loss, providing answers such as, "There are different sorts of bears" or "I don't know; I've seen a black bear—I've never seen any others."[5]

How WEIRD Is Argumentation?

Evolution thrives on diversity. It is only because individuals of the same species vary in their heritable features that species evolve. Natural selection, however, swamps the diversity it feeds on. When a heritable trait allows its bearer to out-reproduce its conspecifics, it spreads through the population until, some generations later, everyone carries it. There are exceptions. Some traits are sex specific, and others are more advantageous when they are not universally shared within a species, but we do not see why reason would be one of these exceptions. If we are right that reason is an adaptation that helps solve problems of coordination, reputation management, and communication encountered by all, it should be shared by normally developing humans and not just by a minority of them, or just by men or women.

In every individual, reason needs, to develop normally, some input—conversation, arguments. Different cultures and milieus may provide different input in this respect, both in terms of quantity—argumentation is strongly encouraged in some cultures, somewhat inhibited in others—and in terms of quality—various forms of argumentation may be favored or disfavored. All human societies, however, rely on a richness of communication not found in other species, and this reliance provided the selection pressures for the emergence of reason. Hence, if our approach is right, reason could not be the cultural product of institutions that only spread in the last centuries, such as schooling.

The very idea that reason is a historically situated cultural invention has been a commonplace in the social sciences. Before Luria's expedition, the French theoretical anthropologist Lucien Lévy-Bruhl had painted a picture of a "primitive mentality" "uncultivated in following a chain of reasoning which is in the slightest degree abstract."[6] He and others had argued that people in other cultures may reason, but on the basis of an altogether different logic. Both views—that reason is a relatively recent historical development and that it takes radically different forms across cultures—are incompatible with the evolutionary approach we defend.

Historical, anthropological, and linguistic evidence points to a potentially damning flaw in our argument so far: the focus on examples and experiments from Western cultures. As a group of cross-cultural psychologists and anthropologists recently put it, these are WEIRD people—people coming from Western, Educated, Industrialized, Rich, Democratic countries. The acronym is well deserved, for this sample often sits at the extreme range of the variability observed in human populations. For instance, American undergraduates—by far the largest pool of participants in psychology experiments—are more individualistic[7] than their noncollege peers, who are more individualistic than Americans from the previous generation, who were already more individualistic than just about any other people on earth.[8]

The importance given to argumentation among WEIRD people could be another freak trait inspired by the ancient Greeks' reliance on argumentation in science, politics, and legal institutions. In most Western cultures, the existence of disagreements is seen as a normal aspect of human interaction, one that should be organized rather than suppressed and that can have positive effects. Universities in particular are supposed to encourage the practice of debate. Couldn't it be, then, that seeing argumentation as beneficial is a cultural bias, and that reason—at least reason as we have described it with its justificatory and argumentative functions—is a culturally acquired skill rather than an evolved, universal trait? Could, for instance, people in other cultures be excellent solitary reasoners but terrible arguers? Or not be reasoners at all?

How to Avoid Looking Like a Fool

The conclusion "the bears in Novaya Zemlya are white" seems so inescapable that it strains credulity to believe some people incapable of drawing it. Still, more recent research validates Luria's results. His experiments were successfully replicated with several unschooled populations,[9] and other experiments remind us that even the most taken-for-granted skills can need to be culturally acquired.

Imagine someone putting three coins, one by one, into an opaque container. Your task is to retrieve the coins. Can't everybody do this? As a matter of fact, no. Only people who have learned to count can. Had you been born a Pirahã, a member of a small Amazonian tribe with no words for numbers, you would be at a loss to retrieve the coins. When Peter Gordon performed this simple experiment with Pirahã participants, only two-thirds stopped exactly at three coins—and the performance quickly deteriorated as the number of coins increased.[10] The Pirahã were not simply wary of this foreigner asking them to play weird games. They did very well on other tasks. They genuinely lacked the ability to count to three.

Is it the case, then, that some people are really unable to produce and evaluate arguments simply because they have unfamiliar premises? No, in the case of reasoning, the problem, it turns out, is merely one of motivation and, more specifically, of social propriety. For one thing, in all the populations tested, some people—a third of the participants, perhaps—easily provided the right answer. They hadn't developed on their own a new cognitive ability. Some people were just more willing to play the experimenter's game.

Why would anyone be reluctant to answer such simple questions as "What color are bears there"? In small-scale populations, people are very cautious with their assertions, only stating a position when they have a good reason to (unlike, say, pundits).[11] Clearly, these conditions are not met when a stranger tells a weird story about some absurdly colored bears in a far-off place. Only a fool would dare make such a statement, a statement she could not appropriately defend. As we saw in Chapter 14, people try to avoid doing things they cannot justify. This is exactly what happened in this exchange, captured as the experimenter asked the white bear question to an unschooled adult Uzbek: "If a man was sixty or eighty and had seen a white bear and had told about it, he could be believed, but I've never seen one and hence I can't say. That's my last word. Those who saw can tell, and those who didn't see can't say anything!" At this point, a young Uzbek volunteered: "From your words it means that bears there are white." But the older man concluded, "What the cock knows how to do, he does. What I know, I say, and nothing beyond that!"[12] They both knew what the experimenter wanted, but only the young man was willing to wager his credit in this weird game.

In order to let people be more comfortable engaging with unfamiliar premises, psychologist Paul Harris and his collaborators gave unschooled participants a richer context. Instead of happening in a far-off but real place, the problems—otherwise similar to those used by Luria—were set on a distant planet. In these conditions, people had less scruple to engage in playful suppositions, and they gave the logical answers more easily.[13] If we learn in school to play reasoning games, drawing weird conclusions from arbitrary premises,[14] basic reasoning skills require no such formal education. All normally developing humans can produce and evaluate arguments. But do they argue or keep these skills for private use?

Argumentation in East Asia

On March 11, 2011, one of the strongest recorded earthquakes wreaked havoc in Japan's Tōhoku region. The Fukushima Daiichi power plant was badly damaged, leading to the worst nuclear accident in the world since the 1986 Chernobyl accident in Ukraine. The Japanese political world was badly shaken, as people discovered glaring gaps in the regulatory process supposed to keep nuclear power plants safe.

The aptly nicknamed "nuclear power village" was one of the culprits.[15] In this gathering of private companies and regulatory agencies, peace reigned. And that was the problem. For the Japanese scholar Takeshi Suzuki, the village was governed by *kotodama*—a belief in the mystical power of words.[16] By making taboo the words that can lead to disputes, *kotodama* "serve[s] a social role to discourage or hinder argumentation." More generally, because of the intense pressure to maintain social harmony, "the Japanese are not trained to argue and reason." Japanese culture would have precluded the members of the nuclear power village to discuss the dangers of nuclear energy, leading them instead to construct a "safety myth of nuclear power plants."[17]

Suzuki is not the first to deplore the "lack of argumentation and debate in the Far-East," to quote the title of an article by Carl Becker.[18] Becker lays the blame for this lacuna on age-old cultural precepts prevalent in Eastern cultures, encapsulated in such dictums as "Eloquence . . . is nothing but hell-producing karma" (from the Rinzai school of Zen). Other authors suggest that East Asian languages are poorly equipped to deal with logic and argument. In his wide-ranging comparison of Eastern cultures, Hajime Nakamura noted the vagueness inherent in the Chinese and the Japanese languages and the obstacle this vagueness sets to "expressing logical conceptions."[19]

Because they are often singled out for the opprobrium they cast on argumentation, East Asian cultures offer a good case study for our claim that some traits of argumentation (and to begin with, obviously, its very presence) are universal. Do people in East Asian cultures argue and benefit from doing so? Do they even have the option of properly arguing, given the alleged vagueness of their languages? The answer is a resounding yes. East Asian languages did not stop the development of logic and rhetoric. In the fifth century BCE, the Chinese Mohists created logical systems as complex and abstruse as those of Western scholastics. Two hundred years later, the rhetoric of the great Han Feizi was so rich that no Roman rhetor could have matched the "subtlety of [its] psychological analysis."[20]

What about the alleged cultural taboos against argumentation, castigated as an egotistical breach of social harmony? Ironically, for all their proclamations against eloquence and argumentation, East Asian intellectuals never stopped confronting one another. Confucius's *Analects* state that "the superior man is slow to speak but quick to act." This did not stop Confucianists from writing treatise after treatise "reacting to criticisms of opponents" and "engaging in philosophical debate with rival doctrines."[21] The history of Daoism is equally full of arguments, in spite of the *Tao Te Ching*'s assertion that "a good man does not argue; he who argues is not a good man."[22]

Berating argumentation didn't stop East Asian intellectuals from arguing. No big surprise here. Arguing is what intellectuals do for a living. Are other people more inclined to respect the precepts erected—and flouted—by the wise men? Hardly. Recent historical work has revealed that even a country with such tight social control as Japan "is not the 'relatively peaceful' arhetorical society described by Becker and others, but a country whose past three hundred years have been marked by great ideological and often physical conflict, and whose disputes have often been conducted and recorded in the form of debates."[23]

Even early Japanese tradition reflected an understanding of the benefits of group discussion. At the dawn of the seventh century, Prince Umayado set out a new constitution. Its last article reads:

> Decisions on important matters should not be made by one person alone. They should be discussed with many. But small matters are of less consequence. It is unnecessary to consult a number of people. It is only in the case of the discussion of weighty affairs, when there is a suspicion that they may miscarry, that one should arrange matters in concert with others, so as to arrive at the right conclusion.[24]

What little experimental evidence there is regarding the benefits of argumentation in East Asian cultures converges with the results obtained in the West. When Japanese students are given the Wason four-card selection task to solve on their own, they perform as badly as their American or European counterparts. But when the same students must discuss the answer in small groups, they enjoy the benefits of argumentation, and most groups converge on the right answer.[25]

The scorn that Japanese culture supposedly displays toward argumentation is not to be blamed for what happened in Fukushima. It was a run-of-the-mill case of regulatory capture. Through more or less direct forms of bribery, regulatory agencies can come to serve the private interests they are supposed to regulate. In Japan, civil servants who play along with companies can expect to be offered cushy jobs upon retirement—a practice called *amakudari.* Other countries can do it differently, but they also suffer from regulatory capture. Let's not blame Rinzai Zen for a sadly widespread political practice.

Argumentation in Small-Scale Societies

For millennia, Eastern and Western cultures have relied on writing to develop complex rhetorical traditions and argumentation-centered institutions. The human species did not evolve in such a culturally sophisticated context. The environment of our ancestors was closer to that of Luria's unschooled peasants, and even closer to the conditions in which modern-day hunter-gatherers live. Even if we accept that everyone, schooled or unschooled, can reason, that doesn't mean that everyone argues.

Of the two classic depictions of the original state of humanity—the "noble savage" and Thomas Hobbes's "war of all against all"—neither offers a good context for argumentation. The noble savage would not have had the motivation to engage in argumentation. The war of all against all offers even fewer opportunities for debates to flourish.

These views are nowhere near the truth. Our ancestors were neither living in harmony with one another nor waging constant war against one another. And argumentation may have played at least as important a role in their social lives as in in ours. When a collective decision has to be made in a modern democracy, people go to the voting booth. Our ancestors sat down and argued—at least if present-day small-scale societies are any guide to the past. In most such societies across the globe, when a grave problem threatens the group—ecological crisis, war, protection of common resources—people gather, debate, and work out a solution that most find satisfying.[26]

Even the most egalitarian of our modern societies looks quite hierarchical compared with a typical small-scale society. Hunter-gatherer societies have no king, no general, no manager to boss people around. If some members have more influence, it is not because they are invested with some supernatural or birthright authority but because of the services they render to the community, for instance, in hunting, in war, and in coming through discussion to good collective decisions. This generalization holds even when a rigorous chain of command might be expected—as in societies plagued with constant warfare.

In the many years anthropologist Napoleon Chagnon spent among the Amazonian tribe of the Yanomamö, he witnessed so much conflict that he attributed about 50 percent of adult male deaths to violence.[27] Yet this permanent state of conflict did not give rise to a strong hierarchical structure. When Kąobawä, a village headman, wanted to be heard, he couldn't simply raise his voice or threaten with his club. He had to rely on argumentation. As Chagnon reports, "should someone be planning to do something potentially dangerous, [Kąobawä] simply points out the danger." Despite his position, "he so diplomatically exerts his influence that the others are not offended."[28]

We could pile up examples of sophisticated argumentative practices in small-scale societies. In the Trobriand Islands, Edwin Hutchins reported convoluted legal argumentation and complex chains of reasoning.[29] Among the Lozi of Zambia, Max Gluckman discovered a culture focused on debates with a rich vocabulary to describe the quality (or lack thereof) of someone's argumentation:

kuyungula—to speak on matters without coming to the point
kunjongoloka—to wander away from the subject when speaking
kubulela siweko—to talk without understanding
muyauluki—a judge who speaks without touching on the important points at issue
siswasiwa—a person who gets entangled in words
siyambutuki—a talker at random[30]

To show that argumentation can be as effective in traditional, small-scale societies as it is in ours, Thomas Castelain visited remote groups of K'iché Maya in rural Guatemala, people who practice subsistence farming and, in most cases, can neither read nor write, and only speak their native language.[31]

With local help, he asked K'iché participants to solve so-called conservation tasks on their own. These tasks require understanding that a given property of an object is conserved across some changes. In the present case, participants were presented with two glasses containing the same amount of water and asked in which glass the water would rise more: in the first glass, in which a ball of Play-Doh was about to be plunged, or in the second glass, in which the two halves of a ball of Play-Doh of identical size would be plunged.

Only a third of the participants answered that the water would rise as much in both glasses. But when they had to discuss the task in small groups, over 70 percent got the right answer. In fact, as soon as a participant had the correct answer, she was nearly always able to convince other group members to change their mind—exactly what had been repeatedly observed in WEIRD cultures.

How to Reconcile Evolutionary, Cognitive, and Anthropological Perspectives on Reasoning

We have focused on the universal traits of reasoning and argumentation, debunking the idea that schooling is necessary to understand simple reasons or that, in some cultures, argumentation might be suppressed or even never develop. Reasoning and argumentation are found everywhere, as we should expect if reason is an evolved module and if the production and evaluation of argument is one of its two main functions.

To say that reason is a universal mechanism does not imply that it works in exactly the same manner in all places or that specific societies can affect reason only in superficial, merely cosmetic ways. Reason is deployed in a variety of cultural practices and institutions that may inhibit some of its uses, enhance others, and provide socially developed cognitive tools that complement and extend naturally evolved abilities.

In all human societies, for instance, there are rules and institutions for resolving issues of rights, going from private disputes to criminal cases. The way these issues are being argued varies greatly across cultures. In many small-scale societies, there is no court system; the parties make their case without the help of lawyers. Elders, local assemblies, or political leaders play the role of arbiter or judge. Still, institutional forms of epistemic vigilance may play a role in these proceedings. There may be, for instance, culturally developed forms of vigilance toward the source such as oaths or even ordeals, which are believed to deter lying. Vigilance toward the content, on the other hand, typically relies just on commonsense production and evaluation of arguments. There are typically no rules regarding admissible evidence and no standard of proof. In larger societies with a state organization, by contrast, arbitration, litigation, and criminal justice are in the hands of complex institutions and obey a whole range of precise rules (with much cultural variation).

One of the most famous criminal trials in recent history, that of the American football star O. J. Simpson, provides a striking illustration of the degree to which legally regimented argumentation may depart from commonsense reasoning. Simpson was accused of having murdered his ex-wife and a friend of hers. In October 1995, at the end of an eleven-month trial where his lawyers argued that he was not involved in any way in the murders, he was acquitted.

The families of the two victims, however, filed a civil suit against him for "wrongful death." Simpson lost this civil trial, and in February 1997, he was ordered to pay $33.5 million in damages for the two deaths the jury concluded he had caused.

To people unfamiliar with the American legal system, this may look like blatant incoherence: how can the same man be acquitted of a double murder and yet made to pay compensation for it? The key consideration here is the difference in the standards of proof that apply in a criminal and in a civil trial. In the criminal trial, the standard being "proof beyond reasonable doubt," Simpson's lawyers had argued that there was indeed reasonable doubt and won. In the civil trial, the standard being "clear and convincing evidence," the civil parties pleaded that there was such evidence and won against Simpson. Even so, most people, even in the United States, found it hard to make any intuitive sense of this double verdict. Legal scholars, on the other hand, could easily find reasons for these two divergent decisions by reasoning at a more abstract level: there are good reasons for different standards of proofs in a criminal trial (where you might sentence someone to prison or even death) and a civil trial (where what is at stake is merely money). Legal arguments in O. J. Simpson's case may well have been at odds with commonsense reasoning. Still, like all arguments, these arguments were ultimately rooted in intuitions about higher-order reasons.

The use of argumentation for resolving conflicting interpretations of events and rights fits quite well the interactionist approach to reason and the argumentative theory of reasoning. Not all culturally developed forms of reasoning provide such obvious illustration of the approach. We mentioned in Chapter 9 the puzzles, riddles, paradoxes, and other brain twisters that in some cultures are produced as stimuli for the pleasure of solitary reasoners. How is this supposed to fit with the interactionist approach? Some cases of argumentative interaction do not fit the approach in any obvious way, either—consider the debating societies such as the Oxford Union, founded in 1823 and currently the world champion of competitive debating. In competitive debates, two or more teams try to best each other through arguments, arguing for a point of view that they have been arbitrarily assigned to defend, and on a topic the audience might not care much about, such as "Should states construct false historical narratives that promote social cohesion?" and "There is a potion which can stop you falling in love. As an eighteen-year-old, should you take the potion?"[32] Why should people bother to argue for opinions they do not hold on issues indifferent to them? Why should people pay any attention to such exchanges of arguments?

We argued in Chapter 10 that reasoning has a double argumentative function: for a communicator, reasoning is a means to produce arguments in order to convince a vigilant audience; for the audience, reasoning is a means to evaluate these arguments and accept them when good, or reject them when bad. In the case of solitary reasoning on puzzles and enigmas pursued as a leisure activity, producing arguments to convince others or evaluating others' arguments needn't play any role. In the case of competitive debating, arguments are produced not in order to convince an audience of the truth of their conclusion but in order to convince a jury of one's argumentative skills.

These interesting cases, however, do not present to the argumentative theory of reasoning a greater challenge than do recreational sex, masturbation, and pornography to the claim that the main function of sex is reproduction. Some evolutionary considerations should help make the point.

Any evolved mechanism is adapted to the environmental conditions in which it has evolved and may malfunction or produce nonfunctional effects in different conditions. Breathing more carbon dioxide than is found in normal air may lead to suffocation. Breathing more oxygen may cause euphoria. These abnormal effects do not challenge the standard view that the breathing mechanisms of mammals are well adapted to their function of delivering oxygen to the body and removing carbon dioxide as needed.

Evolved cognitive modules are typically adapted to processing information belonging to a given domain and to drawing specific inferences from it. We called such a domain the "proper domain" of a cognitive module.[33] In the proper domain of a snake avoidance module, for instance, are snakes present in the environment. In the proper domain of a mindreading module are the mental states of people with whom the individual is or might be interacting. The operations of such cognitive modules are triggered by input information provided by other modules, lower-level perception modules in the case of the snake module, modules processing information about, for instance, the behavior, speech, or visual expression of others for the mindreading module.

Inputs that trigger the operation of a module are imperfect. They do not pick out all and only cases at hand that fall within the proper domain of the module. There is a cost in time and energy to detecting what belongs to the proper domain of a module. This cost makes it more efficient to accept a certain rate of detection errors. The operations of most modules are, in fact, triggered by simple diagnostic criteria rather than by a complex pondering of a variety of factors (a pondering that might need a cognitive mechanism of its own and might not be that efficient anyhow).[34] Often, the trigger is oversensitive—think of the trigger of jealousy or of danger detection—but this oversensitivity may well be adaptive.

The range of inputs that actually activates a module is its actual domain. There is no way the actual domain would exactly correspond to the proper domain. Mistakes in detection are unavoidable and result in either false positives (false alarms) or false negatives. As we noted in Chapter 11, depending on the relative cost of false negatives (such as mistaking a snake for a piece of wood) and false positives (mistaking a piece of wood for a snake), an efficient module may well be biased toward making the less costly type of error so as to avoid as much as possible ever making the more costly one. In such cases, the mismatch between the proper and the actual domain of a module is in fact advantageous.

The mismatch between the proper and the actual domain of cognitive modules can be exploited in interaction across species (such as in prey-predator interaction, where mimicry in one species results in false positive for another species, or camouflage, which results in false negatives). It can also be exploited in social interaction within a species. Many aspects of culture are based on such exploitation. Belief in supernatural agents, for instance, may be rooted in a disposition to overdetect agency and intentionality to the point of attributing mental states to nonagents such as the sun or a mountain, and to seeing in natural patterns the effect of the actions of an imaginary agent.[35] Mismatches between the actual and proper domains of modules are a bonanza for the development of cultural ideas, practices, and artifacts.

The proper domain of reasoning is disagreements between oneself and others—clashes of ideas. Reasoning aims at reducing these disagreements by producing arguments to convince others or by evaluating the arguments of others and possibly be convinced oneself. The actual domain of reasoning, the kind of input that triggers its operations, is, we have argued in Chapter 9, the detection of a clash of ideas. Most clashes of ideas are detected in discussion with others. This is not surprising; most clashes of ideas emerge in clashes of people. In these cases, the proper and the actual domains of reasoning overlap.

Notwithstanding, the overlap between the proper and the actual domains of reasoning remains partial. There are typical false negatives: people in a dominant position or in the vocal majority might pay little attention to the opinion of subordinates or minorities and fail to detect disagreements. There are also false positives: either clashes of ideas that occur between third parties with whom we are not even in a position to interact, as when we read ancient polemics or watch a debate on television, or clashes of ideas within oneself. Clashes of ideas within oneself may be internally generated dilemmas. We cannot make up our minds between two opinions or two possible decisions that have comparable intuitive support. Clashes of ideas within oneself may also consist of adopted artifacts: culturally produced brainteasers for entertainment, or paradoxes for flexing one's philosophical muscle.

Why should solving puzzles, thinking about paradoxes, or watching other people argue be, for some people at least, an enjoyable experience? Here again, evolutionary considerations are relevant. The operation of many evolved mechanisms requires energy, time, and effort to fulfill their function, but the benefit achieved may not be perceived by the organism, or at least not sufficiently to motivate the effort. The main biological benefit of sex is reproduction, but animals (including, for most of their history, humans) are not aware that sex produces offspring. Strong desires and sexual pleasure have evolved to motivate animals to mate.

Once sex comes with hedonistic rewards, these rewards may be achieved or enhanced not just by basic mating but by means of various forms of sexual activity that may or may not contribute to reproductive success. Moreover, among humans, the procurement of these rewards to others may bring economic and social benefits and develop in a creative variety of sexual practices.

Reasoning is a relatively high-investment cognitive activity bringing indirect fitness benefits. It is ultimately beneficial to one's fitness, we claim, to overcome the limits to communication imposed by insufficient trust and, in so doing, to be better able to influence others and to accept the influence of others wisely. Such perspectives, however, are unlikely to motivate much investment in reasoning here and now. So, we suggest, the very performance of reasoning provides some hedonistic reward. This may not be true for all humans to the same degree or in the same way. Those who, because of their social position or their personal disposition, are reluctant to argue may nevertheless enjoy watching others do so, and they may use reasoning for evaluation rather than for production of arguments.

Competitive debates, for instance, spread because they hijack reasoning, putting it to uses it did not evolve for. The audience listens to arguments not to acquire sounder beliefs but for the sheer pleasure of watching a competition and of processing very good arguments. The debaters produce arguments not to convince an audience but to enjoy and display their reasoning skills. Ancient philosophical quarrels, controversies between bloggers on abstruse topics, and formal debates do not belong to reasoning's proper domain—reasoning did not evolve to process such stimuli—but they fall within its actual domain, and this is what explains their cultural success.

An understanding of reasoning as a function of an evolved reason module need not conflict with a historical and anthropological interest in the remarkable variety of the use of reasons and arguments across cultures. In fact, we suggest, it helps explain it better and formulate testable hypotheses not just about the varieties of reasoning but also about its very variability.

Early Reasoners

Work with adult reasoners suggests that some traits of reasoning are universal. Striking results from developmental and educational psychology show that these traits emerge very early on. The acquisition of reason doesn't depend on teaching institutions and a reason-oriented cultural environment—children spontaneously use reason to defend their ideas and their actions, and to evaluate the reasons offered by others.

We start with a fictional character, the heroine of the Countess of Ségur's *Les Malheurs de Sophie* (The Misfortunes of Sophie), a bestseller of nineteenth-century French children's literature. Sophie's very argumentative character aggravated her mother but delighted readers. Here is an example:

Sophie: Mother, why don't you want me to go see the stonemasons without you? And when we go, why do you always want me to stay with you?

Mother: Because stonemasons throw stones and bricks that could hit you, and because there is sand and lime which could make you slip or hurt you.

Sophie: Oh! Mother, first I'll be very careful, and also sand and lime can't hurt me.

Mother: You believe this because you are a little girl. But I, being older, know that lime burns.

Sophie: But Mother . . .

Mother, interrupting: So, do not reason so much, and be quiet. I know better than you what can hurt you or not. I do not want you to go in the yard without me.

Sophie lowered her head and said nothing more; but she took on a morose air and said to herself: I'll go anyway; this amuses me, I'll go.[36]

Although Sophie is only four years old, she is well able to understand her mother's arguments and to reply to them. Is this a figment of a writer's imagination? No, the Countess had been able to observe her eight children, and psychological studies have confirmed that children start to give arguments extremely early, sometimes as early as eighteen months. Three-year-olds exchange arguments routinely, using them in one-third of their (numerous) disputes.[37]

Children's reasoning shares two basic features of adult reasoning: myside bias and low evaluation criteria for one's own reasons. Judy Dunn and Penny Munn observed that three-year-olds were fifteen times more likely to invoke their own feelings rather than their interlocutor's feelings in their arguments.[38] A three-year-old boy (Hugo's son Christopher) who wanted to be left alone to climb some big stairs argued that he was a big boy. When a few minutes later he got tired and wanted to be picked up, he pointed out that he was a little boy. The inconsistency didn't seem to bother him. He also once argued that he shouldn't go to bed right away because he was too tired. Clearly, there is some room for improvement in how children use arguments (although, to be fair, his parents sometimes used similar arguments without much more regard for consistency).

Parents will not be shocked to hear that children spontaneously produce many arguments—indeed, like Sophie's mother, they might even find this skill a tad exasperating. But children do not only use reasoning to explain why they shouldn't go to bed or why they are entitled to take their sibling's toys. They also pay attention to others' arguments and evaluate them.

Psychologists have conducted many observations of parents interacting with their children in an effort to document differences in parenting styles. When it comes to telling children what they should or shouldn't do, some parents mostly rely on authority. Others have a more reasoned approach, explaining to the children why they should go to bed, take their bath, pick up their toys, stop bothering their sister, and so forth. The psychologists measured various traits of the children to see whether parenting style had an effect on children's cognition and behavior.

These studies showed a clear advantage of the use of reasons: the reasoned approach was "successful in promoting resistance to temptation, guilt over antisocial behavior, reparation after deviation, altruism, and high levels of moral reasoning."[39] It seems that the children had, at least in part, made the reasons given their own.

An issue with these studies is that they only show a correlation. Parents with a reasoned style have children who do better on a variety of measures. This does not mean that the reasoned style caused the children to do better. Perhaps the same causal factors explain both why some parents are inclined to give reasons for their requests and why children born of these parents are better at resisting temptation or at feeling guilty over antisocial behavior. Even if this concern were dispelled, it could still be that children are influenced in their attitudes by the fact that they are given reasons, and not at all by the quality of the reasons given.

To more properly test children's sensitivity to argument quality, with Stéphane Bernard and Fabrice Clément we conducted a simple experiment with small children.[40] Using Playmobil toys, the experimenter told a story with a little girl, Anna, who has a dog. One day, the dog has strayed away and the children are invited to help Anna look for it. To help the children find the dog, two new characters are introduced. The first points to the left and says, "The dog went this way because I saw it go in this direction"—a pretty strong argument. The second points to the right and says, "The dog went this way because it went in this direction"—a circular argument. Even three-year-olds were sensitive to the quality of the arguments: they were more likely to be swayed by the stronger one (and two-year-olds already start to display the same skills).[41]

Conservation tasks—like the task used earlier in this chapter in which we dropped Play-Doh balls in water—were invented nearly a century ago by Jean Piaget to test children's understanding of elementary physics and mathematics. You can find online many videos of young children performing these tasks. One such video pits a candid little girl against a wily experimenter. Having placed two crackers on her side and only one on the girl's side, the experimenter asks, "Do you think that we share those fairly?" "No!" emphatically replies the little girl. The psychologist then breaks the girl's cracker in two pieces and asks, "Now is it fair?" "Yeah." "Why?" "Because we both have two!"[42]

Conservation tasks can be used to con young children out of their crackers but also to study various aspects of developing cognition, including the importance of social interaction in fostering understanding. In the 1970s, several groups of psychologists—in Switzerland,[43] in the United States, and in England, all influenced by Piaget—started asking six- and seven-year-olds to solve conservation tasks in pairs. To make things more interesting, the psychologists created as many pairs as possible in which one child could solve the conservation tasks on her own ("conservers") and the other one couldn't ("nonconservers"). The conservers proved very persuasive, being three times more likely to convince the nonconservers than the other way around.[44]

The nonconservers changed their mind because the conservers had good arguments. They were not simply following the lead of more confident or smarter-sounding peers. In fact, when conservers and nonconservers discussed other topics, such as "What is the best TV show?," they were equally likely to win the argument.[45]

Typically, when two nonconservers discuss a conservation task, they don't go anywhere, as they have no grounds for arguing. However, if they fail the task in different ways, a real understanding can emerge from their discussion. In one variant of conservation tasks, children are shown the same amount of water being poured into a thin tall glass on the one hand, and a short wide glass on the other. Fatima might believe there is more water in the thin tall glass, while Mariam believes there's more in the short wide glass. Fatima can try to show Mariam the error of her ways, and vice versa. Through argumentation, they can come to accept that they were both wrong, and understand the correct answer: there is as much water in one glass as in the other.[46]

From very early on, children are influenced by good reasons—not only adults' reasons but also their peers' reasons. Indeed, in some cases they seem to pay more attention to their peers' reasons than to adults'. When a teacher spells out the solution to a math problem, students believe her on trust. They don't need to pay much attention to her explanations, since they are willing to accept the conclusion anyhow. By contrast, if students disagree among themselves, they mostly change their minds when confronted with good reasons. So, when they do change their minds, they are more likely to understand why they did so. Students can be each other's best teachers.

Educators haven't missed the pedagogical potential of argumentation. Beginning in the early twentieth century and gathering strength in the 1970s, research into cooperative, or collaborative, learning has become "one of the greatest success stories in the history of educational research."[47] Hundreds of studies have shown that when students discuss a task in small groups, they often reach a deeper understanding of the materials.[48] Implementing cooperative learning in the classroom is not always easy. There must be disagreement, but not to the point of generating conflict. Letting the students talk things through takes time. Yet in spite of these practical obstacles, by the 1990s more than two-thirds of elementary and middle school teachers in the United States relied on cooperative learning, generally to good effect.[49]

Learning to Argue Better

Claiming that basic traits of reason are universal does not mean denying cultural variation. Similarly, drawing attention to the early developing character of argumentation does not mean denying the difference between the argumentation of a three-year-old and of an adult. If the basic skills that enable producing and evaluating reasons do not require learning, people can certainly get better—much, much better—at argumentation.

The most basic way in which people become better arguers is by acquiring local knowledge relevant to persuading their audience. For instance, as you get to know your friends' tastes in movies, you can convince them more effectively to go see a movie you think they'll like. People can also learn what kinds of arguments are appropriate for a given audience. For instance, scientists know that arguments from authority carry little weight in academic articles. Learning when to argue is also critical. Figuring out the contexts in which argumentation is frowned upon or encouraged is at least as important as learning how to argue. But none of this really amounts to acquiring new reasoning skills or becoming a better reasoner.

We saw in Chapter 12 that it often makes sense not to bother searching for the strongest arguments at the beginning of a discussion. Finding strong arguments is effortful and is not always necessary; a weak argument might convince the audience. Moreover, the counterarguments offered by our interlocutor help us understand better the opposition and find more appropriate arguments. As the conversation unfolds, people come up with arguments that do a better job at taking the audience's point of view into account.

Children offer an exaggerated version of this dynamic. When young children lie, they do a poor job at taking their interlocutors' point of view into account—for instance ignoring the fact that their mother can see a trail of crumbs leading to the cookie thief. Similarly, young children's reasons are too blatantly egocentric. Young children are apt to respond to their mother asking why they took away their sibling's toy with a "Because I want it." In doing this, they ignore the point of view of the sibling, who wanted the toy as well (and of the mother, who had already figured out as much). They expose themselves to easy counterarguments: "He wants it too," "He had it first," and so on.

When the child adjusts well to his mother's counterarguments—by accepting them or refuting them with new arguments—this dynamic is the mark of a well-functioning reasoning mechanism. Some counterarguments, however, are deemed so easy to anticipate that people resent having to produce them. For instance, if the interaction between the child and the mother repeats itself, the mother is likely to become annoyed: "I've already told you, that's not a good reason!" The mother expects the child to learn that his argument is a poor one, and to either admit he is wrong right away or move straight on to a better argument.

In the preceding example, the child has two things to learn. The first is that there are counterarguments that make his "Because I want it" weak. The second is that failing to anticipate these counterarguments has adverse social consequences—in the form of an annoyed mother. It might take a few iterations, but the child will likely learn to argue better.

Experiments have demonstrated how people learn to anticipate counterarguments on more complex topics. We reported in Chapter 12 a study by Deanna Kuhn and her colleagues in which discussion between peers enabled adolescents and adults to produce better arguments about capital punishment.[50] Among the improvements was an ability to see both sides of the argument. This does not mean that people had necessarily changed their mind to adopt a middle-of-the-road position. Instead, they were able to anticipate some of the arguments for the other side and counter them. One of the participants, who was clearly opposed to capital punishment, cited an argument for the death penalty: "I could understand killing a repeat offender to stop the chain [of murders]." However, he mentioned this argument only to refute it in the next sentence, pointing out that life imprisonment would achieve the same goal.

While this experiment reports a real improvement of argument quality, its scope is limited. Arguing about capital punishment made people better at arguing about capital punishment. To make people argue better in a more general way, researchers and educators have had more often recourse to other tools, such as teaching critical thinking. This typically involves lessons about the many (supposed) argumentative fallacies—the ad hominem, the slippery slope, and so on—and cognitive biases—such as the myside bias. Overall, such programs have had weak effects.[51] If people are very good at spotting fallacies and biases in others, they find it much harder to turn the same critical eye on themselves.[52]

If learning to reason is, to a large extent, learning to anticipate counterarguments, then the best solution might be to expose people to more counterarguments—to make people argue more. When someone's first argument is easily shot down and when that person encounters strong resistance in the form of valid arguments for the other side, he or she might learn not only about the content of these specific arguments but about the challenge presented by counterarguments more generally. That person might also learn that some anticipation might be a good thing in order to avoid putting forward indefensible points of view or arguments that are so easily shot down that they hurt their credibility.

Deanna Kuhn and Amanda Crowell set out to test whether making students argue in groups would make the students better reasoners even on topics they had no experience arguing about.[53] The researchers compared the effects of two interventions, both heavy-handed: twice-weekly meetings of fifty minutes over three school years. One intervention consisted of philosophy classes covering a range of social issues. It included a lot of essay writing as well as some whole-class discussions. In the other intervention, the students had to exchange arguments on topics such as home schooling or China's one-child policy. This intervention let the students spend time honing their arguments before sparring with their peers in several argumentation sessions.

At the end of each year, the students had to defend their position on a novel issue in writing. The students who had followed the more standard philosophy classes had had more extensive experience in essay writing, yet they developed simpler arguments. The students trained in argumentation offered more complex arguments that often incorporated both sides of the issue.

Arguing, it seems, makes one a better reasoner across the board. By being confronted with counterarguments on a specific topic, one learns to anticipate their presence in other contexts. While the argumentative theory of reasoning does not predict the exact scope with which people generalize their anticipation of counterarguments, it is in a better position to account for these findings than the intellectualist theory. In the experiment of Kuhn and her colleagues, the students receiving the standard philosophy classes showed very little improvement in spite of intensive training in individual reasoning. By contrast, those asked to argue with each other interiorized the dynamic of argumentation and wrote better essays. By learning to argue together, they had learned to reason better on their own.

17

Reasoning about Moral and Political Topics

You are taking part in a psychology experiment about moral judgment. The experimenter makes you sit in a small room and fill in a questionnaire containing several short stories. One of them tells of a documentary film that used dubiously acquired footage: some of the people in the movie claim they didn't realize they were being filmed when they were interviewed. Asked whether you approve of the decision of the studio to release the movie anyway, you voice a rather strong disapproval. Why are you so severe? Perhaps it's because the complaints came from Mexican immigrants to the United States, a population that doesn't need to be portrayed in a bad light, or perhaps because you worry about recent assaults on privacy, or perhaps it's because of a terrible smell in the room.[1]

Psychologists can be creative when it comes to surreptitiously manipulating people's behavior. To study the impact of disgust on moral judgments, they have had recourse to hypnosis, video clips of nauseating toilets, and trashcans overflowing with old pizza boxes and dirty tissues. In this case, they used fart spray: some of the participants filled in their questionnaires after the foul smell had been sprayed around. Those smelling the unpleasant odor were more severe in their moral judgments than those breathing a cleaner air. Reason wasn't driving moral judgment. Fart spray was.

Other unwanted influences are even more unsettling, as Israeli prisoners might discover if they read the scientific literature. In 2011, three researchers reported a strange pattern in the decisions of Israeli judges sitting on parole commissions.[2] The judges would start the day relatively lenient, granting about two-thirds of the parole requests. Then the rate would drop to zero by

10:00 AM. At 10:30 AM a strong rebound brought the rate of parole back to 65 percent, only to see it plunge back after a couple of hours. There was another shot back to over 60 percent of requests granted at 2:00 PM and then a quick decline back to very low rates for the end of the day.

No rational factor could explain this pattern. What was happening? Breaks were happening. The judges were served a snack around 10:00 AM and took lunch at 1:00 PM. Those breaks brought them back to the same good mood and energy they had started the day with. But their motivation quickly waned, and since more paperwork is required to accept a parole request than to deny it, so did the prisoners' hopes of getting out. We do not know if prisoners' associations have bought a snack vending machine for the courtroom. What we do know is that the judges never gave as a reason to deny parole that they were getting tired.

So far we have mostly looked at issues that admit of a more or less right answer, whether it is a logical task, making predictions, or even delivering a verdict. However, reasoning is also used in domains in which what is the right answer or even whether there is one is much less clear, such as esthetics or morality.

Moral reason has often been treated quite independently from other types of reason. We can still discern, though, the equivalent in the moral realm of the intellectualist approach to reason. This intellectualist view of moral reason—a simplistic version of Kant's position, for instance—suggests that reason can be and should be the ultimate arbiter in moral matters. Through reason, people should reach sound moral principles and act or judge in line with these principles. For most of the twentieth century, moral psychologists such as Jean Piaget and Lawrence Kohlberg have adopted a version of the intellectualist view, postulating that better use of reason is what makes people behave more morally.

However, reflections on morality have also led some thinkers—from Paul to Kierkegaard—to view morality as being rightfully dominated by emotions and intuitions. They, too, have found allies among psychologists—such as the experimenters who conducted the ingenious studies described in the preceding paragraphs.

While we have built a solid case against the individualist approach in Chapters 11 through 15, the moral domain offers a fresh challenge. Perhaps in this domain solitary reason is in fact able to overcome intuitions and guide the lone reasoner toward more enlightened decisions. Or, on the contrary, perhaps reason is so impotent in the moral realm that even sound arguments fail to change people's minds.

How Reasoning Lets Us Behave Immorally

In 2001 Jonathan Haidt published a groundbreaking article called "The Emotional Dog and Its Rational Tail."[3] For Haidt, reasoning is here only to "wag the dog," to create post-hoc justifications that cover the tracks of the intuitions and emotions secretly running the show. The studies mentioned earlier in this chapter fit well with Haidt's theory, as they show moral judgments being driven by irrelevant factors—a bad smell or tiredness-induced bad mood—rather than reason. But Haidt went further, suggesting that instead of making us do the right thing, reason may give us excuses *not* to do the right thing.

In the 1970s, Melvin Snyder and his colleagues performed a clever experiment showing that students are ready to jump on the flimsiest excuse to avoid sitting next to someone with a disability.[4] Participants were told they would have to evaluate old comedies. The movies were showing on two TV screens, in a single room separated by a partition. In front of each TV screen were two chairs, an empty one and one occupied by a confederate—an experimenter pretending to be just another participant. While one of the confederates had no distinguishable signs, the other confederate's heavy metal braces signaled a motor handicap.

Participants were told that each TV would play a different type of movie—a slapstick comedy or a sad clown comedy. Which movie did the participants prefer? It turned out that they consistently wanted to see the movie that would make them sit close to the confederate without a disability—whichever movie that was. They were making up on the fly preferences for old comedies in order to avoid sitting next to someone with a disability.

Similar demonstrations have piled up since. For instance, male participants adjust their preferences in order to pick the sports magazine with the swimsuit issue: if it's the one that has more sports cover, then sports cover is the decisive factor; if it's the one that has more feature articles, then feature articles become the decisive factor. As the old excuse goes, "I read *Playboy* for the articles."[5]

The philosopher Eric Schwitzgebel took this logic to the extreme and looked at the behavior of expert moral reasoners, people whose job it is to read about, think about, and talk about moral reason: ethics professors. It turns out that for all their moral reflection, the ethicists are not more likely to vote, to pay conference registration fees, to reply to students' emails, or to abstain from rude behavior than other philosophy professors.[6]

These examples support Haidt's model and demonstrate the pettiness of moral reason, whether it helps undergrads avoid people who make them feel uncomfortable, lets men look at scantily clad models, or allows ethics professors to skip voting.[7] In none of these cases are the rationalizations produced likely to cause any further harm. The undergrads' newfound passion for slapstick comedies will hurt neither them nor people with disabilities. But for moral violations of a different scale, more powerful rationalizations are needed, and these can take on a ghastly life of their own.

Great Reasoner, Awful Rationalizations

A few years ago, one of us, Hugo, was invited by Jon Haidt to share our ideas at the University of Virginia. No trip to Charlottesville is complete without a tour of Monticello, the home of Thomas Jefferson. There is much to be learned about the founding father in this "little mountain." His love of books, which used to fill two big rooms. His admiration for the thinkers of the French enlightenment, immortalized in marble busts. His ingenuity, on display with a giant clock of his making. His architectural acumen, which gave birth to this neoclassical marvel.

Yet none of this house's wonders should make us forget who built it and who operated the five-thousand-acre plantation it dominated. Slaves. Nearly two hundred of them.[8] Slaves who were sold like chattel when Jefferson needed to pay for these fancy busts and other frivolous expenses.[9] Slaves who were whipped into submission.[10] Slaves who were sold away to distant quarters "to make an example . . . in terrorem to others."[11]

As many of Jefferson's biographers have pointed out,[12] there is nothing extraordinary about this behavior for a Virginia planter of the revolutionary era. Jefferson, however, was anything but a typical Virginia planter of the revolutionary era. He was a proponent of universal education, the founder of a major university, a fighter of cruel punishment, the architect of religious toleration in Virginia, and the writer of these words: "We hold these truths to be self-evident, that all men are created equal, that they are endowed by their Creator with certain unalienable Rights, that among these are Life, Liberty and the pursuit of Happiness." Talk about cognitive dissonance.

Because he was such a brilliant reasoner, Jefferson offers one the most dramatic illustrations of Haidt's model. When Jefferson reflects on what is to be done about slavery, he has no trouble finding reasons to oppose emancipation.

In his *Notes on the State of Virginia,*[13] Jefferson laid down his fear that emancipation would only lead to "the extermination of the one or the other race." He could have stopped there, but he really wanted to bolster his point, and so "to these objections, which are political," he "added others, which are physical and moral." Blacks and whites can't live in harmony together because of the many defects in black people's physique and spirit. "Are not the fine mixtures of red and white, the expressions of every passion by greater or less suffusions of colour in the one, preferable to that eternal monotony, which reigns in the countenances, that immoveable veil of black which covers all the emotions of the other race?" The blacks may be "more adventuresome" but only "from a want of forethought." Their love is but "an eager desire." "Their griefs are transient." "Their existence appears to participate more of sensation than reflection." To conclude, "this unfortunate difference of colour, and perhaps of faculty, is a powerful obstacle to the emancipation of these people."

This is reason at its worst. Patently biased, it turns the most subjective evaluation—"a more elegant symmetry of form"—into an objective assessment—"the real distinctions which nature has made." It makes of a scientific mind a dunce ready to accept that the orangutan has a preference "for the black women over those of his own species." It pushes a sharp intellect to say that blacks both "seem to require less sleep"—when it comes to "sit up till midnight" for the "slightest amusements"—and have a "disposition to sleep"—after all, "an animal whose body is at rest, and who does not reflect, must be disposed to sleep of course." It lets a master rhetorician argue, in effect, that there's nothing to be done right now about blacks being reduced to the most abject submission, *because they don't have flowing hair.*[14]

One cannot tell whether Jefferson's fear of a race war or his racist beliefs drove his rejection of immediate emancipation.[15] Instead of emancipation, he favored long-term and far-fetched plans for educating young blacks, separating them from their parents and sending them back to Africa.[16] But it can only be his views on the inferiority of the black race that made him so fearful of interracial encounters (except when it came to sleeping with his mistress, his slave Sally Hemings). Why send emancipated slaves as far away as Africa? Because "when freed, [they are] to be removed beyond the reach of mixture."[17] After all "their amalgamation with the other color produces a degradation to which no lover of his country, no lover of excellence in the human character can innocently consent."[18]

It is already difficult to figure out why people hold such and such beliefs when they can be asked; reconstructing the thought process of a dead man is an even more speculative business. Yet we know that Jefferson didn't become a slave owner because of his racist beliefs. Rather, he inherited a plantation along with its slaves and, presumably, the attitude of ordinary slave owners at the time. Later he also adopted the enlightenment ideals of his intellectual peers. This massive contradiction could be reconciled only by a creative reasoner, and the sad way Jefferson rose to the challenge is also part of his legacy.

Clearly, reason is not behaving as we would like it to, helping people pass more enlightened judgments and make fairer decisions. Jefferson, armed with a brilliant intellect, all the knowledge of his time, and the noblest ideals, should have reasoned his way to the right creed and the just behavior. Instead, reason provided him with convenient rationalizations, allowing him to keep his slaves and his wealth. Sadly, these rationalizations proved far from inert, turning him into the "intellectual godfather of the racist pseudo-science of the American school of anthropology."[19]

Such examples might prompt us to safely lock up moral reason and throw away the key. Yet we should also consider that if reason's power of rationalization is immense, it is not limitless. Sometimes no excuse is to be found, and people have the choice of either behaving immorally without any justification or behaving morally after all. We have described how students invented a taste for one kind of movies to avoid sitting next to someone with a disability. In the same study, another group of participants was denied that opportunity: the two TV screens showed the same movie. These participants could not use even a bogus preference to justify sitting away from someone with a disability—as a result, they were much less likely to do so. Likewise, some of Jefferson's contemporaries found it beyond their ability to justify owning people. George Washington freed his slaves and provided for them in his will. Benjamin Franklin freed his slaves in his lifetime. It seems that one cannot, in fact, "find or make a reason for *everything* one has a mind to do"[20]—unless perhaps one is as smart as Thomas Jefferson.[21]

Can Reasoning Change People's Moral Opinions?

The picture of moral reasoning painted so far, the one stressed by Haidt, fits one side of the interactionist approach to reason perfectly. Instead of proceeding to a careful assessment of the moral value of a judgment or a decision, reason looks for justifications that may be mere excuses for what people wanted to do all along, moral or not—the myside bias at work. Being content with shallow reasons and flimsy rationalizations reflects another pitfall of solitary reason: the lack of critical examination of one's own justifications and arguments.

But Haidt's theory has another component, the "wag-the-other-dog's-tail illusion." As the argumentative theory of reasoning might predict, "in a moral argument, we expect the successful rebuttal of an opponent's arguments to change the opponent's mind." For Haidt, "such a belief is like thinking that forcing a dog's tail to wag by moving it with your hand will make the dog happy."[22] In other words, however strong your arguments might seem, they won't change other people's position on moral issues. People will keep being driven by their intuitions and emotions instead.[23]

Haidt's famous "Emotional Dog" article begins with an example of reasoning's powerlessness to affect moral judgments:

> Julie and Mark are brother and sister. They are traveling together in France on summer vacation from college. One night they are staying alone in a cabin near the beach. They decide that it would be

> interesting and fun if they tried making love. At the very least it would be a new experience for each of them. Julie was already taking birth control pills, but Mark uses a condom too, just to be safe. They both enjoy making love, but they decide not to do it again. They keep that night as a special secret, which makes them feel even closer to each other. What do you think about that? Was it OK for them to make love?[24]

Among the participants Haidt and his colleagues interviewed, most said it was not acceptable for Julie and Mark to make love. When the experimenter asked them why, they had many reasons. Nine reasons each, on average. All of them were shot down by the experimenter. "They might have children with problems." No, they used two forms of contraception, so there will be no children. "They'll be screwed up psychologically." On the contrary, they grow closer after the experience. "They'll be shunned when people find out." They keep it a secret; no one finds out. And so forth. But the participants didn't say, "I cannot articulate a rational basis for my moral condemnation, and therefore I retract it." Most held fast to a judgment they could not support anymore.

The interactionist approach, however, predicts that good reasons should carry some weight. Why, then, don't some of the people who find themselves unable to answer Haidt's argument change their minds?[25] More generally, if it is true that people don't change their minds in response to moral arguments, why the reluctance?

While it can be infuriating and depressing to fail to change people's minds, especially on important moral matters, that doesn't mean that those who won't budge are being irrational. According to the interactionist approach to reason, people should be sensitive to strong reasons, but even seemingly strong reasons shouldn't overwhelm every other concern. For instance, we might have a strong intuitive reluctance to accept a given conclusion. Some intuitions are difficult to make explicit, so that we can be at a loss when explaining why we reject an apparently strong argument. That doesn't mean that the intuitions are irrational—although failing to defend our point of view in the face of strong arguments might make us look so. Some of the most important intuitions that stop us from accepting even arguments we cannot effectively counter have to do with deference to experts.

For instance, when Moana tries to convince Teiki, his more liberal friend, that climate change is a hoax, they both defer to experts—but not the same experts. Deferring to experts is rational. If we didn't, we would be clueless about a wide variety of important issues about which we have no personal experience and no competent reflection. Once we defer to some experts, it makes sense to put relatively little weight on challenging arguments from third parties. Even though we might not be able to come up with counterarguments, we believe that the experts we defer to would. For instance, Moana could give Teiki many arguments that he cannot refute on the spot, since he does not know exactly why the experts he trusts believe in climate change. Still, Teiki would likely not change his mind, thinking that his experts would be able to counter Moana's arguments.

When beliefs are not readily testable, it is quite rational to accept them on the basis of trust, and it is quite rational for people who trust different authorities to stubbornly disagree. We don't mean that these are the most rational attitudes possible. An intellectually more demanding approach asks for clarity and for a willingness to revise one's idea in the light of evidence and dissenting arguments. This approach, which has become more common with the development of the sciences, is epistemically preferable—but no one has the time and resources to apply it to every topic.

How Argumentation Helps Get Moral Problems Right

Should we keep reasoning about moral issues? Solitary reasoning has dubious effects, and even argumentation faces many obstacles. Yet our answer is a resounding yes. In fact, we suspect that most moral beliefs are more amenable to arguments than, say, gut feelings about incest. Beliefs about what the police can do to fix the crime problem in the neighborhood or beliefs about how wrong Ross was to cheat on Rachel don't have a preset consensual answer in one's community; they don't have the same power to signify whether we are a friend or a foe. When the overriding concern of people who disagree is to get things right, argumentation should not only make them change their mind, it should make them change their mind for the best.

An obvious problem for testing this prediction is the lack of a clear moral benchmark to tell whether argumentation leads to better moral beliefs—by definition, if there is a clear moral benchmark, then there should be no reason to argue. However, it is possible to look at cases in which adults agree and see what happens during child development. If children of a certain age differ or are confused about a given issue, it is possible to see which children are more convincing: those who share the adults' judgment or those who defend less mature points of view.

Jean Piaget made an art of confusing children. For instance, he would give children—for example, nine-year-olds—the following two stories:

> Story 1
> Once there was a little boy called John. He was in his room and his mother called him to dinner. He opened the door to the dining room, but behind the door there was a tray with six cups on it. John couldn't have known that the tray was behind the door. He opened the door, knocked the tray and all six cups were smashed.

> Story 2
> Once there was a little boy called David. One day when his mother was out he tried to get some sweets from the cupboard. He climbed on a chair and stretched out his arm. But the sweets were too high and he couldn't reach, and while he was trying to reach [them] he knocked over a cup and it fell and broke.[26]

Piaget would ask the children: Which of the two boys, John or David, is naughtier? When Patrick Leman and Gerard Duveen replicated Piaget's experiment, they found that most nine-year-olds thought John was naughtier.[27] Now, as adults, we can presumably all agree that this answer is wrong. Exactly how naughty David was is a matter for discussion, but clearly John did nothing wrong. His breaking the cups was purely accidental, not even the result of negligence. Reassuringly, when pairs of children who had different views on the matter were left to discuss with one another, they were five times more likely to end up thinking that David was naughtier. Thanks to argumentation, their moral judgments had gotten more accurate.

The Surprising Efficacy of Political Debates among Citizens

Among adults, some moral debates become political debates—debates not only about what is right or wrong but also about what the community should do to fix the problem. It is tempting to have a dim view of political debates. In some democratic countries, the most publicized of those debates occur between contenders for the presidency. These are somewhat unnatural spectacles in which the debaters know they have no chance of convincing each other and mostly seek to strengthen the support of their base. Fortunately, debates about political matters don't only occur between presidential contenders; they also occur between citizens.

Samuel Huntington expressed a common opinion when he argued that "elections, open, free and fair, are the essence of democracy."[28] But voting is not the only way to aggregate opinions in a (potentially) fair manner. Indeed, it is neither the oldest nor the most common.[29] As we noted in Chapter 16, bands of hunter-gatherers make group decisions based on public deliberation. To the extent that life in these bands bears a resemblance to that of our Paleolithic ancestors, this suggests that deliberation has a far greater antiquity than voting. In *Democracy and Its Global Roots,* Amartya Sen takes the reader on a brief tour of non-Western democratic traditions—many of which were deliberative.[30] From the great interreligious debates sponsored by the emperor Akbar in sixteenth-century India to the Thembu's open meetings that left a young Nelson Mandela with the impression of "democracy in its purest form,"[31] deliberation throughout the world carries the hope of reaching better beliefs and making better decisions.

In the early 1980s, political scientists started paying more attention to the role played by deliberation in a healthy democracy.[32] At first, the new field of *deliberative democracy* focused on lofty ideals, on the potential of deliberation to promote rational discourse, civility, public engagement, and mutual respect. Then political scientists confronted these lofty ideals to the reality of deliberation between divided, misinformed, sometimes irate citizens. To the surprise of many, the lofty ideals won. When a sample of citizens is brought together, divided in small groups, and, with the soft prodding of a moderator, made to discuss policy, good things happen.[33] The participants in these discussions end up better informed, with more articulate positions but also a deeper understanding of other people's point of view. Their opinions tend to converge toward a reasonable compromise. They are more likely to participate in public life in the future. Deliberation among citizens works.

One of the most successful deliberative democracy experiments was launched by Robert Luskin and James Fishkin. In dozens of cities, they conducted deliberative polls in which citizens discussing among themselves reached more informed positions on various policy matters. One of these cities was Omagh, Northern Ireland.

On August 15, 1998, a bomb had exploded in Omagh, killing twenty-nine people and injuring more than two hundred. Claimed by a splinter group of the Irish Republican Army—creatively called the Real Irish Republican Army—the attack is remembered as one of the worst atrocities in the long and bloody conflict over the control of Northern Ireland. In Omagh, Catholics and Protestants have plenty of reasons to distrust each other and to stick to their group's beliefs—not the best place for deliberation to work.

Yet when Luskin, Fishkin, and two colleagues asked a sample of the local population that included both Catholics and Protestants to discuss education policy, the debates proved constructive, even on highly loaded topics.[34] When questions related to mixed religious schools emerged in the debate, the participants didn't fight and polarize. After the discussions, participants had changed their minds on several points, and they were much more knowledgeable about education policy. They also found that their interlocutors were more trustworthy and open to reason than they expected.

Critics of deliberative democracy have pointed out its scaling-up problem: debates work well with a handful of people, not so well with several millions. Fishkin, joined by the American constitutional scholar Bruce Ackerman, has proposed a Deliberation Day, a national holiday in which citizens would be invited to debate upcoming elections. While such institutions would further boost the role of discussion in public life, argumentation has already proven its ability to effect large-scale moral and political change.

Abolitionism: Not Such an Easy Argument to Make

By the end of the eighteenth century, the British dominated the transatlantic slave trade,[35] and they had just acquired huge swaths of territory in the Americas, bearing the promise of untold wealth. Economic logic dictated that they capture and ship hundreds of thousands of slaves to exploit these lands.[36] Instead they chose to abolish the slave trade. How did the abolitionists manage such a complete reversal?

From our modern vantage point, it seems like an easy argument to make. Why would it be necessary to convince someone that slavery is so wrong that it should be banished? Unfortunately, the evil of slavery hasn't always been a moral truism. Indeed, for most of history slavery was part of the fabric of life. Practiced by the Greeks and the Romans, sanctioned by Judaism, Christianity, and Islam, slavery hadn't been an issue for most of European history. The tide turned when the Enlightenment's heralds, such as Diderot, staunchly denounced the practice. At the same time, new religious movements—most notably the Quakers—offered a new reading of the Bible that made of slavery a very un-Christian institution. At long last slavers and slave owners had to offer justifications for their practice.

Apologists of slavery obliged and, for a while, even tried to take the moral high ground. They argued that life in Africa was so tough as to be practically unbearable. By comparison, during the Middle Passage, the slaves were treated as VIPs, provided with "Cordial . . . Pipes and Tobacco," and "amused with Instruments of Music." The contrast was such that "Nine out of Ten [slaves] rejoice at falling into our Hands," the slavers claimed.[37] Moreover, the whole slave trade was only necessary because slaves, having reached their destination, failed to have enough children to maintain the population. That was due to female slaves' being "prostitutes" who must have frequent "abortions, in order that they may continue their trade without loss of time." "Such promiscuous embraces," continues Edward Long in his *History of Jamaica,* "must necessarily hinder, or destroy, conception."[38] Slaves should be thankful for the slave trade yet also blamed for it.

These arguments sound not just abominable but also preposterous. At the time, though, British citizens lacked reliable information about what was going on in Africa, the West Indies, or America. And British lives weren't exactly cushy, either. The industrial revolution generated great wealth but also its share of misery. In ports across Britain, thousands of men were "impressed," kidnapped and brought onboard navy ships for "several years of floggings, scurvy, and malaria."[39] Given the picture painted by the slavers, common people might have thought the slaves weren't much worse off.

Still, the anti-abolitionists' strongest arguments weren't moral, but economic. Entire cities, such as Liverpool, relied on the slave trade. Even inland cities like Manchester were dependent on a constant supply of raw material gathered by slaves in the colonies to employ textile workers. Slavers' mouthpieces never tired of mentioning the "widows and orphans" that abolition would leave in its trail all over Great Britain.[40] The anti-abolitionists didn't hesitate to make up numbers—seventy million pounds were at stake![41]—or to invoke the British's favorite beverage—the lack of "Sugar and Rum[!]" would "render the Use of Tea insupportable."

Yet by the mid-1780s, the Quakers and other early abolitionists had managed to reclaim the moral high ground. They had done so by using an essential argumentative tool: displaying inconsistencies in the audience's position. In this case, the inconsistencies were glaring enough: Christianity and the English spirit on the one hand, slavery on the other. "The very idea of trading the persons of men should kindle detestations in the breasts of MEN—especially of BRITONS—and above all of CHRISTIANS," pleaded James Dore in a 1788 sermon.[42] Historian Seymour Drescher pointed out that the strength of this inconsistency was the main propeller of popular abolitionism: "How could the world's most secure, free, religious, just, prosperous, and moral nation allow itself to remain the premier perpetrator of the world's most deadly, brutal, unjust, immoral offenses to humanity?"[43] Still, the economic considerations put forward by the slavers held fast. The moral arguments were too abstract, the immensity of the suffering wrought by slavery not plain enough. The abolitionists needed more evidence for their arguments to carry their full weight.

Convincing a Country

For years, the abolitionist Thomas Clarkson crisscrossed England, accumulating the greatest wealth of evidence ever gathered on the slave trade. The fruits of his labors—*An Abstract of the Evidence Delivered before a Select Committee of the House of Commons in the Years 1790, and 1791; on the Part of the Petitioners for the Abolition of the Slave-Trade*—became the main weapon in the abolitionists' growing arsenal. This arsenal was completed by the men and women who developed a "rhetoric of sensitivity," composing poems meant to restore the slaves' full humanity;[44] by the freed slaves who wrote widely successful autobiographies, putting a face on the numbers of the *Abstract;* by the former slavers who attested to the horrors they had witnessed, lending credibility to the cause. Yet it was the *Abstract* that remained "the central document of British mass mobilization."[45] But the abolitionists needed something more than popular clamor outside the walls of Parliament. They needed an insider's voice.

William Wilberforce, member of Parliament, was conservative on many issues, but by the mid-1780s he had become an evangelical, a conversion that seemingly made him more responsive to the abolitionists' arguments. Wilberforce was lobbied by the movement and finally convinced to lend his voice to the cause. Like other abolitionists, Wilberforce pointed out the inconsistency between slavery and belonging to a nation "which besides the unequalled degree of true civil liberty, had been favored with an unprecedented measure of religious light, with its long train of attendant blessings."[46] But Wilberforce didn't simply rehearse the standard arguments. He mastered the evidence, familiarized himself with the anti-abolitionists' arguments, and fought them on their own ground. Slavers claimed it would make no economic sense to mistreat their most precious cargo. Wilberforce pointed out that on the contrary, "the Merchants profit depends upon the number that can be crouded together, and upon the shortness of their allowance."[47] The anti-abolitionists relied on Long's supposedly well-informed *History of Jamaica* for many of their arguments, so Wilberforce decided to use Long's own assertions as premises. "Those Negroes breed the best, whose labour is least, or easiest,"[48] claimed Long. Well, added Wilberforce, if only slave owners exerted a less brutal dominion, the slave population would be self-sustaining, and trade unnecessary.

The overwhelming mass of reasons and evidence gathered by the abolitionists ended up convincing most members of Parliament—directly or through the popular support the arguments had gathered. In 1792, three-quarters of the House of Commons voted for a gradual abolition of the slave trade. The House of Lords, closer to the slavers' interests, asked for more time to ponder the case. Awkward timing: for years, the threats posed by the French revolution, and then by Napoleon, would quash all radical movements—which, at the time, included abolition. But as soon as an opportunity arose, the abolitionists whetted their arguments, popular clamor rekindled, Parliament was flooded with new petitions, and Wilberforce again handily carried the debate in the Commons. In 1807, both houses voted to abolish the slave trade.

The British abolitionists didn't invent most of the arguments against slavery. But they refined them, backed them with masses of evidence, increased their credibility by relying on trustworthy witnesses, and made them more accessible by allowing people to see life through a slave's eyes. Debates, public meetings, and newspapers brought these strengthened arguments to a booming urban population. And it worked. People were convinced not only of the evils of slavery but also of the necessity of doing something about it. They petitioned, gave money, and—with the help of other factors, from economy to international politics—had first the slave trade and then slavery itself banned.

The Best and Worst of Reason

The interactionist approach is in a unique position to account for the range of effects reason has on moral judgments and decisions. Many experiments and, before them, countless personal and historical observations have rendered the intellectualist view of moral reason implausible. Moral judgments and decisions are quite commonly dominated by intuitions and emotions with reason providing, at best, inert rationalizations and, at worst, excuses that allow the reasoner to engage in morally dubious behavior—from sitting away from someone with a disability to keeping one's slaves. Reason does what it is expected to do as a biased and lazy producer of justifications.

Yet we do not quite share the pessimism regarding the ability of reason to change people's minds. People do not just provide their own justifications and arguments; they also evaluate those of others. As evaluators, people should be able to recognize strong arguments and be swayed by them in all domains, including the moral realm. Clearly, arguments that challenge the moral values of one's community can be met with disbelief, distrust of motives, even downright hostility. Still, on many moral issues, people have been influenced by good arguments, from local politics—for example, how to organize the local school curriculums—to major societal issues—such as the abolition of the slave trade.

18

Solitary Geniuses?

In the 1920s, the small community of particle physicists faced a dilemma of epic consequences. The mathematics they used to understand elementary particles conflicted with their standard representations of space. While mathematics enabled very accurate predictions of the particles' behavior, the equations could not ascribe to these particles a trajectory in space. There seemed to be a deep flaw in quantum theory. Would the physicists have to reject this extraordinarily successful framework? Or would they be forced to accept its limitations? Werner Heisenberg was already a well-recognized figure in this community. By developing the mathematical framework used to understand elementary particles, he had contributed to creating this dilemma. So he set out to resolve it.

At the time, most of the action in particle physics was happening in Niels Bohr's laboratory in Copenhagen. Bohr not only was a great scientist; he also had a flair for obtaining funding, which he used wisely, promoting the best young researchers—such as Heisenberg. However, Bohr's larger-than-life persona clashed with Heisenberg's ambition and independence. In search of a more peaceful atmosphere, Heisenberg retreated to the small island of Heligoland, a three hours' boat ride off the coast of Germany. After months of isolation, he finally found a solution to the dilemma that was haunting quantum physics: a mathematical formulation of what would become the uncertainty principle. Heisenberg summarized the principle in layman's terms: "One can never know with perfect accuracy both of those two important factors which determine the movement of one of the smallest particles—its position and its velocity."[1] This is a third route: quantum physics does not have to be replaced by a new framework, and physicists do not have to resign themselves to an imperfect understanding of the world. If elementary particles cannot be ascribed a precise trajectory, it is not because quantum physics is flawed; it is because position and velocity do not simultaneously exist before being measured and cannot be measured simultaneously.

A stunningly brilliant mind, isolated on an island, reaches a deeper understanding of the nature of the world—a perfect illustration of the solitary genius view of science. In this popular view, scientific breakthroughs often come from great minds working in isolation. The image of the lonely genius is a figment of the romantic imagination of the eighteenth century, as it can be found in the verses of the poet Wordsworth. To him, a statue of Newton was

> The marble index of a mind for ever
> Voyaging through strange seas of thought, alone[2]

In the solitary genius view of science, geniuses are fed data by underlings—students, assistants, lesser scientists—they think very hard about a problem, and come up with a beautiful theory. This is very close to the view of science that had been advocated by Francis Bacon.

Bacon was a visionary statesman and philosopher living in England at the turn of the sixteenth century. He urged his colleagues to abandon their specious arguments and adopt a more empirical approach, to stop adding new commentaries on Aristotle and conduct experiments instead—ideas that would inspire the scientific revolution. While Bacon emphasized the collaborative character of science, he held a very hierarchical notion of collaboration. In his description of a utopian New Atlantis, Bacon makes a long list of people in charge of "collect[ing] the experiments which are in all books . . . [and] of all mechanical arts," and "try[ing] new experiments." But the theorizing is reserved to a very select group: "Lastly, we have three that raise the former discoveries by experiments into greater observations, axioms, and aphorisms. These we call Interpreters of Nature."

As one of the two foremost lawyers of his day, Bacon was intimately familiar with argumentation, yet there is little place for it in his grand scheme. In such a hierarchical view of science, if argumentation is needed at all, it is to enable recognized geniuses to convince lesser minds of their discoveries. And even that may not be working so well.

Dispirited by what he perceived to be the slow acceptance of his ideas, Max Planck, one of the founders of quantum physics, quipped, "A new scientific truth does not triumph by convincing its opponents and making them see the light, but rather because its opponents eventually die, and a new generation grows up that is familiar with it."[3] Quoted in millions of web pages, thousands of books, and hundreds of scientific articles, Planck's aphorism encapsulates a deeply held belief about scientific change (or lack of it).

If scientific progress is achieved by lone geniuses building revolutionary theories, if even these geniuses fail to convince their colleagues to share in their vision, then science, the most successful of our epistemic endeavors, violates all the predictions of our interactive approach to reason. That would be a problem.

Scientists Are Biased, Too

Scientists rely on diverse cognitive skills. As many researchers, from Poincaré to Einstein, have pointed out, intuitions play a crucial role in the emergence of new insights. No one denies, of course, that reason plays an important role in science. Is it that individual scientists can successfully answer very complex questions on their own? If so, they must be exceptionally good reasoners capable of overcoming the two problems that plague laypeople in their solitary reasoning: myside bias and low evaluation criteria for one's own arguments—that is, laziness. Perhaps scientists are a special breed. Perhaps they have been drilled about falsification to such an extent that it has become second nature, allowing them to reason impartially about their own ideas. The example of Linus Pauling, discussed in Chapter 11, gave us a glimpse of the answer, but less anecdotal evidence also speaks to this question.

In the 1970s, Michael Mahoney and his colleagues conducted a series of experiments to discover whether scientists are also prey to the myside bias. One of their studies compared the answers of scientists from two different fields, psychology and physics, to the answers of Protestant ministers on a task designed to assess the myside bias. The three groups showed a bias, the scientists being as biased as the ministers or the participants of previous experiments.[4] When Mahoney conducted an experiment in which scientists had to reason about their own domain of expertise, he again observed a strong myside bias.[5]

The same pattern emerges from observations of scientists in their natural environment. Kevin Dunbar conducted an extraordinary study of how scientists think in real life and not just in experimental settings.[6] He interviewed scores of researchers, in an effort to understand how they made sense of their data, developed new hypotheses, and solved new problems. His observations showed that scientists reason to write off inconvenient results. When an experiment has a disappointing outcome, researchers do not reason impartially, questioning their initial hypothesis and trying to come up with a new one. Instead, they are satisfied with weak arguments rescuing the initial hypothesis: a technical problem occurred; the experiment was flawed; someone made a mistake.

Yet Scientists Are Sensitive to Good Arguments

If scientists' reasoning shares the biases and limitations of laypeople's reasoning, it should also share its strengths: be much more objective in the evaluation than in the production of reasons. Scientists should pay attention to each other's arguments and change their minds when given good reasons to. But, following Planck and the common wisdom, it seems they fail to do even that, resisting new theories to their last breath.

To be fair to scientists, they've been asked to swallow some crazy ideas: that we are the descendants of unicellular organisms, the product of billions of years of random mutations, barreling around the sun at over 100,000 kilometers per hour, glued to the ground by the same force that keeps the earth in its orbit, seeing this page thanks to light particles for which time does not exist, our consciousness the product of a three-pound slab of gray matter. When Planck was complaining about the sluggish diffusion of his ideas, he was trying to convince physicists that energy is not a continuous variable but that it comes instead in quanta of a specified size. In giving birth to quantum physics, Planck was violating one of classical physics' most fundamental assumptions. That he was met with skepticism should not come as a surprise. The counterintuitiveness of scientific ideas is not the only factor slowing their spread. Thomas Kuhn, the author of the revolutionary book *The Structure of Scientific Revolutions,* argued that much of the resistance to new theories comes from scientists who have staked their entire careers on one paradigm. One could hardly blame them for being somewhat skeptical of the upstarts' apparently silly yet threatening ideas.

Although he was more conscious than anyone before him of the challenge involved in getting revolutionary ideas accepted, Kuhn did not share Planck's bleak assessment: "Though some scientists, particularly the older and more experienced ones, may resist indefinitely, most of them can be reached in one way or another."[7] Ironically, Bernard Cohen, another scholar of scientific revolutions, noted how "Planck himself actually witnessed the acceptance, modification, and application of his fundamental concept by his fellow scientists."[8] Planck and Kuhn also seem to have been wrong about the deleterious effects of age on the capacity to take new ideas in stride. Quantitative historical studies suggest that older scientists are only barely less likely to accept novel theories than their younger peers.[9]

In his review of scientific revolutions, Cohen went further. Not only does change of mind happen, but "whoever reads the literature of scientific revolution cannot help being struck by the ubiquity of references to conversion."[10] Being confronted with good arguments sometimes leads to epiphanies, as that recounted by a nineteenth-century chemist who had been given a pamphlet at a conference: "I . . . read it over and over at home and was astonished at the light which the little work shed upon the most important points of controversy. Scales as it were fell from my eyes, doubts evaporated, and a feeling of the most tranquil certainty took their place."[11]

Revolutions in science proceed in a gentle manner. There is no need to chop off senior scientists' heads; they can change their minds through "unforced agreement."[12] Ten years after Darwin had published *The Origin of Species,* three-quarters of British scientists had been at least partly convinced.[13] Barely more than ten years after a convincing theory of plate tectonics was worked out, it had been integrated into textbooks.[14]

Science Makes the Best of Argumentation

Scientists' reasoning is not different in kind from that of laypeople. Science doesn't work by recruiting a special breed of super-reasoners but by making the best of reasoning's strengths: fostering discussions, providing people with tools to argue, giving them the latitude to change their minds.

The Royal Society, founded in England in 1660, was one of the first scientific societies in the world. Several of its founders would become important figures in the scientific revolution. Yet, however brilliant these founders might have been, they still got a few things wrong. Christopher Wren believed that "a true astrology [could] be found by the inquiring philosopher." The hand of a freshly hanged man was, according to Robert Boyle, a cure for goiter. John Wilkins championed the miraculous powers of Valentine Greatrakes, "the most famous occult healer of the seventeenth century,"[15] who could supposedly cure tuberculosis and other ailments. But what they got right was much more important than any particular belief they got wrong. It was a way of acquiring new, better beliefs. These pioneers thought that real experiments should replace thought experiments,[16] that scholars ought to engage in open-ended dialogue and not in the sterile mind games of medieval *disputationes,* that more knowledge could be gained from tradesmen and travelers than from centuries-old books.

Wren, Boyle, Wilkins, and their colleagues aimed high. To illustrate the power of discussion, Boyle wrote *The Sceptical Chymist,* the report of an imaginary symposium of chemists. The dialogue style was hardly new; Galileo had used it with great power to expose his revolutionary theories. But in the *Dialogue Concerning the Two Chief World Systems,* it is all too clear that Salviati—Galileo's mouthpiece—has it all figured out from the beginning.

Boyle's use of dialogue was different. In *The Sceptical Chymist,* none of the characters know the answer from the start. It is a "piece of theatre that exhibit[s] how persuasion, dissensus and, ultimately, conversion to truth ought to be conducted."[17] Carneades, the protagonist whose views most closely mirror Boyle's, does not inculcate the truth to his interlocutors; "rather [truth] is dramatized as emerging through the conversation."[18]

Boyle's grand vision might not have materialized, but he wouldn't be entirely disappointed by the everyday workings of contemporary science.

Experiments have become common currency. And Boyle's constructive dialogues can be observed in thousands of lab meetings every day. What happened to the scientists that Dunbar had found so prompt to rationalize away disappointing results? When they brought their excuses to the lab meetings, they were rebuffed. The rationalizations were only good enough to convince those who offered them, not the other lab members. The researchers were forced to come up with better explanations of their results, assisted by other group members who provided alternative hypotheses and explanations.[19]

The lab meetings Dunbar observed offer a perfect demonstration of how discussion can rein in the biases that mislead individual reasoners. And this is only one of the many forms that the exchange of arguments takes in science, from informal chats to peer review and international symposia.

The importance of discussions and arguments for science has not escaped contemporary scientists—as opposed to the poets and other external observers who created the myth of the lone genius. When interviewed by Mihaly Csikszentmihalyi and Keith Sawyer about creativity, a mathematical physicist described science as

> a very gregarious business, it's essentially the difference between having this door open and having it shut. If I'm doing science, I have the door open. That's kind of symbolic, but it's true. You want to be all the time talking with people. . . . It's only by interaction with other people in the building that you get anything interesting done; it's essentially a communal enterprise.[20]

Even Daniel Kahneman, who with Amos Tversky has made some of the most important contributions to the development of an individualist view of reasoning, is well aware of the power of discussion: "I did the best thinking of my life on leisurely walks with Amos."[21]

How Solitary Are Solitary Geniuses?

Sociologists and historians of science concur: science is an intrinsically collective enterprise. Not in Bacon's sense of a division of labor between the lowly data gatherer and the high-minded theoretician, but as a more integrated collaborative endeavor, shot through with discussions and arguments. Yet it seems that there are exceptions. What of those scientists who break new ground on their own? What of Heisenberg on his island?

The historian of science Mara Beller has looked in detail at the process that led Heisenberg to formulate the uncertainty principle, and she doesn't buy the solitary genius view. Drawing from a range of sources—scientific papers, letters, interviews, and autobiographies—she reconstructed the dialogical nature of Heisenberg's achievement. "Not the magisterial unfolding of a single argument, but the creative coalescence of different arguments, each reinforcing and illuminating the others, resulted in Heisenberg's monumental contribution to physics."[22] Heisenberg's insights were a reaction to Schrödinger's position, built by engaging with the thought of Bohr and Dirac, recycling ideas of less famous physicists such as Norman Campbell and H. A. Sentfleben.[23]

Most importantly, Heisenberg shared a dense correspondence with Wolfgang Pauli. It is by confronting Pauli's arguments and rising to his challenges that Heisenberg was able to push his ideas to their full potential. Indeed, the first draft of Heisenberg's uncertainty paper appears in a letter to Pauli. Even geniuses need people to argue with to develop their best ideas.

The exchange of arguments between Heisenberg and Pauli was essential to the formulation of the uncertainty principle, but it was far from an everyday conversation. Instead of short statements briskly flying back and forth, we find long, well-structured arguments separated by days of solitary thinking. From what we can piece together of it, the thought process that leads to scientific breakthroughs doesn't look like the type of solitary reasoning depicted by psychology experiments, no unexamined arguments piling up endlessly in support of the scientist's views. There is a myside bias, yes, but it is tempered by a more demanding quality control that weeds out the weakest arguments.

In Chapter 12, we argued that in an everyday conversation, it makes sense for reasoners not to spend much energy anticipating potential counterarguments. Anticipating counterarguments is both difficult and not that useful, since failing to convince one's interlocutor right away carries little cost. It is improbable, however, that throughout our evolution reasoning was only ever used in these conditions. The stakes were bound to be higher at times, and in some situations counterarguments could be more easily anticipated. So we should expect some flexibility in how much and how well people scrutinize their own arguments.

Some institutions put a premium on well-crafted arguments. Lawyers only get one final plea. Politicians must reduce their arguments to efficient sound bites. Scientists compete for the attention of their peers: only those who make the best arguments have a chance of being heard. To some extent, reasoning rises to these challenges. With training, professionals manage to impose relatively high criteria on their own arguments. In most cases this improvement transparently aims at conviction. Lawyers must persuade a judge or a jury. If there are other people, including the lawyers themselves, whom arguments fail to convince, so be it.

By contrast, scientists—at their best—seem to be striving for the truth, not for the approval of a particular audience. We suggest that in fact their reasoning still looks for arguments intended to convince but that the target audience is large and is particularly demanding.[24] If anyone finds a counterargument that convinces the scientific community, the scientist's argument is shot down. She cannot afford to appeal to a judge's inclinations or to play on a jury's ignorance of the law. Her arguments must be aimed at universal validity—in a universe composed of her peers.

At least as important as the incentive to strive for universal assent are the means to achieve it. In Chapter 12, when Hélène was trying to convince Marjorie to dine at Isami, she could hardly have anticipated Marjorie's counterarguments, since they were based on idiosyncratic pieces of information such as where Marjorie had dined the week before or how much she could afford to spend. By contrast, experts in a given scientific field are likely to share much of the information relevant to settle a disagreement. This makes actual argumentation much more efficient. This also improves the power of individual reasoning. If a scientist, while evaluating her own arguments, finds a counterargument, then others could have discovered it as well; and if she doesn't find any, it's a decent indication that others won't find any either, or at least none that will be recognized as compelling by the members of the relevant community. The more overlap there is between one's beliefs and those of the relevant audience, the more useful individual ratiocination can be. Sitting at the end of this spectrum, mathematicians from Newton to Perelman have been known to produce remarkable new results in complete isolation.

Of all the scientific communities, mathematicians are those who are most likely to recognize the same facts and to be convinced by the same arguments—they share the same axioms, the same body of already established theorems, and the arguments they are aiming at are proofs. From a formal point of view, a proof is a formal derivation where the conclusion necessarily follows from the premises. From a sociology of science point of view, a proof is an argument that is considered, in a scientific community, as conclusive once and for all. In logic and mathematics, the formal and the sociological notions of proof tend to be coextensive.[25] In 1930, Kurt Gödel presented his first incompleteness theorem, destroying the dreams of building a consistent and complete set of axioms for mathematics, shattering the most ambitious project mathematicians had ever devised. It was a truly revolutionary and threatening result. Yet it was promptly accepted by every mathematician who read Gödel's proof.[26]

If mathematicians are able to reach a consensus so quickly, if they all agree on the rules of the games, then they should also be able to anticipate each other's counterarguments very efficiently. More than any other group, they have both the incentive and the means to evaluate their own arguments thoroughly. The community's many minds are bound to uncover flaws overlooked by a solitary mind, but a lone mathematician still has a chance of achieving great results. Grigory Perelman solved the Poincaré conjecture in a mostly empty, decrepit Russian institute and in his mother's apartment.[27] Andrew Wiles worked for six years in near-total secrecy to prove Fermat's conjecture.

Dialogue can still bring great benefits to mathematicians—Paul Erdős became one of the twentieth century's great mathematicians through hundreds of collaborations. But it is not as necessary as in other disciplines where, even in the so-called hard sciences, a researcher can hardly hope to achieve the same degree of exigency toward her arguments as a mathematician, leaving more room for improvements through discussion.

The Social Context of Science Drives Improvement in Solitary Reasoning

Because science offers its practitioners the incentives and the capacity to engage in productive solitary reasoning, the most brilliant scientists seem endowed with a preternatural ability to generate great insights—and none more so than Isaac Newton. The historian Richard Westfall, Newton's famous biographer, tells us:

> I have never, however, met one [man] against whom I was unwilling to measure myself, so that it seemed reasonable to say that I was half as able as the person in question, or a third, or a fourth, but in every case a finite fraction. The end result of my study of Newton has served to convince me that with him there is no measure. He has become for me wholly other, one of the tiny handful of supreme geniuses who have shaped the categories of human intellect, a man not finally reducible to the criteria by which we comprehend our fellow beings.[28]

Newton's deification began long before Westfall,[29] but there were also some early skeptics, such as Joseph Priestley, who was careful to bring Sir Isaac's achievements back to human scale: "Could we have entered into the mind of Sir Isaac Newton, and have traced all the steps by which he produced his great works, we might see nothing very extraordinary in the process."[30] Priestley was, among many things, a chemist, and Sir Isaac's work in that domain seems to justify this irreverent assessment.

Prying in Newton's notes, one does encounter some surprising passages:

> The Dragon kild by Cadmus is the subject of our work, & his teeth are the matter purified.
>
> Democritus (a Grecian Adeptist) said there were certain birds (volatile substances) from whose blood mixt together a certain kind of Serpent ([symbol for mercury]) was generated which being eaten (by digestion) would make a man understand the voice of birds (the nature of volatiles how they may be fixed).
>
> St John the Apostle & Homer were Adeptists.
>
> Sacra Bacchi (vel Dionysiaca) [the rites of Bacchus (or Dionysus)] instituted by Orpheus were of a Chymicall meaning.[31]

These are hardly unique. Newton wrote hundreds of pages on chemistry and alchemy, some describing experiments, others trying to understand the deep meaning of such allegories. While to many the passages from Newton's greatest physics work—the *Principia*—would sound equally obscure, they come from different operations of reasoning. The most relevant difference is not that Newton happened to be right in one case and wrong in the other but that the quality of his arguments varies widely. It would take an extreme relativist to argue that the tight mathematical arguments of the *Principia* are not any sounder than, say, this: "Neptune with his trident leads philosophers into the academic garden. Therefore Neptune is a mineral, watery solvent and the trident is a water ferment like the Caduceus of Mercury, with which Mercury is fermented, namely, two dry Doves with dry ferrous copper."[32] Yet both are the product of the same undeniably brilliant mind.

An obvious difference between Newton's reasoning about astronomy and about alchemy is the quality of the data he had access to. On the one hand Tycho Brahe's precise recording of stellar and planetary positions. On the other hand a mix of hermeneutical texts, vague rumors about people able to transmute metal, and bogus recipes. But there is another difference.

When reasoning about astronomy, Newton knew he would have to convince the most brilliant minds of his time, and he could try to anticipate many of their counterarguments. Even when his academic colleagues weren't there to talk with him, they were influencing the way he thought. This preoccupation is reflected in Newton's publication choices. When he published his revolutionary ideas, Newton made sure his colleagues would be convinced, even if that meant not reaching a broader audience. While the first version of the *Principia* was written "in a popular method, that it might be read by many," Newton then realized that "such as had not sufficiently entered into the principles could not easily discern the strength of the consequences, nor lay aside the prejudices to which they had been many years accustomed" and so, "to prevent the disputes which might be raised upon such accounts," he "chose to reduce the substance of this book into the form of Propositions (in the mathematical way)."[33]

By contrast, in his alchemical pursuits, Newton lacked serious interlocutors: at the time there were only a "few 'chemical philosophers' besides J. B. van Helmont, long dead, Robert Boyle, and Isaac Newton."[34] Moreover, the whole topic was shrouded in secrecy. Newton "intend[ed] not to publish anything"[35] on this subject, and he complained about Boyle keeping his recipes to himself.[36] At Boyle's death, Newton asked John Locke, who was one of Boyle's executors, to share some of Boyle's previously hidden notes, while forcefully denying that they contained anything valuable.[37] Such a social context put no pressure on Newton to produce strong arguments, and provided him with little possibility to anticipate counterarguments anyway.

When reasoning about gravity, Newton had to convince a community of well-informed and skeptical peers. He was forced to develop better arguments. When reasoning about alchemy, there were no such checks. The same brilliant mind reasoning on its own went nowhere.

Conclusion: In Praise of Reason after All

Philosophers have depicted reason as a superior power of the human mind. Experimental psychologists have suggested that this superpower is, alas, badly flawed. From an evolutionary point of view, the idea of a flawed superpower makes little sense. So, we set out to rethink what reason is and what it is for.

What Reason Is (and Isn't)

Reason is, we argued, one *module of inference* among many. Inferential modules are specialized: they each have a narrow domain of competence and they use procedures adapted to their narrow domain. This contrasts with the old and still dominant view that all inference is done by means of the same logic (or the same probability calculus, or logic plus probabilities).

But isn't reason characterized by the fact that it is general? How, then, could it be a specialized inference module? The first part of our answer—let's not rush—has been to insist that reason is indeed specialized; it draws intuitive inferences just about reasons.

Reason draws inferences about reasons? This may look like a vague truism or a cheap play on words (at least in English and in Romance languages, where a single word of Latin origin refers both to the faculty of reason and to reasons as motives). Yet in the history of philosophy and of psychology, reason and reasons have been studied as two quite distinct topics. So the hypothesis that what reason does is draw inference about reasons, far from being a truism, is a serious challenge to dominant views. But how does it help?

True, humans can reason about any topic whatsoever. How can a mechanism that draws intuitive inferences just about reasons be the mechanism of reason itself? The second part of our answer has been that the reason module produces not only *intuitive* conclusions about reasons—and indeed only about reasons. In doing so, it also indirectly produces *reflective* conclusions about the things reasons are themselves about. Since reasons may be about anything—rabbits, rain, boats, people, law, or numbers—reason may, indirectly, produce reflective conclusions about all kinds of topics.

So, for instance, if there are dark clouds in the sky and you want to go out, you might intuit, "It might rain. This is a strong reason to take my umbrella." When, on the basis of that intuition, you decide to take your umbrella, your decision is an indirect, reflective conclusion of your intuitive inference. Your intuition was about the strength of a reason; your reflective decision is about taking your umbrella.

Instead of assuming that intuition and reasoning must be produced by two quite different types of mechanisms—an old idea currently refurbished in dual process theories—we show how reasoning itself can be achieved by an intuitive inference mechanism. The mechanism is highly specialized but it indirectly contributes to our ideas in all domains: it has what we called "virtual domain-generality."

Compare: the mechanism of visual perception is highly specialized. It processes patterns of stimulation on the retina caused by photons and draws unconscious inferences on the things in the environment that may have emitted or reflected these photons. Not in spite of its high specialization but thanks to it, vision indirectly contributes to most of our thoughts and decisions. Vision and reason are specialized in very different ways, but they both exemplify virtual domain-generality.

The procedures of vision exploit regularities in the way objects reflect light in a normal environment, such as the fact that light generally come from above, and more specific regularities concerning types of objects, such as the fact that faces are generally seen upright. Similarly, the procedures of reason exploit properties of reasons in general, for example, relevance, clarity, or strength; they also exploit properties of specific types of reasons, for example, the force of precedent in reasons concerning coordination, from parent-children relationships to legal matters.

We show, in other terms, how reason fits among other modules of intuitive inference rather than being a towering superpower. Notwithstanding its virtual domain generality, reason is not a broad-use adaptation that would be advantageous to all kinds of animal species. Reasons, we argued, are for social consumption. Reason is an adaptation to the hypersocial niche humans have built for themselves. First part of the enigma of reason solved.

What Reason Is (and Isn't) For

We have been working together on reason for more than ten years. While our account of the mechanisms of reason is developed for the first time in this book, we have been presenting our earlier work on the function of reason—the "argumentative theory of reasoning"—in a number of publications and conferences.

Most of the philosophers and psychologists we talked to endorse some version of the dominant intellectualist view: they see reason as a means to improve individual cognition and arrive on one's own at better beliefs and decisions. Reason, they take for granted, should be objective and demanding. Still, when we present evidence that, on the contrary, reason is hopelessly biased and lazy, they accept it without a hitch. Indeed, many of them are familiar with this evidence—some have even contributed to collecting it.

Reading this book, you might have felt somewhat the same way. When you discovered Bertillon's system or read about Pauling's obsession with vitamin C, you might have been more entertained than surprised. Psychology books make their hay showcasing human biases—as we did in Chapters 11 to 14. This has become part of pop scientific culture. In a way, we all know how biased and limited reason is—well, other people's reason at least.

We are as good at recognizing biases in others as we are bad at acknowledging our own.[1] Perhaps this explains why many people can both hold onto an intellectualist position (for themselves and some kindred spirits) and firmly believe that reason is biased and lazy (particularly in individuals who disagree with them).

Actually, the usual defenses of the intellectualist approach to reason are themselves good examples of biased and lazy reasoning. It is an undisputed fact that individual reasoning is rarely if ever objective and impartial as it should be if the intellectualist approach were right. In discussing what to do with this mismatch between theory and evidence, the possibility that the approach itself might be mistaken is rarely considered. Failures of reasoning are lazily explained by various interfering factors and by weaknesses of reason itself. Again, this doesn't make much evolutionary sense. A genuine adaptation is adaptive; a genuine function functions.

In our interactionist account, reason's bias and laziness aren't flaws; they are features that help reason fulfill its function. People are biased to find reasons that support their point of view because this is how they can justify their actions and convince others to share their beliefs. You cannot justify yourself by presenting reasons that undermine your justification. You cannot convince others to change their minds by giving them arguments for the view you want them to abandon or against the view you want them to adopt. And if people reason lazily, it is because, in typical interactions, this is the most efficient way to proceed. Instead of laboring hard to anticipate counterarguments, it is generally more efficient to wait for your interlocutors to provide them (if they ever do).

Reason properly understood as a tool for social interaction is certainly not perfect, but flawed it is not. Second part of the enigma of reason solved.

While the argumentative theory of reasoning has been generally well received, it has often been misunderstood in two ways not just by people critical of the theory but also—and this has been more worrying—by people who were attracted to it.

A first misunderstanding that we encountered again and again consists in attributing to us the view that argumentation is just a way to manipulate and deceive others and that it has no real intellectual merit. This very cynical view of reasoning and argumentation must have some appeal—possibly that of making one feel superior to naïve ordinary folks. To the risk of disappointing some of our readers, this is a view we do not hold and a cynicism we do not share.

Of course, people are sometimes deceived by an argument. This can happen, however, only because most arguments are not deceitful and are easier to evaluate than just would-be authoritative pronouncements. Without the possibility to objectively evaluate arguments, why would anybody ever take an argument seriously? Reasoning is not only a tool for producing arguments to convince others; it is also, and no less importantly, a tool for evaluating the arguments others produce to convince us. The capacity to produce arguments could evolve only in tandem with the capacity to evaluate them.

Intellectualists are committed to the view that reason should be demanding and objective both in the production and in the evaluation of arguments. They cannot but observe with resignation that human reason actually is not up to what it should be.

The interactionist approach, on the other hand, makes two contrasting predictions. In the production of arguments, we should be biased and lazy; in the evaluation of arguments, we should be demanding and objective—demanding so as not to be deceived by poor or fallacious arguments into accepting false ideas, objective so as to be ready to revise our ideas when presented with good reasons why we should.

The first prediction—that the production of reasons is lazy and biased—is not, strictly speaking, a prediction at all. The data we "predicted," or rather retrodicted, were already there in full view. What the interactionist approach does (and the intellectualist approach fails to do) is make sense of this evidence.

The second prediction—that evaluation is demanding and objective—is a genuine prediction. There is hardly any direct evidence on the issue in the literature, and the little there is is inconclusive. This second prediction is original to the interactionist approach. Ask an intellectualist psychologist of reasoning to predict whether people will be better at producing or evaluating arguments, and chances are your interlocutor won't predict any difference or won't even see the rationale of the question.

Just as widespread as the view that people—at least other people—are biased and lazy is the view that they are gullible: they accept the most blatantly fallacious arguments; and that they are pigheaded: they reject perfectly valid arguments. If people were both gullible and pigheaded, it would be all too easy to spread false new ideas and all too difficult to dispel old mistaken views. The exchange of ideas would, if anything, favor bad ideas. This pessimistic view is widely shared. Group discussion, in particular, is often reviled,[2] an attitude well expressed by the saying, "A camel is a horse designed by a committee."

Actually, camels are a marvel of nature, group discussions often work quite well, and the study of these discussions provides good indirect evidence in favor of our second prediction. In problem solving, the performance of a group tends to be much better than the average individual performance of the group's members and, in some cases, even better than the individual performance of any of its members: people can find together solutions that none of them could find individually. We reviewed some of this evidence in Chapter 15.[3] There is much further evidence in the literature supporting our prediction that people are more demanding and objective in evaluation than in production. This evidence is, alas, indirect. We have begun, however, testing our prediction directly. Stay tuned!

In this book, we have highlighted the remarkable success of reasoning in a group. This, however, had often given rise to a second misunderstanding of our approach.

According to the interactionist approach, reason didn't evolve to enhance thinking on one's own but as a tool for social interaction. We produce reasons to justify ourselves and to convince others. This exchange of reasons may benefit every interlocutor individually. It may also, on some occasions, benefit a group. Why not envisage, then, that the exchange of reasons and the mechanism of reason itself could have evolved for the benefit of the group rather than for the benefit of individuals?

The idea that Darwinian selection works at several levels and in particular at the level of groups has been much developed and discussed lately. It has been argued in particular that group-level selection has played a major evolutionary role in making human cooperation and morality possible.[4] Couldn't the evolution of reason, then, be a case of group-level rather than individual-level selection for cognitive cooperation? No, ours is definitely not a group-level selection hypothesis. In fact, it would be inconsistent with the interactionist approach to reason to think of it as a group-level adaptation.

Group-level selection favors the pursuit of collective benefits over that of individual benefits. Reason as we have described it is, by contrast, a mechanism for the pursuit of individual benefits. An individual stands to benefit from having her justifications accepted by others and from producing arguments that influence others. She also stands to benefit from evaluating objectively the justifications and arguments presented by others and from accepting or rejecting them on the basis of such an evaluation. These benefits are achieved in social interaction, but they are individual benefits all the same.

In interactions where reasons play a role, the people interacting may have converging or diverging interests. The exchange of reasons may play an important role in either case. Argumentation, for instance, plays a major role in negotiations, where interests diverge, often quite strongly. To the extent that members of a group share their interests, they can trust one another, and people who trust one another have a very reduced use or no use at all for justifications and arguments. Group selection would favor systematic trust and trustworthiness in a group. Reason as we describe it is an adaptation to social life where trust has to be earned and remains limited and fragile.

Group discussion is not always efficient. When people have their ideas closely aligned to start with, it leads to polarization. When people start with conflicting ideas and no common goal, it tends to exacerbate differences. Group discussion is typically beneficial when participants have different ideas and a common goal. The collective benefits reaped in such cases should be seen, we suggest, as a side effect of a mechanism that serves individual interests.

To say as we do that reason is an individual-level rather than a group-level adaptation doesn't mean that it has consequences only for individuals and not for social groups and networks. In Chapters 17 and 18, we have evoked various ways in which the individual dispositions of many reasoners can be harnessed for moral, political, or scientific goals. More generally, an issue well worth investigating is that of the population-scale effects of this individual disposition. What role does reason play in the success or failure of different cultural ideas and practices? Conversely, while we have shown that reason is a human universal, much more must be done to find out to what extent and in which ways it can be harnessed, enriched, and codified differently in various cultural traditions (the cultural history of logics being only one quite interesting aspect of the question).

And now to conclude this conclusion: reason has stood for far too long on a broken pedestal, overhanging other faculties but with an awkward tilt. What we hope to have done is put it back where it belongs, level with other cognitive mechanisms but quite upright, and, as other evolved mechanisms, powerful in complex and subtle ways—and endlessly fascinating.

NOTES

REFERENCES

ACKNOWLEDGMENTS

ILLUSTRATION CREDITS

INDEX

Notes

Introduction

1. For a more detailed answer, see Dawkins 1996.
2. Kahneman 2011.
3. Mercier 2016a; Mercier and Sperber 2011.

1. Reason on Trial

1. Descartes 2006.
2. Ibid., p. 13.
3. Ibid., p. 18.
4. Ibid., p. 5.
5. Ibid.
6. Cited, translated, and discussed in Galler 2007.
7. Luther 1536, LW 34:137 (nos. 4–8).
8. The story is well told in Nicastro 2008. Dan Sperber used it in answering John Brockman's question "What Is Your Favorite Deep, Elegant, or Beautiful Explanation?" (Brockman 2013).
9. Actually, there was no universally recognized value of the stade at the time: several states each had their own version. What was the length of the stade Eratosthenes used? There is some doubt on the issue, and some doubt therefore on the accuracy of his measurement of the circumference of the earth. What is not in doubt are the inventiveness and rationality of his method.
10. Kaczynski 2005.
11. Chase 2003, p. 81.
12. A theme well developed in Piattelli-Palmarini 1994.
13. Tversky and Kahneman 1983.
14. Mercier, Politzer, and Sperber in press.

15. As shown, for instance, in Prado et al. 2015.

16. Braine and O'Brien 1998; Piaget and Inhelder 1967; Rips 1994.

17. E.g., Johnson-Laird and Byrne 1991.

18. See, e.g., Evans 1989.

19. Byrne 1989. In Byrne's experiments, the character who may study late in the library is not named. For ease of exposition, we call her Mary.

20. E.g., Bonnefon and Hilton 2002; Politzer 2005.

21. The role of relevance in comprehension is emphasized in Sperber and Wilson 1995.

2. Psychologists' Travails

1. Interpreting Aristotle on the issue is not easy. See, for instance, Frede 1996.

2. Kant 1998, p. 106.

3. For different perspectives on the experimental psychology of reasoning, see Adler and Rips 2008; Evans 2013; Holyoak and Morrison 2012; Johnson-Laird 2006; Kahneman 2011; Manktelow 2012; Manktelow and Chung 2004; Nickerson 2012.

4. Exceptions include Evans 2002 and Oaksford and Chater 1991.

5. These and related issues have been highlighted by the philosopher Gilbert Harman in his book *Change in View* (1986), but this hasn't had the impact we believe it deserved on the psychology of reasoning.

6. Khemlani and Johnson-Laird 2012, 2013.

7. F. X. Rocca, "Why Not Women Priests? The Papal Theologian Explains," Catholic News Service, January 31, 2013, available at http://www.catholicnews.com/services/englishnews/2013/why-not-women-priests-the-papal-theologian-explains.cfm.

8. Personal communication, November 21, 2012.

9. Our preferred explanation of the task is Sperber, Cara, and Girotto 1995.

10. Evans 1972; Evans and Lynch 1973.

11. For further striking evidence that, in solving the selection task, people do not reason but just follow intuitions of relevance (of a richer kind than Evans suggested), see Girotto et al. 2001.

12. Evans and Wason 1976, p. 485; Wason and Evans 1975.

13. Evans and Over 1996.

14. Sloman 1996.

15. Stanovich 1999.

16. Kahneman 2003a.

17. For some strong objections to dual system approaches, see Gigerenzer and Regier 1996; Keren and Schul 2009; Kruglanski et al. 2006; Osman 2004.

3. From Unconscious Inferences to Intuitions

1. Hume 1999, p. 166.

2. When we say of an organism that it has "information" about some state of affairs, we mean roughly that it is in a cognitive state normally produced only if the state of affairs in question obtains. You have, for instance, information that it is raining when you come to believe that it is raining through mechanisms that evolved and developed so as to cause such a belief only if it is raining. This understanding and use of "information" is inspired by the work of Fred Dretske (1981), Ruth Millikan (1987, 1993), and other authors who have developed similar ideas. We are aware of the complex issues such an approach raises. While we don't need to go into them for our present purpose, we do develop the idea a bit deeper in Chapter 5, when we discuss the notion of representation (see also Floridi 2011; Jacob 1997).

3. Darwin 1938–1939, p. 101.

4. Steck, Hansson, and Knaden 2009; Wehner 2003; Wittlinger, Wehner, and Wolf 2006.

5. Wehner 1997, p. 2. For the figure, see Wehner 2003.

6. On the understanding of perception as unconscious inference, see Hatfield 2002. On Helmholtz, see Meulders 2010. On Ptolemy, see Smith 1996.

7. Shepard 1990.

8. Bartlett 1932.

9. Ibid., p. 204.

10. Strickland and Keil 2011.

11. Miller and Gazzaniga 1998.

12. Grice 1989.

13. Kahneman 2003a.

14. See, for instance, Proust 2013; Schwartz 2015.

15. Thompson 2014.

4. Modularity

1. Vouloumanos and Werker 2007.

2. Pinker 1994.

3. Marler 1991.

4. To use Frank Keil's apt expression (Keil 1992).

5. Kanwisher, McDermott, and Chun 1997.

6. Csibra and Gergely 2009; Gergely, Bekkering, and Király 2002; see also Nielsen and Tomaselli 2010. These studies were based on earlier experiments of Meltzoff 1988.

7. Rakoczy, Warneken, and Tomasello 2008; Schmidt, Rakoczy, and Tomasello 2011; Schmidt and Tomasello 2012.

8. Dehaene and Cohen 2011.

9. Schlosser and Wagner 2004.

10. Fodor 1983.

11. As is well illustrated in the work of, for instance, Clark Barrett, Peter Carruthers, Leda Cosmides and John Tooby, Rob Kurzban, Steven Pinker, and Dan Sperber. For an in-depth discussion, see Barrett 2015.

5. Cognitive Opportunism

1. Smith 2001.

2. For instance, Oaksford and Chater 2007; Tenenbaum et al. 2011.

3. Needham and Baillargeon 1993. We thank Renée Baillargeon for kindly providing us with an improved version of the original figure.

4. Hespos and Baillargeon 2006; see also Luo, Kaufman, and Baillargeon 2009.

5. Fodor 1981, p. 121.

6. Winograd 1975.

7. We are using a broad, naturalistic notion of representation inspired in particular by the work of Fred Dretske (1981, 1997), Pierre Jacob (1997), and Ruth Millikan (1987, 2004). For a more restricted view of representation that, while addressing philosophical concerns, pays close attention to empirical evidence, see Burge 2010.

8. On the other hand, we are not committing here to any particular view on the metaphysics of representations and on the causal role, if any, of representational content. For a review and useful discussion, see Egan 2012.

9. Another possibility is that some transmission of information is done not through direct module-to-module links but by broadcasting a module's output on a "global workspace" where other modules can access it, as envisaged in theories of consciousness such as Baars's and Dehaene's (Baars 1993; Dehaene 2014).

10. Tenenbaum, Griffiths, and Kemp 2006.

11. Gallistel and Gibbon 2000; Rescorla 1988.

12. As suggested long ago by Daniel Dennett (1971).

13. For instance, Sperber 2005.

6. Metarepresentations

1. See, for instance, Carey 2009; Hirschfeld and Gelman 1994; Sperber, Premack, and Premack 1995.

2. See Sperber 2000.

3. Premack and Woodruff 1978.

4. Gopnik and Wellman 1992; Perner 1991.

5. Baillargeon, Scott, and He 2010; Leslie 1987.

6. Baron-Cohen et al. 1985; Wimmer and Perner 1983.

7. Onishi and Baillargeon 2005.

8. Surian, Caldi, and Sperber 2007.

9. For various uses of the notion of mental file, see Kovács 2016; Perner, Huemer, and Leahy 2015; Recanati 2012. Our approach is closer to Kovács's.

10. Frith and Frith 2012; Kovács, Téglás, and Endress 2010; Samson et al. 2010.

11. The distinction between personal and subpersonal levels was introduced by Dennett 1969. It has been interpreted in several ways, including by Dennett himself; see Hornsby 2000.

12. There is an influential theory defended, in particular, by the philosopher Alvin Goldman and the neuroscientist Vittorio Gallese, according to which we understand what happens in other people's minds by simulating their mental processes (e.g., Gallese and Goldman 1998). The approach that we suggest is at odds with standard versions of "simulation theory" and more compatible with others, where, however, the notion of "simulation" is understood in a technical sense, so much broader than the ordinary sense of the word that this becomes a source of misunderstandings.

13. Some psychologists and philosophers (e.g., Apperly and Butterfill 2009; Heyes and Frith 2014) have argued that the mechanism that tracks the mental states of others is more rudimentary than we have suggested and doesn't even detect beliefs or intentions as such. According to them, the more sophisticated understanding exhibited by four-year-old children who "pass" the standard false belief task (or by adults who enjoy reading a Jane Austen novel) is based on an altogether different "system 2" mindreading mechanism. The main advantage of this dualist hypothesis is to accommodate the new evidence regarding mindreading in infants with minimal revisions of earlier standard views of mindreading. Others—and we agree with them—argue in favor of revising these standard views and question the alleged evidence in favor of a dual-system understanding of mindreading (see Carruthers 2016; see also Brent Strickland and Pierre Jacob, "Why Reading Minds Is Not like Reading Words," January 22, 2015, available at http://cognitionandculture.net/blog/pierre-jacobs-blog/why-reading-minds-is-not-like-reading-words).

14. Dehaene 1999.

15. Gilmore, McCarthy, and Spelke 2007.

16. Rozenblit and Keil 2002.

7. How We Use Reasons

1. The word "reason" is also used in a much broader sense as a synonym of cause, as in "The reason why the night is dark is that today is the new moon." In this book, we have avoided using "reason" in this broad sense when there was any risk of confusion.

2. For a review of recent philosophical work on reasons relevant to our concerns, see Alvarez 2016.

3. Nagel 1976; Pritchard 2005; Williams 1981.

4. See, for instance, Baars 1993; Block 2007; Carruthers 2005; Chalmers 1996; Dehaene 2014; Dennett 1991; Graziano 2013; Rosenthal 2005.

5. Nisbett and Wilson 1977; see also Greenwald and Banaji 1995.

6. Carruthers 2011.

7. Darley and Latané 1968.

8. Latané and Rodin 1969.

9. Hannes Rakoczy (2012) underscores the confusion brought about by the variety of uses of the explicit / implicit distinction. Zoltan Dienes and Josef Perner (1999) do define "implicit" in a precise but somewhat idiosyncratic way, which would deserve a detailed discussion in another context.

10. See Jessen 2001.

11. Sperber and Wilson 2002.

8. Could Reason Be a Module?

1. Raz 2000, 2010; see also Wallace et al. 2004.

2. Carroll 1895. For a richer modern discussion, with some attention to psychology, see Kornblith 2012; Railton 2004.

3. See Köhler 1930; Ramachandran and Hubbard 2001. For recent relevant work on the issue, see Monaghan et al. 2014.

4. Green and Williams 2007.

5. As we argued in Chapter 3, see also Thompson 2014.

6. For an excellent example of this, read Robert Zaretsky and John Scott's *The Philosopher's Quarrel* (2009), where they describe how Jean-Jacques Rousseau stubbornly resisted David Hume's sensible arguments in the name of his own intuition, bolstered by his own philosophical arguments in favor of intuition.

7. A point argued in detail by Proust 2013.

8. Mercier 2012.

9. On reputation in general, see Origgi 2015. On moral reputation, see Baumard, André, and Sperber 2013; Sperber and Baumard 2012.

10. The fit between this approach and Jürgen Habermas's theory of communicative action (1987) would be worth exploring.

11. Frederick 2005.

9. Reasoning

1. Mercier and Sperber 2009; Sperber 1997.

2. It is common to call "intuitive," in an extended sense, a reflective conclusion for which one has an intuitive argument. This is a harmless extension of meaning, and we do

not object to it in general. In this book, however, where we discuss the relationship between intuitive reasons and their reflective conclusions, we use "intuition" and "intuitive" in their strict, nonextended sense.

3. Whitehead and Russell 1912.

4. Sometimes reasoning also resorts to pictures or gestures, which may provide better insights than words, but typically these insights are verbally invoked in support of a verbal conclusion. See, for instance, Stenning and Oberlander 1995.

5. It can be argued that a prior ability to metarepresent representations is what made language possible in the evolution of the species and what makes language acquisition possible in individual development (Origgi and Sperber 2000; Scott-Phillips 2014; Sperber and Wilson 2002). This is not a universally accepted view. For the opposite view that language is what makes metarepresentations possible and hence comes before metarepresentation both in evolution and in development, see Dennett 2000; De Villiers 2000.

6. Blakemore 2002; Ducrot 1984; Ducrot and Anscombre 1983; Hall 2007; Wilson 2016.

7. As stressed by Harman 1986.

8. On the reductio ad absurdum in a dialogical perspective, see Dutilh Novaes 2016.

9. See, for example, Evans and Over 2004.

10. For a detailed discussion, see Carston 2002; Wilson and Sperber 2002.

11. See Chapter 12 and Mercier et al. in press.

12. Carruthers 2006.

13. See Oaksford and Chater 2007; Stenning and Van Lambalgen 2012. Both books, by the way, propose a reinterpretation of Byrne's "suppression effect," that is, the fact that *modus ponens* inference can be suppressed by adding a logically irrelevant but pragmatically relevant premise, as we saw in Chapter 1.

14. We are not denying, of course, that there are questions that can be resolved by using syllogism or other forms of deduction in a strict and literal manner. As the case of the Wason four-card selection task we discussed in Chapter 2 illustrates, even when a problem can be solved deductively, people often fail to realize this. Nevertheless, especially when people argue about such problems, logical considerations may end up being invoked normatively.

15. This is an issue that is raised in Louis Lee, Goodwin, and Johnson-Laird 2008.

16. Stanovich 2012.

10. Reason

1. Darwin 1981, p. 46.

2. Ibid., pp. 188–189.

3. See, for instance, Einhorn and Hogarth 1981; Nisbett et al. 1983.

4. Cosmides and Tooby 1987, 1989; Tooby and Cosmides 1989.

5. Cosmides 1989.

6. Girotto et al. 2001; Sperber, Cara, and Girotto 1995.

7. For an argument to the same effect via a different route, see Kruglanski and Gigerenzer 2011.

8. Heyes 2003; Sterelny 2003.

9. Mercier and Sperber 2011.

10. Bicchieri 2006.

11. Malle 2004.

12. Reid 2000, pp. 193–194.

13. Mascaro and Morin 2014.

14. For recent ideas on the relationship between cooperation and communication, see Godfrey-Smith and Martínez 2013; Skyrms 1996, 2010; Sterelny 2012; Tomasello 2010.

15. Axelrod 1984; Blais 1987.

16. Baumard, André, and Sperber 2013.

17. Sperber et al. 2010.

18. Clément 2010; Harris 2012; Mascaro and Sperber 2009.

19. Montefiore 2003, p. 360.

20. For an interpretation of the history of classical logic that focuses on its argumentative roots, see Dutilh Novaes 2015.

21. We are not the first to develop an argumentative theory of reasoning. Others (for instance, Billig 1996; Perelman and Olbrechts-Tyteca 1958; Toulmin 1958) have maintained that reasoning is primarily about producing and evaluating arguments. They have done so mostly on introspective grounds and in a philosophical perspective. We may be more original in having done so (Mercier 2016a; Mercier in press; Mercier and Sperber 2011) on empirical grounds and in a naturalistic and evolutionary perspective (see also Dessalles 2007, who develops a different evolutionary perspective where argumentation plays an important role).

22. On the distinction between adaptation and beneficial side effects, see Sober 1993; for more on the evolution of sea turtles' limbs, see Wyneken, Godfrey, and Bels 2007, pp. 97–138.

11. Why Is Reasoning Biased?

1. "Linus C. Pauling Dies at 93; Chemist and Voice for Peace," *New York Times*, August 21, 1994, available at http://www.nytimes.com/learning/general/onthisday/bday/0228.html.

2. Watson 1997, Kindle location 395.

3. Maurice Wilkins, BBC radio 4 interview, 1997, Oregon State University Libraries, available at http://oregondigital.org/cdm4/item_viewer.php?CISOROOT=/dna&CISOPTR=166&CISOBOX=1&REC=1.

4. "Linus Pauling Rebuts New Mayo Study on Vitamin C," September 28, 1979, p. 2, available at https://profiles.nlm.nih.gov/ps/retrieve/ResourceMetadata/MMBBKK (see also Collins and Pinch 2005).

5. Creagan et al. 1979.

6. Moertel et al. 1985.

7. Collins and Pinch 2005.

8. Pauling and Herman 1989.

9. Ibid., p. 6837.

10. Ibid., p. 6835.

11. Ibid., p. 6835; emphasis added on the derogatory "described."

12. Mansel Davies, "Obituary: Professor Linus Pauling," *Independent,* August 21, 1994, available at http://www.independent.co.uk/news/people/obituary-professor-linus-pauling-1377923.html.

13. *Oxford English Dictionary,* https://en.oxforddictionaries.com/definition/bias.

14. Tversky and Kahneman 1973.

15. E.g., Gigerenzer 2007.

16. Funder 1987; Haselton and Buss 2000.

17. Ings and Chittka 2008.

18. Liszkowski et al. 2004.

19. Crombie 1971, pp. 84ff. For another perspective on Grosseteste and falsification, see Serene 1979.

20. Both citations are from Bacon 1620, bk. 1, p. 105. Crombie 1971 links Grosseteste and Bacon, and Urbach 1982 points out the commonalities between Bacon and Popper on this issue.

21. Popper 1963, p. 6.

22. Mynatt, Doherty, and Tweney 1977.

23. Wason 1968, p. 142.

24. Ball et al. 2003; for related methodology, see Evans 1996.

25. Lucas and Ball 2005.

26. Nickerson 1998, p. 175.

27. Kuhn 1991, computed from table 5.4, p. 142, with 160 participants.

28. Sanitioso, Kunda, and Fong 1990.

29. Taber and Lodge 2006.

30. Johnson-Laird and Byrne 2002; Stanovich and West 2008; Stanovich, West, and Toplak 2013.

31. Evans 1989, p. 41; although not all demonstrations are equally convincing, see Mercier 2016b.

32. "Julie Burchill and Psychopaths," n.d., available at http://jonathanronson.tumblr.com/post/43312919747/julie-burchill-and-psychopaths.

33. Evans and Over 1996, p. 20.

34. Stanovich 2004, p. 134.
35. Lilienfeld, Ammirati, and Landfield 2009, p. 391.
36. Nickerson 1998, p. 205.
37. Bacon 1620.
38. Thaxter et al. 2010.
39. Evans 2007; Kahneman 2003b; Stanovich 2004.
40. Stanovich 2004, p. 134.
41. Evans 1989, p. 42.
42. Pyke 1984.
43. For foraging, see, e.g., Hill et al. 1987.
44. Allen et al. 2009.
45. Ibid., p. 1083.
46. Shaw 1996, p. 80.
47. Mercier and Sperber 2011.
48. The term has been previously used with slightly different meanings by, e.g., Baron 1995 and Perkins 1989.
49. Cicero, *De Inventione,* bk. 1, chap. 52, trans. C. D. Yonge, available at http://classicpersuasion.org/pw/cicero/dnv1-4.htm#97.
50. Indeed, being defended by a lawyer who does not display a consistent myside bias could be grounds for appeal because the counsel has "undermined the proper functioning of the adversarial process," *Strickland v. Washington,* 466 U.S. 668 (1984), available at http://supreme.justia.com/cases/federal/us/466/668/case.html (ineffective assistance of counsel).
51. For previous uses of the analogy, see, e.g., Knobe 2010; Tetlock 2002.

12. Quality Control

1. Cicero, *De Oratore,* bk. 1, chap. 2, p. 7, trans. J. S. Watson, available at http://archive.org/details/ciceroonoratorya00ciceuoft.
2. Ibid., bk. 3, chap. 21, p. 214.
3. Kuhn 1991, p. 87.
4. Ibid.
5. Nisbett and Ross 1980, p. 119.
6. Perkins 1985, p. 568.
7. Levinson 2006; see also Enfield 2013.
8. Loosely based on an example provided by Levinson 2005.
9. Dingemanse and Enfield 2015; Dingemanse et al. 2015.
10. Resnick et al. 1993, pp. 362–363.
11. Kuhn, Shaw, and Felton 1997.
12. Boudry, Paglieri, and Pigliucci 2015.

13. Hahn and Hornikx 2016; Hahn and Oaksford 2007; Oaksford and Hahn 2004.

14. For more evidence that people, at least when they are motivated, are good at evaluating arguments, see the literature on persuasion and attitude change. A good review is provided by Petty and Wegener 1998.

15. Trouche et al. 2016.

16. Trouche, Shao, and Mercier submitted; see also Trouche, Sander, and Mercier 2014.

13. The Dark Side of Reason

1. Arthur Conan Doyle, *The Hound of the Baskervilles,* Project Gutenberg, available at http://www.gutenberg.org/files/3070/3070-h/3070-h.htm.

2. "Bertillonnage," Wikipedia, available at http://fr.wikipedia.org/wiki/Bertillonnage.

3. Blair 1901, p. 395.

4. Reinach 1903, Kindle locations 2911–2924.

5. Whyte 2008, p. 78.

6. Anonymous 1900, p. 330.

7. Ibid., p. 344.

8. Ibid., p. 362.

9. Ibid., p. 369.

10. Anonymous 1904.

11. Ibid.

12. Koriat, Lichtenstein, and Fischhoff 1980. Note that overconfidence is a complex phenomenon, and reasoning is not its only cause.

13. Sadler and Tesser 1973; for review, see Tesser 1978.

14. Ross, Lepper, and Hubbard 1975. On the role of reasoning, see Anderson, Lepper, and Ross 1980; Anderson, New, and Speer 1985.

15. Keynes 1989, p. 252.

16. See, for instance, Kay 2011.

17. Janis 1982.

18. Myers and Bishop 1970.

19. Myers and Bach 1974.

20. Sunstein 2002.

21. E.g., Kay 2011.

22. Preface to Halberstam 2001, p. xi, written by John McCain.

23. Kunda 1990, p. 480.

24. Nickerson 1998, p. 197.

25. Dunning, Meyerowitz, and Holzberg 1989.

26. Kay 2011, p. 217.

27. See Scott-Kakures 2001.

28. Scott-Phillips, Dickins, and West 2011; Tinbergen 1963.

29. Millikan 1987, p. 34.
30. Dehaene 2009.
31. Keynes 1936, p. vii.

14. A Reason for Everything

1. Hall, Johansson, and Strandberg 2012.
2. Nisbett and Wilson 1977.
3. Franklin 1799.
4. Wilson et al. 1993.
5. Simonson 1989.
6. Dijksterhuis 2004.
7. Hsee 1999.
8. Simonson 1989.
9. Hsee et al. 1999.
10. Thompson and Norton 2011.
11. Rozin, Millman, and Nemeroff 1986.
12. See Thompson, Hamilton, and Rust 2005.
13. Tversky and Shafir 1992, p. 305.
14. Or at least the participants thought so; see Simonson 1990.
15. Thompson and Norton 2011.
16. Morabia 2006.
17. Simonson and Nye 1992, p. 442. We've adjusted the dollar amounts to broadly take inflation up to publication date into account.
18. Arkes and Ayton 1999.
19. McAfee, Mialon, and Mialon 2010 make a similar point.

15. The Bright Side of Reason

1. Quotes from *12 Angry Men,* directed by Sidney Lumet, original story by Reginald Rose (1957; Beverly Hills, CA: MGM, 2008), DVD.
2. Habermas 1975, p. 108.
3. Twenty-seven percent for the Harvard students of Cosmides 1989; 18 percent for the top one-quarter students of Stanovich and West 2000.
4. Johnson-Laird and Byrne 2002.
5. Oaksford and Chater 2003, p. 305.
6. Open Science Collaboration 2015.
7. Mercier, Trouche, et al. 2015.
8. Trouche, Sander, and Mercier 2014.
9. See ibid. for references.

10. Koriat 2012; Levin and Druyan 1993.

11. Moshman and Geil 1998; Trognon 1993.

12. Trouche, Sander, and Mercier 2014.

13. See Laughlin 2011.

14. See the references in Kravitz and Martin 1986.

15. Ringelmann 1913, p. 10, translated by Kravitz and Martin 1986.

16. Latané, Williams, and Harkins 1979.

17. E.g., Mullen, Johnson, and Salas 1991.

18. Nemeth et al. 2004. This article shows that dissent increases only the quantity, not the quality, of idea generation, but further evidence suggests that quality of idea generation is also improved by dissent; see, for instance, Nemeth and Ormiston 2007; Smith 2008. For more evidence of the positive role played by dissension in group discussion, see, e.g., Schulz-Hardt et al. 2006.

19. Montaigne 1870, p. 540.

20. Barry Popik, "Never Make Forecasts, Especially about the Future," October 30, 2010, available at http://www.barrypopik.com/index.php/new_york_city/entry/never_make_forecasts_especially_about_the_future/.

21. Tetlock 2005.

22. Ibid., p. 51.

23. See Silver 2012.

24. Tetlock 2005, p. 61.

25. Ibid., p. 123.

26. Ibid., p. 62.

27. Ibid., pp. 121ff.

28. Ibid., p. 128.

29. Dalkey and Helmer 1963, p. 461.

30. E.g., Linstone and Turoff 1976; Rowe and Wright 1999.

31. Yaniv and Kleinberger 2000.

32. Rowe and Wright 1996.

33. True story . . . "Paul the Octopus," Wikipedia, available at http://en.wikipedia.org/wiki/Paul_the_Octopus.

34. Liberman et al. 2012; Minson, Liberman, and Ross 2011.

35. Mellers et al. 2014.

36. Sunstein, Kahneman, and Schkade 1998, p. 2129.

37. Caenegem 1987, p. 14.

38. Blackstone 1979, p. 380.

39. *Ballew v. Georgia,* 435 U.S. (1978), p. 234, cited in Ellsworth 1989, full text available at http://www.law.cornell.edu/supct/html/historics/USSC_CR_0435_0223_ZO.html.

40. Braman 2009; Sunstein et al. 2007.

41. E.g., Sandys and Dillehay 1995.
42. Hastie, Penrod, and Pennington 1983.
43. Ibid., p. 60.
44. Ellsworth 1989, p. 217.
45. Ellsworth 2003, p. 1406.
46. Mercier, Trouche, et al. 2015.
47. Virginia Statute on Religious Freedom, 1786, Digital History, available at http://www.digitalhistory.uh.edu/disp_textbook.cfm?smtID=3&psid=1357.

16. Is Human Reason Universal?

1. "Geography of Uzbekistan," Wikipedia, available at http://en.wikipedia.org/wiki/Geography_of_Uzbekistan; "Uzbekistan," Wikipedia, available at http://en.wikipedia.org/wiki/Uzbekistan.
2. The first part of this chapter relies on Mercier 2011a and the second on Mercier 2011b.
3. Luria 1934.
4. Luria 1976, p. 107.
5. Ibid., pp. 108–109.
6. Lévy-Bruhl 1910, p. 22.
7. "Individualistic" in the relatively coarse sense in which this term is used in cross-cultural psychology and in particular in East-West comparisons.
8. Henrich, Heine, and Norenzayan 2010.
9. Cole 1971; Scribner 1977.
10. Gordon 2004.
11. Blurton Jones and Konner 1976.
12. Luria 1976, p. 109.
13. Dias, Roazzi, and Harris 2005.
14. Dutilh Novaes 2015.
15. "Japan's Well-Placed Nuclear Power Advocates Swat Away Opponents," *NBC News,* March 12, 2014, available at http://www.nbcnews.com/storyline/fukushima-anniversary/japans-well-placed-nuclear-power-advocates-swat-away-opponents-n50396.
16. Suzuki 2012.
17. All three citations from ibid., p. 179; see also Suzuki 2008.
18. Becker 1986. The citation that follows is drawn from this article.
19. Nakamura 1964, p. 534.
20. Lloyd 2007, p. 10.
21. Liu 1996.
22. Combs 2004.

23. Branham 1994, p.131.

24. "Prince Shotoku's Seventeen-Article Constitution" [Jushichijo Kenpo], SaruDama, available at http://www.sarudama.com/japanese_history/jushichijokenpo.shtml.

25. Mercier, Deguchi, et al. 2016.

26. Boehm et al. 1996; Ostrom 1991.

27. Chagnon 1992. For the record, The Yanomamö are not hunter-gatherers but horticulturalists.

28. Ibid., p. 134.

29. Hutchins 1980.

30. Gluckman 1967, p. 277.

31. Castelain et al. 2016.

32. These topics were actually debated at the 2014 Durham Open debating competition: http://www.debate-motions.info/other-tournament-motions/442-durham-open-2014.

33. For the notions of proper domain, actual domain, and their relevance to the study of culture, see Sperber 1994; Sperber and Hirschfeld 2004.

34. Gigerenzer 2007.

35. Barrett 2000; Boyer 2001.

36. Comtesse de Ségur, *Les Malheurs de Sophie,* 1858, available at http://fr.wikisource.org/wiki/Les_Malheurs_de_Sophie/3; our translation.

37. Dunn and Munn 1987.

38. Ibid., p. 795.

39. Grusec and Goodnow 1994, p. 5.

40. Mercier, Bernard, and Clément 2014; see also Corriveau and Kurkul 2014; Koenig 2012.

41. Castelain, Bernard, and Mercier submitted.

42. "Conservation Task," YouTube, February 10, 2007, available at http://www.youtube.com/watch?v=YtLEWVu815o.

43. Doise, Mugny, and Perret-Clermont 1975; Doise and Mugny 1984; Perret-Clermont 1980.

44. Silverman and Geiringer 1973.

45. Miller and Brownell 1975.

46. Ames and Murray 1982.

47. Slavin 1996, p. 43.

48. Slavin 1995.

49. Ibid.

50. Kuhn, Shaw, and Felton 1997.

51. Mercier et al. in press; Willingham 2008.

52. Pronin, Gilovich, and Ross 2004.

53. Kuhn and Crowell 2011.

17. Reasoning about Moral and Political Topics

1. Schnall et al. 2008.

2. Danziger, Levav, and Avnaim-Pesso 2011; but see Glöckner 2016 for a skeptical interpretation of these results.

3. Haidt 2001.

4. Snyder et al. 1979.

5. Chance and Norton 2008.

6. Rust and Schwitzgebel 2013.

7. And there are many others: see, e.g., Uhlmann et al. 2009; Valdesolo and DeSteno 2008.

8. "Property" on the Jefferson Monticello website, available at http://www.monticello.org/site/plantation-and-slavery/property.

9. Thomas Jefferson to John Wayles Eppes, June 30, 1820, *Founders Early Access,* available at http://rotunda.upress.virginia.edu/founders/default.xqy?keys=FOEA-print-04-02-02-1352.

10. Stanton 1993, pp. 158–159.

11. Letter to Thomas Mann Randolph, June 8, 1803, in Betts 1953, p. 19. This specific citation only refers to one slave.

12. On their distortion of the historical record, see Finkelman 1993.

13. All quotes are from Query XIV, which is available from many online sources; see, for instance, http://xroads.virginia.edu/~hyper/jefferson/ch14.html.

14. All quotes from Query XIV (including the flowing hair).

15. This was a rejection that led both to his own failure to free more than a handful of slaves and to the advice given to other people not to free their slaves either. That was, at any rate, the advice given by Thomas Jefferson to Edward Coles in a letter written on August 25, 1814; see *American History: From Revolution to Reconstruction and Beyond,* available at http://www.let.rug.nl/usa/presidents/thomas-jefferson/letters-of-thomas-jefferson/jefl232.php.

16. Ibid.; see also Finkelman 1993.

17. From Query XIV again.

18. Letter from Thomas Jefferson to Edward Coles, August 25, 1814.

19. Finkelman 1993, p. 186. More generally, on the development of racism as a justification for slavery when confronted with egalitarian ideals, see Davis 1999; Fields 1990.

20. Franklin 1799; emphasis added.

21. Jefferson is also another example of how awareness of a bias doesn't reduce it. As he said, "the moment a person forms a theory, his imagination sees, in every object, only the traits which favor that theory" (Jefferson 1829, p. 235).

22. Haidt 2001, p. 823.

23. Although see Haidt and Bjorklund 2007.

24. Haidt 2001, p. 814; see Haidt, Bjorklund, and Murphy 2000.

25. On the limits of the moral dumbfounding effect, see Royzman, Kim, and Leeman 2015.

26. Drawn from Piaget 1932, adapted by Leman and Duveen 1999, p. 575.

27. Leman and Duveen 1999.

28. Huntington 1993, p. 9.

29. Unless one counts voting with one's feet; see Conradt and Roper 2005.

30. Sen 2003.

31. Mandela 1994, p. 21.

32. E.g., Bessette 1980; Cohen 1989.

33. E.g., Fishkin 2009; for more references, see Landemore 2013; Mercier and Landemore 2012.

34. Luskin et al. 2014.

35. Drescher 2009, p. 207.

36. Ibid., p. 224.

37. Richard Miles, quoted in Sanderson 1972, pp. 68, 71; see also Hochschild 2006.

38. Long 1774; see also Perry 2012.

39. Hochschild 2006, p. 223.

40. Perry 2012, p. 95.

41. Ibid., p. 90.

42. Cited in Swaminathan 2016.

43. Drescher 2009, p. 213.

44. See Carey 2005, pp. 85ff.

45. Drescher 1990, p. 566.

46. Wilberforce 1789.

47. Ibid., p. 52, cited in Perry 2012.

48. Long 1744, p. 437, cited in Perry 2012.

18. Solitary Geniuses?

1. Heisenberg 1952, p. 30.

2. Cited in Shapin 1991, p. 194.

3. Planck 1968, pp. 33–34, cited in Kuhn 1962.

4. Mahoney and DeMonbreun 1977.

5. Mahoney 1977.

6. Dunbar 1995.

7. Kuhn 1962, p. 152.

8. Cohen 1985, p. 468.

9. Studies reviewed by Wray 2011, who concludes that "Planck's principle is a myth. Older scientists are not especially resistant to change" (p. 190).

10. Cohen 1985, p. 468.

11. Cited in Cohen 1985, p. 472.

12. Rorty 1991.

13. Hull, Tessner, and Diamond 1978; Kitcher 1993; Levin, Stephan, and Walker 1995.

14. Oreskes 1988; for other examples, see Kitcher 1993; Wootton 2006.

15. Citations from Thomas 1971.

16. King 1991.

17. Cited in Steven and Schaffer 1985.

18. Ibid., p. 75.

19. Dunbar 1995, p. 380.

20. Csikszentmihalyi and Sawyer 1995, p. 347.

21. Kahneman 2011, p. 40.

22. Beller 2001, pp. 103–104.

23. Ibid., pp 65–102.

24. See also Zamora Bonilla 2006.

25. Azzouni 2007.

26. Mancosu 1999.

27. Gessen 2009.

28. Westfall 1983, p. x.

29. See Fara 2002.

30. Priestley 1786, p. 346.

31. Cited in Hall 1996, p. 188.

32. Cited in ibid., p. 187.

33. Here Newton talks about book 3 of his *Principia*. The quotes are from the introduction to book 3.

34. Hall 1996, p. 196.

35. At least that's what he seems to have admitted to a colleague; cited in Hall 1996, p. 184.

36. Hall 1996, p. 199.

37. Principe 2004.

Conclusion

1. Pronin 2007.

2. Mercier, Trouche, et al. 2015.

3. For instance, the data show that group discussion allows the perfect or nearly perfect spread of the correct solution to manageable logical or mathematical problems

(Claidière, Trouche, and Mercier submitted; Laughlin 2011; Trouche, Sander, and Mercier 2014). It allows group members to reach a better collective answer than that reached by the best group member individually for some inductive problems (Laughlin 2011). It allows pupils and students to do better on a wide variety of school tasks (Johnson and Johnson 2009; Slavin 1995). It allows jurors to deliver better verdicts (Hastie, Penrod, and Pennington 1983). It allows scientists to discard mistaken hypotheses and form new ones (Dunbar 1997). It allows citizens to reach better-informed opinions (Fishkin 2009). It allows doctors and medical students to make better diagnoses (Kesson et al. 2012; Reimer, Russell, and Roland 2016). It allows forecasters to make better political and economic forecasts (Mellers et al. 2014). It allows group members to make more rational strategic decisions (Kugler, Kausel, and Kocher 2012). It allows investors to make better investments (Cheung and Palan 2012). And it allows good ideas to spread among scientists and in the general public (Chanel et al. 2011; Kitcher 1993; Wootton 2015).

4. For instance, Boyd et al. 2003; Sober and Wilson 1998. For an alternative view, see Baumard, André, and Sperber 2013.

References

Adler, J. E., and L. J. Rips, eds. 2008. *Reasoning: Studies of human inference and its foundations.* Cambridge: Cambridge University Press.

Allen, T. J., et al. 2009. Stereotype strength and attentional bias: Preference for confirming versus disconfirming information depends on processing capacity. *Journal of Experimental Social Psychology,* 45(5): 1081–1087.

Alvarez, M. 2016. Reasons for action: Justification, motivation, explanation. *The Stanford encyclopedia of philosophy.* Available at http://plato.stanford.edu/archives/sum2016/entries/reasons-just-vs-expl/.

Ames, G. J., and F. B. Murray. 1982. When two wrongs make a right: Promoting cognitive change by social conflict. *Developmental Psychology,* 18: 894–897.

Anderson, C. A., M. R. Lepper, and L. Ross. 1980. Perseverance of social theories: The role of explanation in the persistence of discredited information. *Journal of Personality and Social Psychology,* 39(6): 1037–1049.

Anderson, C. A., B. L. New, and J. R. Speer. 1985. Argument availability as a mediator of social theory perseverance. *Social Cognition,* 3(3): 235–249.

Anonymous. 1900. *Le procès Dreyfus devant le conseil de guerre de Rennes.* Paris: P-V Stock.

———. 1904. *Rapport de MM les experts (Poincaré etc) [portant sur le mémoire présenté à la Cour de Cassation en 1899 présentant le système de M. Bertillon].* Available at http://www.maths.ed.ac.uk/~aar/dreyfus/dreyfusfrench.pdf.

Apperly, I. A., and S. A. Butterfill. 2009. Do humans have two systems to track beliefs and belief-like states? *Psychological Review,* 116(4): 953–970.

Arkes, H. R., and P. Ayton. 1999. The sunk cost and Concorde effects: Are humans less rational than lower animals? *Psychological Bulletin,* 125(5): 591–600.

Axelrod, R. 1984. *The evolution of cooperation.* New York: Basic Books.

Azzouni, J. 2007. How and why mathematics is unique as a social practice. In B. Van Kerkhove and J. P. van Bendegem, eds., *Perspectives on mathematical practices.* Dordrecht: Springer, pp. 3–23.

Baars, B. J. 1993. *A cognitive theory of consciousness.* Cambridge: Cambridge University Press.

Bacon, F. 1620. *Novum Organum,* 1st ed.

Baillargeon, R., R. M. Scott, and Z. He. 2010. False-belief understanding in infants. *Trends in Cognitive Sciences,* 14(3): 110–118.

Ball, L. J., et al. 2003. Inspection times and the selection task: What do eye-movements reveal about relevance effects? *Quarterly Journal of Experimental Psychology,* 56(6): 1053–1077.

Baron, J. 1995. Myside bias in thinking about abortion. *Thinking and Reasoning,* 1: 221–235.

Baron-Cohen, S., et al. 1985. Does the autistic child have a "theory of mind"? *Cognition,* 21(1): 37–46.

Barrett, H. C. 2015. *The shape of thought: How mental adaptations evolve.* New York: Oxford University Press.

Barrett, J. L. 2000. Exploring the natural foundations of religion. *Trends in Cognitive Sciences,* 4(1): 29–34.

Bartlett, S. F. C. 1932. *Remembering: A study in experimental and social psychology.* Cambridge: Cambridge University Press.

Baumard, N., J. B. André, and D. Sperber. 2013. A mutualistic approach to morality: The evolution of fairness by partner choice. *Behavioral and Brain Sciences,* 36(1): 59–78.

Becker, C. B. 1986. Reasons for the lack of argumentation and debate in the Far East. *International Journal of Intercultural Relations,* 10(1): 75–92.

Beller, M. 2001. *Quantum dialogue: The making of a revolution.* Chicago: University of Chicago Press.

Bessette, J. 1980. Deliberative democracy: The majority principle in republican government. In R. Goldwin and W. Schambra, eds., *How democratic is the Constitution?* Washington, DC: American Enterprise Institute, pp. 102–116.

Betts, E. M. 1953. *Thomas Jefferson's farm book.* Princeton, NJ: Princeton University Press.

Bicchieri, C. 2006. *The grammar of society: The nature and dynamics of social norms.* Cambridge: Cambridge University Press.

Billig, M. 1996. *Arguing and thinking: A rhetorical approach to social psychology.* Cambridge: Cambridge University Press.

Blackstone, W. 1768 / 1979. Of private wrongs. Vol. 3 of *Commentaries on the laws of England.* Facsimile of the first edition. Chicago: University of Chicago Press.

Blair, F. P. 1901. A review of Bertillon's testimony in the Dreyfus case. *American Law Review,* 35: 389–401.

Blais, M. J. 1987. Epistemic tit for tat. *Journal of Philosophy,* 84(7): 363–375.

Blakemore, D. 2002. *Relevance and linguistic meaning: The semantics and pragmatics of discourse markers.* Cambridge: Cambridge University Press.

Block, N. 2007. Consciousness, accessibility, and the mesh between psychology and neuroscience. *Behavioral and Brain Sciences,* 30(5): 481–548.

Blurton Jones, N., and M. J. Konner. 1976. !Kung knowledge of animal behavior. In R. Lee and I. DeVore, eds., *Studies of the !Kung San and their neighbors.* Cambridge, MA: Harvard University Press, pp. 325–348.

Boehm, C., et al. 1996. Emergency decisions, cultural-selection mechanics, and group selection (and comments and reply). *Current Anthropology,* 37(5): 763–793.

Bonnefon, J.-F., and D. J. Hilton. 2002. The suppression of modus ponens as a case of pragmatic preconditional reasoning. *Thinking & Reasoning,* 8(1): 21–40.

Boudry, M., F. Paglieri, and M. Pigliucci. 2015. The fake, the flimsy, and the fallacious: Demarcating arguments in real life. *Argumentation,* 29(4): 431–456.

Boyd, R., et al. 2003. The evolution of altruistic punishment. *Proceedings of the National Academy of Sciences,* 100(6): 3531–3535.

Boyer, P. 2001. *Religion explained: The evolutionary origins of religious thought.* London: Heinemann.

Braine, M. D. S., and D. P. O'Brien. 1998. *Mental logic.* Mahwah, NJ: Lawrence Erlbaum Associates.

Braman, E. 2009. *Law, politics, and perception: How policy preferences influence legal reasoning.* Charlottesville: University of Virginia Press.

Branham, R. J. 1994. Debate and dissent in late Tokugawa and Meiji Japan. *Argumentation and Advocacy,* 30(3): 131–149.

Brockman, J., ed. 2013. *This explains everything: Deep, beautiful, and elegant theories of how the world works.* New York: Harper.

Burge, T. 2010. *Origins of objectivity.* New York: Oxford University Press.

Byrne, R. M. 1989. Suppressing valid inferences with conditionals. *Cognition,* 31(1): 61–83.

Caenegem, R. C. 1987. *Judges, legislators and professors: Chapters in European legal history.* Cambridge: Cambridge University Press.

Carey, B. 2005. *British abolitionism and the rhetoric of sensibility: Writing, sentiment and slavery, 1760–1807.* New York: Springer.

Carey, S. 2009. *The origin of concepts.* New York: Oxford University Press.

Carroll, L. 1895. What the tortoise said to Achilles. *Mind,* 4(14): 278–280.

Carruthers, P. 2005. *Consciousness: Essays from a higher-order perspective.* New York: Clarendon Press.

———. 2006. *The architecture of the mind: Massive modularity and the flexibility of thought.* Oxford: Oxford University Press.

———. 2011. *The opacity of mind: An integrative theory of self-knowledge.* New York: Oxford University Press.

———. 2016. Two systems for mindreading? *Review of Philosophy and Psychology,* 7(1): 141–162.

Carston, R., 2002. Linguistic meaning, communicated meaning and cognitive pragmatics. *Mind & Language,* 17(1–2): 127–148.

Castelain, T., S. Bernard, and H. Mercier. Submitted. Evidence that 2-year-old children are sensitive to argument strength.

Castelain, T., et al. 2016. Evidence for benefits of argumentation in a Mayan indigenous population. *Evolution and Human Behavior,* 37(5): 337–342.

Chagnon, N. A. 1992. *Yanomamö: The fierce people,* 4th ed. New York: Holt, Rinehart & Winston.

Chalmers, D. J. 1996. *The conscious mind: In search of a fundamental theory.* New York: Oxford University Press.

Chance, Z., and M. I. Norton. 2008. I read *Playboy* for the articles. In M. S. McGlone and M. L. Knapp, eds., *The interplay of truth and deception: New agendas in theory and research.* New York: Routledge, pp. 136–148.

Chanel, O., et al. 2011. Impact of information on intentions to vaccinate in a potential epidemic: Swine-origin Influenza A (H1N1). *Social Science & Medicine,* 72(2): 142–148.

Chase, A. 2003. *Harvard and the Unabomber: The education of an American terrorist.* New York: W. W. Norton.

Cheung, S. L., and S. Palan. 2012. Two heads are less bubbly than one: Team decision-making in an experimental asset market. *Experimental Economics,* 15(3): 373–397.

Claidière, N., E. Trouche, and H. Mercier. Submitted. Argumentation and the diffusion of counter-intuitive beliefs.

Clément, F. 2010. To trust or not to trust? Children's social epistemology. *Review of Philosophy and Psychology* 1, no. 4: 531–549.

Cohen, I. B. 1985. *Revolution in science.* Cambridge, MA: Harvard University Press.

Cohen, J. 1989. Deliberation and democratic legitimacy. In A. Hamlin and P. Pettit, eds., *The good polity.* New York: Basil Blackwell, pp. 17–34.

Cole, M. 1971. *The cultural context of learning and thinking: An exploration in experimental anthropology.* London: Methuen.

Collins, H., and T. Pinch. 2005. *Dr. Golem: How to think about medicine.* Chicago: University of Chicago Press.

Combs, S. C. 2004. The useless-/usefulness of argumentation: The Dao of disputation. *Argumentation and Advocacy,* 41(2): 58–71.

Conradt, L., and T. J. Roper. 2005. Consensus decision making in animals. *Trends in Ecology & Evolution,* 20(8): 449–456.

Corriveau, K. H., and K. E. Kurkul. 2014. "Why does rain fall?" Children prefer to learn from an informant who uses noncircular explanations. *Child Development,* 85(5): 1827–1835.

Cosmides, L. 1989. The logic of social exchange: Has natural selection shaped how humans reason? Studies with the Wason selection task. *Cognition,* 31(3): 187–276.

Cosmides, L., and J. Tooby. 1987. From evolution to behavior: Evolutionary psychology as the missing link. In J. Dupré, ed., *The latest on the best: Essays on evolution and optimality.* Cambridge, MA: MIT Press, pp. 277–306.

———. 1989. Evolutionary psychology and the generation of culture, part II: Case study: A computational theory of social exchange. *Ethology and Sociobiology,* 10(1): 51–97.

Creagan, E. T., et al. 1979. Failure of high-dose vitamin C (ascorbic acid) therapy to benefit patients with advanced cancer: A controlled trial. *New England Journal of Medicine,* 301(13): 687–690.

Crombie, A. C. 1971. *Robert Grosseteste and the origins of experimental science, 1100–1700.* Oxford: Clarendon Press.

Csibra, G., and G. Gergely. 2009. Natural pedagogy. *Trends in Cognitive Sciences,* 13(4): 148–153.

Csikszentmihalyi, M., and R. K. Sawyer. 1995. Creative insight: The social dimension of a solitary moment. In R. J. Sternberg and J. E. Davidson, eds., *The nature of insight.* Cambridge, MA: MIT Press, pp. 329–363.

Dalkey, N., and O. Helmer. 1963. An experimental application of the Delphi method to the use of experts. *Management Science,* 9(3): 458–467.

Danziger, S., J. Levav, and L. Avnaim-Pesso. 2011. Extraneous factors in judicial decisions. *Proceedings of the National Academy of Sciences,* 108(17): 6889–6892.

Darley, J. M., and B. Latané. 1968. Bystander intervention in emergencies: Diffusion of responsibility. *Journal of Personality and Social Psychology,* 8(4): 377–383.

Darwin, C. 1838–1839. *Notebook N, metaphysics & expression.* Available at http://darwin-online.org.uk/content/frameset?itemID=CUL-DAR126.-&viewtype=text&pageseq=1.

———. 1871/1981. *The descent of man.* Princeton, NJ: Princeton University Press.

Davis, D. B. 1999. *The problem of slavery in the age of revolution, 1770–1823.* New York: Oxford University Press.

Dawkins, R. 1996. Why don't animals have wheels? *Sunday Times,* November 24.

Dehaene, S. 1999. *The number sense: How the mind creates mathematics.* Oxford: Oxford University Press.

———. 2009. *Reading in the brain: The new science of how we read.* London: Penguin.

———. 2014. *Consciousness and the brain: Deciphering how the brain codes our thoughts.* London: Penguin.

Dehaene, S., and L. Cohen. 2011. The unique role of the visual word form area in reading. *Trends in Cognitive Sciences,* 15(6): 254–262.

Dennett, D. C. 1969. *Content and consciousness.* London: Routledge and Kegan Paul.

———. 1971. Intentional systems. *Journal of Philosophy,* 68(4): 87–106.

———. 1991. *Consciousness explained.* Boston: Little, Brown.

———. 2000. Making tools for thinking. In D. Sperber, ed. *Metarepresentations: A multidisciplinary perspective.* Oxford: Oxford University Press, pp. 17–29.

Descartes, R. 1637/2006. *A discourse on method.* Trans. and with an intro. by Ian Mclean. New York: Oxford University Press.

Dessalles, J.-L. 2007. *Why we talk: The evolutionary origins of language.* Trans. J. Grieve. Oxford: Oxford University Press.

De Villiers, J. 2000. Language and theory of mind: What are the developmental relationships? In S. Baron-Cohen, H. Tager-Flusberg, and D. J. Cohen, eds., *Understanding other minds: Perspectives from developmental cognitive neuroscience.* Oxford: Oxford University Press, pp. 83–123.

Dias, M., A. Roazzi, and P. L. Harris. 2005. Reasoning from unfamiliar premises: A study with unschooled adults. *Psychological Science,* 16(7): 550–554.

Dienes, Z., and J. Perner. 1999. A theory of implicit and explicit knowledge. *Behavioral and Brain Sciences,* 22(5): 735–808.

Dijksterhuis, A. 2004. Think different: The merits of unconscious thought in preference development and decision making. *Journal of Personality and Social Psychology,* 87(5): 586–598.

Dingemanse, M., and N. J. Enfield. 2015. Other-initiated repair across languages: Towards a typology of conversational structures. *Open Linguistics,* 1(1): 96–118.

Dingemanse, M., et al. 2015. Universal principles in the repair of communication problems. *PloS One,* 10(9): e0136100.

Doise, W., and G. Mugny. 1984. *The social development of the intellect.* Oxford: Pergamon.

Doise, W., G. Mugny, and A.-N. Perret-Clermont. 1975. Social interaction and the development of cognitive operations. *European Journal of Social Psychology,* 5(3): 367–383.

Drescher, S. 1990. People and parliament: The rhetoric of the British slave trade. *Journal of Interdisciplinary History,* 20(4): 561–580.

———. 2009. *Abolition: A history of slavery and antislavery.* Cambridge: Cambridge University Press.

Dretske, F. 1981. *Knowledge and the flow of information.* Cambridge, MA: MIT Press.

———. 1997. *Naturalizing the mind.* Cambridge, MA: MIT Press.

Ducrot, O. 1984. *Le dire et le dit.* Paris: Minuit.

Ducrot, O., and J. C. Anscombre. 1983. *L'argumentation dans la langue.* Brussels: Mardaga.

Dunbar, K. 1995. How scientists really reason: Scientific reasoning in real-world laboratories. In R. J. Sternberg and J. E. Davidson, eds., *The nature of insight.* Cambridge, MA: MIT Press, pp. 365–395.

———. 1997. How scientists think: On-line creativity and conceptual change in science. In T. B. Ward, S. M. Smith, J. Vaid, and D. N. Perkins, eds., *Creative thought: An investigation of conceptual structures and processes.* Washington, DC: American Psychological Association, pp. 461–493.

Dunn, J., and P. Munn. 1987. Development of justification in disputes with mother and sibling. *Developmental Psychology,* 23: 791–798.

Dunning, D., J. A. Meyerowitz, and A. D. Holzberg. 1989. Ambiguity and self-evaluation: The role of idiosyncratic trait definitions in self-serving assessments of ability. *Journal of Personality and Social Psychology,* 57(6): 1082–1090.

Dutilh Novaes, C. 2015. A dialogical, multi-agent account of the normativity of logic. *Dialectica,* 69(4): 587–609.

———. 2016. Reductio ad absurdum from a dialogical perspective. *Philosophical Studies,* 173(10): 2605–2628.

Egan, F. 2012. Representationalism. In E. Margolis, R. Samuels, S. Stich, and F. Egan, eds., *Oxford handbook of philosophy of cognitive science.* New York: Oxford University Press, pp. 250–272.

Einhorn, H. J., and R. M. Hogarth. 1981. Behavioral decision theory: Processes of judgment and choice. *Annual Review of Psychology,* 32: 53–88.

Ellsworth, P. C. 1989. Are twelve heads better than one? *Law and Contemporary Problems,* 52(4): 205–224.

———. 2003. One inspiring jury. *Michigan Law Review,* 101: 1387–1396.

Enfield, N. J. 2013. *Relationship thinking: Agency, enchrony, and human sociality.* New York: Oxford University Press.

Evans, J. St. B. T. 1972. Interpretation and matching bias in a reasoning task. *Quarterly Journal of Experimental Psychology,* 24(2): 193–199.

———. 1989. *Bias in human reasoning: Causes and consequences.* Hillsdale, NJ: Lawrence Erlbaum.

———. 1996. Deciding before you think: Relevance and reasoning in the selection task. *British Journal of Psychology,* 87: 223–240.

———. 2002. Logic and human reasoning: An assessment of the deduction paradigm. *Psychological Bulletin,* 128(6): 978–996.

———. 2007. *Hypothetical thinking: Dual processes in reasoning and judgment.* Hove, UK: Psychology Press.

———. 2013. *Reasoning, rationality and dual processes: Selected works of Jonathan St BT Evans.* London: Psychology Press.

Evans, J. St. B. T., and J. S. Lynch. 1973. Matching bias in the selection task. *British Journal of Psychology,* 64(3): 391–397.

Evans, J. St. B. T., and D. E. Over. 1996. *Rationality and reasoning.* Hove, UK: Psychology Press.

———. 2004. *If: Supposition, pragmatics, and dual processes.* Oxford: Oxford University Press.

Evans, J. St. B. T., and P. C. Wason. 1976. Rationalization in a reasoning task. *British Journal of Psychology,* 67: 479–486.

Fara, P. 2002. *Newton: The making of genius.* New York: Columbia University Press.

Fields, B. J. 1990. Slavery, race and ideology in the United States of America. *New Left Review*, no. 181 (May–June): 95–118.

Finkelman, P. 1993. Jefferson and slavery: "Treason against the Hopes of the World." In P. S. Onuf, ed., *Jeffersonian Legacies*. Charlottesville: University Press of Virginia, pp. 181–223.

Fishkin, J. S. 2009. *When the people speak: Deliberative democracy and public consultation*. Oxford: Oxford University Press.

Floridi, L. 2011. *The philosophy of information*. New York: Oxford University Press.

Fodor, J. A. 1981. *Representations: Philosophical essays on the foundations of cognitive science*. Cambridge MA: MIT Press.

———. 1983. *The modularity of mind*. Cambridge, MA: MIT Press.

Franklin, B. 1799/1950. *The autobiography of Benjamin Franklin*, New York: Pocket Books.

Frede, M. 1996. Aristotle's rationalism. In M. Frede and G. Striker, eds., *Rationality in Greek thought*. New York: Oxford University Press, pp. 157–173.

Frederick, S. 2005. Cognitive reflection and decision making. *Journal of Economic Perspectives*, 19(4): 25–42.

Frith, C. D., and U. Frith. 2012. Mechanisms of social cognition. *Annual Review of Psychology*, 63: 287–313.

Funder, D. C. 1987. Errors and mistakes: Evaluating the accuracy of social judgment. *Psychological Bulletin*, 101(1): 75–90.

Galler, J. S. 2007. Logic and argument in "The Book of Concord." PhD diss., University of Texas at Austin.

Gallese, V., and A. Goldman. 1998. Mirror neurons and the simulation theory of mindreading. *Trends in Cognitive Sciences*, 2(12): 493–501.

Gallistel, C. R., and J. Gibbon. 2000. Time, rate, and conditioning. *Psychological Review*, 107(2): 289–344.

Gergely, G., H. Bekkering, and I. Király. 2002. Rational imitation in preverbal infants. *Nature*, 415(6873): 755.

Gessen, M. 2009. *Perfect rigor: A genius and the mathematical breakthrough of the century*. Boston: Houghton Mifflin Harcourt.

Gigerenzer, G. 2007. Fast and frugal heuristics: The tools of bounded rationality. In D. Koehler and N. Harvey, eds., *Handbook of judgment and decision making*. Oxford: Blackwell, pp. 62–88.

Gigerenzer, G., and T. Regier. 1996. How do we tell an association from a rule? Comment on Sloman (1996). *Psychological Bulletin*, 119(1): 23–26.

Gilmore, C. K., S. E. McCarthy, and E. S. Spelke. 2007. Symbolic arithmetic knowledge without instruction. *Nature*, 447(7144): 589–591.

Girotto, V., et al. 2001. Inept reasoners or pragmatic virtuosos? Relevance and the deontic selection task. *Cognition*, 81(2): 69–76.

Glöckner, A. 2016. The irrational hungry judge effect revisited: Simulations reveal that the magnitude of the effect is overestimated. *Judgment and Decision Making,* 11(6): 601–610.

Gluckman, M. 1967. *The judicial process among the Barotse of Northern Rhodesia (Zambia).* Manchester, UK: Manchester University Press.

Godfrey-Smith, P., and M. Martínez. 2013. Communication and common interest. *PLOS Computational Biology,* 9(11): e1003282.

Gopnik, A., and H. M. Wellman. 1992. Why the child's theory of mind really is a theory. *Mind & Language,* 7(1–2): 145–171.

Gordon, P. 2004. Numerical cognition without words: Evidence from Amazonia. *Science,* 306(5695): 496–499.

Graziano, M. S. 2013. *Consciousness and the social brain.* Oxford: Oxford University Press.

Green, M. S., and J. N. Williams. 2007. *Moore's paradox: New essays on belief, rationality, and the first person.* New York: Oxford University Press.

Greenwald, A. G., and M. R. Banaji. 1995. Implicit social cognition: Attitudes, self-esteem, and stereotypes. *Psychological Review,* 102(1): 4–27.

Grice, P. 1989. *Studies in the way of words.* Cambridge, MA: Harvard University Press.

Grusec, J. E., and J. J. Goodnow. 1994. Impact of parental discipline methods on the child's internalization of values: A reconceptualization of current points of view. *Developmental Psychology,* 30(1): 4–19.

Habermas, J. 1975. *Legitimation crisis.* Boston: Beacon.

———. 1987. *The theory of communicative action.* Boston: Beacon.

Hahn, U., and J. Hornikx. 2016. A normative framework for argument quality: Argumentation schemes with a Bayesian foundation. *Synthese,* 193(6): 1833–1873.

Hahn, U., and M. Oaksford. 2007. The rationality of informal argumentation: A Bayesian approach to reasoning fallacies. *Psychological Review,* 114(3): 704–732.

Haidt, J. 2001. The emotional dog and its rational tail: A social intuitionist approach to moral judgment. *Psychological Review,* 108(4): 814–834.

Haidt, J., and F. Bjorklund. 2007. Social intuitionists reason, in conversation. In W. Sinnott-Armstrong, ed., *Moral psychology.* Cambridge, MA: MIT Press, pp. 241–254.

Haidt, J., F. Bjorklund, and S. Murphy. 2000. Moral dumbfounding: When intuition finds no reason. Unpublished ms, University of Virginia.

Halberstam, D. 2001. *The best and the brightest.* New York: Random House.

Hall, A. 2007. Do discourse connectives encode concepts or procedures? *Lingua,* 117(1): 149–174.

Hall, A. R. 1996. *Isaac Newton: Adventurer in thought.* Cambridge: Cambridge University Press.

Hall, L., P. Johansson, and T. Strandberg. 2012. Lifting the veil of morality: Choice blindness and attitude reversals on a self-transforming survey. *PloS One,* 7(9): e45457.

Harman, G. 1986. *Change in view: Principles of reasoning.* Cambridge, MA: MIT Press.

Harris, P. L. 2012. *Trusting what you're told: How children learn from others,* Cambridge MA: Belknap Press of Harvard University Press.

Haselton, M. G., and D. M. Buss. 2000. Error management theory: A new perspective on biases in cross-sex mind reading. *Journal of Personality and Social Psychology,* 78(1): 81–91.

Hastie, R., S. Penrod, and N. Pennington. 1983. *Inside the jury.* Cambridge, MA: Harvard University Press.

Hatfield, G. 2002. Perception as unconscious inference. In D. Heyer and R. Mausfeld, eds., *Perception and the physical world: Psychological and philosophical issues in perception.* New York: Wiley, pp. 115–143.

Heisenberg, W. 1952. *Die Physik der Atomkerne.* London: Taylor & Francis.

Henrich, J., S. J. Heine, and A. Norenzayan. 2010. The weirdest people in the world. *Behavioral and Brain Sciences,* 33(2–3): 61–83.

Hespos, S. J., and R. Baillargeon. 2006. Décalage in infants' knowledge about occlusion and containment events: Converging evidence from action tasks. *Cognition,* 99(2): B31–B41.

Heyes, C. 2003. Four routes of cognitive evolution. *Psychological Review,* 110(4): 713–727.

Heyes, C. M., and C. D. Frith. 2014. The cultural evolution of mind reading. *Science,* 344(6190): 1243091.

Hill, K., et al. 1987. Foraging decisions among Ache hunter-gatherers: New data and implications for optimal foraging models. *Ethology and Sociobiology,* 8(1): 1–36.

Hirschfeld, L. A., and S. A. Gelman. 1994. *Mapping the mind: Domain specificity in cognition and culture.* Cambridge: Cambridge University Press.

Hochschild, A. 2006. *Bury the chains: Prophets and rebels in the fight to free an empire's slaves.* Boston: Houghton Mifflin.

Holyoak, K. J., and R. G. Morrison, eds. 2012. *The Oxford handbook of thinking and reasoning,* New York: Oxford University Press.

Hornsby, J. 2000. Personal and sub-personal: A defence of Dennett's early distinction. *Philosophical Explorations,* 3(1): 6–24.

Hsee, C. K. 1999. Value seeking and prediction-decision inconsistency: Why don't people take what they predict they'll like the most? *Psychonomic Bulletin and Review,* 6(4): 555–561.

Hsee, C. K., et al. 1999. Preference reversals between joint and separate evaluations of options: A review and theoretical analysis. *Psychological Bulletin,* 125(5): 576–590.

Hull, D. L., P. D. Tessner, and A. M. Diamond. 1978. Planck's principle. *Science,* 202(4369): 717–723.

Hume, D. 1999. *An enquiry concerning human understanding.* Ed. Tom L. Beauchamp. Oxford: Oxford University Press.

Huntington, S. P. 1993. *The third wave: Democratization in the late twentieth century.* Norman: University of Oklahoma Press.

Hutchins, E. 1980. *Culture and inference.* Cambridge, MA: MIT Press.

Ings, T. C., and L. Chittka. 2008. Speed-accuracy tradeoffs and false alarms in bee responses to cryptic predators. *Current Biology,* 18(19): 1520–1524.

Jacob, P. 1997. *What minds can do: Intentionality in a non-intentional world.* Cambridge: Cambridge University Press.

Janis, I. L. 1982. *Groupthink,* 2nd rev. ed. Boston: Houghton Mifflin.

Jefferson, T. 1829. *Memoir, correspondence, and miscellanies, from the papers of T. Jefferson,* vol. 1. Ed. Thomas Jefferson Randolph. Charlottesville, VA: F. Carr, and Co.

Jessen, C. 2001. *Temperature regulation in humans and other mammals,* 2nd ed. New York: Springer.

Johnson, D. W., and R. T. Johnson. 2009. Energizing learning: The instructional power of conflict. *Educational Researcher,* 38(1): 37–51.

Johnson-Laird, P. N. 2006. *How we reason.* Oxford: Oxford University Press.

Johnson-Laird, P. N., and R. M. J. Byrne. 1991. *Deduction.* Hove, UK: Lawrence Erlbaum Associates.

———. 2002. Conditionals: A theory of meaning, pragmatics, and inference. *Psychological Review,* 109: 646–678.

Kaczynski, T. 2005. *The Unabomber manifesto: Industrial society and its future.* Berkeley, CA: Jolly Roger Press.

Kahneman, D. 2003a. Maps of bounded rationality: Psychology for behavioral economics. *American Economic Review,* 93(5): 1449–1475.

———. 2003b. A perspective on judgment and choice: Mapping bounded rationality. *American Psychologist,* 58(9): 697–720.

———. 2011. *Thinking, fast and slow.* New York: Farrar, Straus & Giroux.

Kant, I. 1781/1998. *Critique of pure reason.* Trans. and ed. Paul Guyer and Allen W. Wood. Cambridge: Cambridge University Press.

Kanwisher, N., J. McDermott, and M. M. Chun. 1997. The fusiform face area: A module in human extrastriate cortex specialized for face perception. *Journal of Neuroscience,* 17(11): 4302–4311.

Kay, J. 2011. *Among the Truthers: A journey through America's growing conspiracist underground.* New York: Harper Collins.

Keil, F. C. 1992. The origins of an autonomous biology. In M. R. Gunnar and M. Maratsos, eds., *Modularity and constraints in language and cognition: The Minnesota symposia on child psychology.* Hillsdale, NJ: Erlbaum, pp. 103–137.

Keren, G., and Y. Schul. 2009. Two is not always better than one: A critical evaluation of two-system theories. *Perspectives on Psychological Science,* 4(6): 533–550.

Kesson, E. M., et al. 2012. Effects of multidisciplinary team working on breast cancer survival: Retrospective, comparative, interventional cohort study of 13 722 women. *British Medical Journal,* 344: e2718.

Keynes, J. M. 1936. *The general theory of employment, interest and money.* London: Macmillan.

——. 1989. *Collected writings,* vol. 12. London: Macmillan.

Khemlani, S., and P. N. Johnson-Laird. 2012. Theories of the syllogism: A meta-analysis. *Psychological Bulletin,* 138(3): 427–457.

——. 2013. The processes of inference. *Argument & Computation,* 4(1): 4–20.

King, P. 1991. Mediaeval thought-experiments: The metamethodology of mediaeval science. In T. Horowitz and G. J. Massey, eds., *Thought experiments in science and philosophy.* Savage, MD: Rowman & Littlefield, pp. 43–64.

Kitcher, P. 1993. *The advancement of science: Science without legend, objectivity without illusions.* New York: Oxford University Press.

Knobe, J. 2010. Person as scientist, person as moralist. *Behavioral and Brain Sciences,* 33(4): 315–329.

Koenig, M. A. 2012. Beyond semantic accuracy: Preschoolers evaluate a speaker's reasons. *Child Development,* 83(3): 1051–1063.

Köhler, W. 1930. *Gestalt psychology.* London: Bell.

Koriat, A. 2012. When are two heads better than one and why? *Science,* 336(6079): 360–362.

Koriat, A., S. Lichtenstein, and B. Fischhoff. 1980. Reasons for confidence. *Journal of Experimental Psychology: Human Learning and Memory and Cognition,* 6: 107–118.

Kornblith, H. 2012. *On reflection.* New York: Oxford University Press.

Kovács, Á. M. 2016. Belief files in theory of mind reasoning. *Review of Philosophy and Psychology,* 7(2): 509–527.

Kovács, Á. M., E. Téglás, and A. D. Endress. 2010. The social sense: Susceptibility to others' beliefs in human infants and adults. *Science,* 330(6012): 1830–1834.

Kravitz, D. A., and B. Martin. 1986. Ringelmann rediscovered: The original article. *Journal of Personality and Social Psychology,* 50(5): 936–941.

Kruglanski, A. W., and G. Gigerenzer. 2011. Intuitive and deliberate judgments are based on common principles. *Psychological Review,* 118(1): 97–109.

Kruglanski, A. W., et al. 2006. On parametric continuities in the world of binary either ors. *Psychological Inquiry,* 17(3): 153–165.

Kugler, T., E. E. Kausel, and M. G. Kocher. 2012. Are groups more rational than individuals? A review of interactive decision making in groups. *Wiley Interdisciplinary Reviews: Cognitive Science,* 3(4): 471–482.

Kuhn, D. 1991. *The skills of arguments.* Cambridge: Cambridge University Press.

Kuhn, D., and A. Crowell. 2011. Dialogic argumentation as a vehicle for developing young adolescents' thinking. *Psychological Science,* 22(4): 545–552.

Kuhn, D., V. Shaw, and M. Felton. 1997. Effects of dyadic interaction on argumentative reasoning. *Cognition and Instruction,* 15(3): 287–315.

Kuhn, T. 1962. *The structure of scientific revolutions,* 50th anniv. ed. Chicago: University of Chicago Press.

Kunda, Z. 1990. The case for motivated reasoning. *Psychological Bulletin,* 108: 480–498.

Landemore, H. 2013. Democratic reason: The mechanisms of collective intelligence in politics. In J. Elster and H. Landemore, eds., *Collective wisdom.* Cambridge: Cambridge University Press, pp. 251–289.

Latané, B., and J. Rodin. 1969. A lady in distress: Inhibiting effects of friends and strangers on bystander intervention. *Journal of Experimental Social Psychology,* 5(2): 189–202.

Latané, B., K. Williams, and S. Harkins. 1979. Many hands make light the work: The causes and consequences of social loafing. *Journal of Personality and Social Psychology,* 37(6): 822–832.

Laughlin, P. R. 2011. *Group problem solving.* Princeton, NJ: Princeton University Press.

Leman, P. J., and G. Duveen. 1999. Representations of authority and children's moral reasoning. *European Journal of Social Psychology,* 29(5–6): 557–575.

Leslie, A. M. 1987. Pretense and representation: The origins of a "theory of mind." *Psychological Review,* 94(4): 412–426.

Levin, I., and S. Druyan. 1993. When sociocognitive transaction among peers fails: The case of misconceptions in science. *Child Development,* 64(5): 1571–1591.

Levin, S. G., P. E. Stephan, and M. B. Walker. 1995. Planck's principle revisited: A note. *Social Studies of Science,* 25(2): 275–283.

Levinson, S. C. 2005. Living with Manny's dangerous idea. *Discourse Studies,* 7(4–5): 431–453.

———. 2006. On the human "interaction engine." In N. J. Enfield and S. C. Levinson, eds., *Roots of human sociality.* Oxford: Berg, pp. 39–69.

Lévy-Bruhl, L. 1910. *How natives think.* Princeton, NJ: Princeton University Press.

Liberman, V., et al. 2012. Naïve realism and capturing the "wisdom of dyads." *Journal of Experimental Social Psychology,* 48(2): 507–512.

Lilienfeld, S. O., R. Ammirati, and K. Landfield. 2009. Giving debiasing away: Can psychological research on correcting cognitive errors promote human welfare? *Perspectives on Psychological Science,* 4(4): 390–398.

Linstone, H. A., and M. Turoff, eds. 1976. *The Delphi method: Techniques and applications.* New York: Addison-Wesley.

Liszkowski, U., et al. 2004. Twelve-month-olds point to share attention and interest. *Developmental Science,* 7(3): 297–307.

Liu, Y. 1996. Three issues in the argumentative conception of early Chinese discourse. *Philosophy East and West,* 46(1): 33–58.

Lloyd, G. E. R. 2007. Towards a taxonomy of controversies and controversiality: Ancient Greece and China. In M. Dascal and H. Chang, eds., *Traditions of controversy.* Amsterdam: John Benjamins, pp. 3–16.

Long, E. 1774. *The history of Jamaica or, A general survey of the antient and modern state of that island: With reflexions on its situation, settlements, inhabitants, climate, products, commerce, laws, and Government.* Montreal: McGill–Queen's University Press.

Louis Lee, N. Y., G. P. Goodwin, and P. N. Johnson-Laird. 2008. The psychological puzzle of Sudoku. *Thinking & Reasoning,* 14(4): 342–364.

Lucas, E. J., and L. J. Ball. 2005. Think-aloud protocols and the selection task: Evidence for relevance effects and rationalisation processes. *Thinking & Reasoning,* 11(1): 35–66.

Luo, Y., L. Kaufman, and R. Baillargeon. 2009. Young infants' reasoning about physical events involving inert and self-propelled objects. *Cognitive Psychology,* 58(4): 441–486.

Luria, A. R. 1934. The second psychological expedition to central Asia. *Journal of Genetic Psychology,* 41: 255–259.

———. 1976. *Cognitive development: Its cultural and social foundations.* Cambridge, MA: Harvard University Press.

Luskin, R. C., et al.. 2014. Deliberating across deep divides. *Political Studies*, 62(1): 116–135.

Luther, M. 1536 / 1960. *The disputation concerning man.* Luther's Works, vol. 34. Ed. Helmut T. Lehman and Lewis W. Spitz. Minneapolis: Augsburg Fortress.

Mahoney, M. J. 1977. Publication prejudices: An experimental study of confirmatory bias in the peer review system. *Cognitive Therapy and Research,* 1(2): 161–175.

Mahoney, M. J., and B. G. DeMonbreun. 1977. Psychology of the scientist: An analysis of problem-solving bias. *Cognitive Therapy and Research,* 1(3): 229–238.

Malle, B. F. 2004. *How the mind explains behavior: Folk explanations, meaning, and social interaction.* Cambridge, MA: MIT Press.

Mancosu, P. 1999. Between Vienna and Berlin: The immediate reception of Godel's incompleteness theorems. *History and Philosophy of Logic,* 20(1): 33–45.

Mandela, N. 1994. *Long walk to freedom.* Boston: Little, Brown.

Manktelow, K. 2012. *Thinking and reasoning: An introduction to the psychology of reason, judgment and decision making,* Hove, UK: Psychology Press.

Manktelow, K., and M. C. Chung, eds. 2004. *Psychology of reasoning: Theoretical and historical perspectives.* London: Psychology Press.

Marler, P. 1991. The instinct to learn. In S. Carey and R. Gelman, eds., *The epigenesis of mind: Essays on biology and cognition.* London: Psychology Press, pp. 591–617.

Mascaro, O., and O. Morin. 2014. Gullible's travel: How honest and trustful children become vigilant communicators. In L. Robinson and S. Einav, eds., *Trust and skepticism: Children's selective learning from testimony.* London: Psychology Press, pp. 69–82.

Mascaro, O., and D. Sperber. 2009. The moral, epistemic, and mindreading components of children's vigilance towards deception. *Cognition,* 112: 367–380.

McAfee, R. P., H. M. Mialon, and S. H. Mialon. 2010. Do sunk costs matter? *Economic Inquiry,* 48(2): 323–336.

Mellers, B., et al. 2014. Psychological strategies for winning a geopolitical forecasting tournament. *Psychological Science,* 25(5): 1106–1115.

Meltzoff, A. N. 1988. Infant imitation after a 1-week delay: Long-term memory for novel acts and multiple stimuli. *Developmental Psychology,* 24(4): 470–476.

Mercier, H. 2011a. On the universality of argumentative reasoning. *Journal of Cognition and Culture,* 11: 85–113.

———. 2011b. Reasoning serves argumentation in children. *Cognitive Development,* 26(3): 177–191.

———. 2012. Looking for arguments. *Argumentation,* 26(3): 305–324.

———. 2016a. The argumentative theory: Predictions and empirical evidence. *Trends in Cognitive Sciences,* 20(9): 689–700.

———. 2016b. Confirmation (or myside) bias. In R. Pohl, ed., *Cognitive illusions.* London: Psychology Press, pp. 99–114.

———. In press. Reasoning and argumentation. In V. A. Thompson and L. J. Ball, eds., *International handbook of thinking & reasoning.* Hove, UK: Psychology Press.

Mercier, H., S. Bernard, and F. Clément. 2014. Early sensitivity to arguments: How preschoolers weight circular arguments. *Journal of Experimental Child Psychology,* 125: 102–109.

Mercier, H., M. Deguchi, et al. 2016. The benefits of argumentation are cross-culturally robust: The case of Japan. *Thinking & Reasoning,* 22(1): 1–15.

Mercier, H., and H. Landemore. 2012. Reasoning is for arguing: Understanding the successes and failures of deliberation. *Political Psychology,* 33(2): 243–258.

Mercier, H., G. Politzer, and D. Sperber. In press. What causes failure to apply the Pigeonhole Principle in simple reasoning problems? *Thinking & Reasoning.*

Mercier, H., and D. Sperber. 2009. Intuitive and reflective inferences. In J. S. B. T. Evans and K. Frankish, eds., *In two minds.* New York: Oxford University Press, pp. 149–170.

———. 2011. Why do humans reason? Arguments for an argumentative theory. *Behavioral and Brain Sciences,* 34(2): 57–74.

Mercier, H., E. Trouche, et al. 2015. Experts and laymen grossly underestimate the benefits of argumentation for reasoning. *Thinking & Reasoning,* 21(3): 341–355.

Mercier, H., et al. In press. Natural born arguers: Teaching how to make the best of our reasoning abilities. *Educational Psychologist.*

Meulders, M. 2010. *Helmholtz: From enlightenment to neuroscience.* Cambridge, MA: MIT Press.

Miller, M. B., and M. S. Gazzaniga. 1998. Creating false memories for visual scenes. *Neuropsychologia,* 36(6): 513–520.

Miller, S. A., and C. A. Brownell. 1975. Peers, persuasion, and Piaget: Dyadic interaction between conservers and nonconservers. *Child Development,* 46(4): 992–997.

Millikan, R. G. 1987. *Language, thought, and other biological categories: New foundations for realism.* Cambridge, MA: MIT Press.

———. 1993. *White Queen psychology and other essays for Alice.* Cambridge, MA: MIT Press.

———. 2004. *Varieties of meaning: The 2002 Jean Nicod lectures.* Cambridge, MA : MIT Press.

Minson, J. A., V. Liberman, and L. Ross. 2011. Two to tango. *Personality and Social Psychology Bulletin,* 37(10): 1325–1338.

Moertel, C. G., et al. 1985. High-dose vitamin C versus placebo in the treatment of patients with advanced cancer who have had no prior chemotherapy: A randomized double-blind comparison. *New England Journal of Medicine,* 312(3): 137–141.

Monaghan, P., et al. 2014. How arbitrary is language? *Philosophical Transactions of the Royal Society B,* 369(1651): 20130299.

Montaigne, M. 1870. *The essays of Michael, Seigneur de Montaigne: With notes and quotations, and account of the author's life.* London: Alex Murray & Son.

Montefiore, S. S. 2003. *Stalin: The court of the Red Tsar.* London: Knopf.

Morabia, A. 2006. Pierre-Charles-Alexandre Louis and the evaluation of bloodletting. *Journal of the Royal Society of Medicine,* 99(3): 158–160.

Moshman, D., and M. Geil. 1998. Collaborative reasoning: Evidence for collective rationality. *Thinking and Reasoning,* 4(3): 231–248.

Mullen, B., C. Johnson, and E. Salas. 1991. Productivity loss in brainstorming groups: A meta-analytic integration. *Basic and Applied Social Psychology,* 12(1): 3–23.

Myers, D. G., and P. J. Bach. 1974. Discussion effects on militarism-pacifism: A test of the group polarization hypothesis. *Journal of Personality and Social Psychology,* 30(6): 741–747.

Myers, D. G., and G. D. Bishop. 1970. Discussion effects on racial attitudes. *Science,* 169(3947): 778–779.

Mynatt, C. R., M. E. Doherty, and R. D. Tweney. 1977. Confirmation bias in a simulated research environment: An experimental study of scientific inference. *Quarterly Journal of Experimental Psychology,* 29(1): 85–95.

Nagel, T. 1976. Moral luck. *Proceedings of the Aristotelian Society, supplementary volumes,* 50: 115–151.

Nakamura, H. 1964. *Ways of thinking of Eastern peoples: India, China, Tibet, Japan.* Honolulu: University of Hawaii Press.

Needham, A., and R. Baillargeon. 1993. Intuitions about support in 4.5-month-old infants. *Cognition,* 47(2): 121–148.

Nemeth, C. J., and M. Ormiston. 2007. Creative idea generation: Harmony versus stimulation. *European Journal of Social Psychology,* 37(3): 524–535.

Nemeth, C. J., et al. 2004. The liberating role of conflict in group creativity: A study in two countries. *European Journal of Social Psychology,* 34(4): 365–374.

Nicastro, N. 2008. *Circumference: Eratosthenes and the ancient quest to measure the globe.* New York: Macmillan.

Nickerson, R. S. 1998. Confirmation bias: A ubiquitous phenomena in many guises. *Review of General Psychology,* 2: 175–220.

———. 2012. *Aspects of rationality: Reflections on what it means to be rational and whether we are.* London: Psychology Press.

Nielsen, M., and K. Tomaselli. 2010. Overimitation in Kalahari Bushman children and the origins of human cultural cognition. *Psychological Science,* 21(5): 729–736.

Nisbett, R. E., and L. Ross. 1980. *Human inference: Strategies and shortcomings of social judgment.* Englewood Cliffs, NJ: Prentice-Hall.

Nisbett, R. E., and T. Wilson. 1977. Telling more than we can know. *Psychological Review,* 84(1): 231–259.

Nisbett, R. E., et al. 1983. The use of statistical heuristics in everyday inductive reasoning. *Psychological Review,* 90(4): 339–363.

Oaksford, M., and N. Chater. 1991. Against logicist cognitive science. *Mind & Language,* 6(1): 1–38.

———. 2003. Optimal data selection: Revision, review, and reevaluation. *Psychonomic Bulletin & Review,* 10(2): 289–318.

———. 2007. *Bayesian rationality: The probabilistic approach to human reasoning.* Oxford: Oxford University Press.

Oaksford, M., and U. Hahn. 2004. A Bayesian approach to the argument from ignorance. *Canadian Journal of Experimental Psychology,* 58(2): 75–85.

Onishi, K. H., and R. Baillargeon. 2005. Do 15-month-old infants understand false beliefs? *Science,* 308: 255–258.

Open Science Collaboration. 2015. Estimating the reproducibility of psychological science. *Science,* 349(6251), p. aac4716.

Oreskes, N. 1988. The rejection of continental drift. *Historical Studies in the Physical and Biological Sciences,* 18(2): 311–348.

Origgi, G. 2015. *La réputation.* Paris: Presses Universitaires de France.

Origgi, G., and D. Sperber. 2000. Evolution, communication and the proper function of language. In P. Carruthers and A. Chamberlain, eds., *Evolution and the human mind: Modularity, language and meta-cognition.* Cambridge: Cambridge University Press, pp. 140–169.

Osman, M. 2004. An evaluation of dual-process theories of reasoning. *Psychonomic Bulletin and Review,* 11(6): 988–1010.

Ostrom, E. 1991. *Governing the commons: The evolution of institutions for collective action.* Cambridge: Cambridge University Press.

Pauling, L., and Z. S. Herman. 1989. Criteria for the validity of clinical trials of treatments of cohorts of cancer patients based on the Hardin Jones principle. *Proceedings of the National Academy of Sciences,* 86(18): 6835–6837.

Perelman, C., and L. Olbrechts-Tyteca. 1958. *The new rhetoric: A treatise on argumentation.* Notre Dame, IN: University of Notre Dame Press.

Perkins, D. N. 1985. Postprimary education has little impact on informal reasoning. *Journal of Educational Psychology,* 77: 562–571.

———. 1989. Reasoning as it is and could be: An empirical perspective. In D. M. Topping, D. C. Crowell, and V. N. Kobayashi, eds., *Thinking across cultures: The Third International Conference on Thinking.* Hillsdale, NJ: Erlbaum, pp. 175–194.

Perner, J. 1991. *Understanding the representational mind.* Cambridge, MA: MIT Press.

Perner, J., M. Huemer, and B. Leahy. 2015. Mental files and belief: A cognitive theory of how children represent belief and its intensionality. *Cognition,* 145: 77–88.

Perret-Clermont, A.-N. 1980. *Social interaction and cognitive development in children.* London: Academic Press.

Perry, A. T. 2012. A traffic in numbers: The ethics, effects, and affect of mortality statistics in the British abolition debates. *Journal for Early Modern Cultural Studies,* 12(4): 78–104.

Petty, R. E., and D. T. Wegener. 1998. Attitude change: Multiple roles for persuasion variables. In D. T. Gilbert, S. Fiske, and G. Lindzey, eds., *The handbook of social psychology.* Boston: McGraw-Hill, pp. 323–390.

Piaget, J. 1932. *The moral judgment of the child.* London: Routledge & Kegan Paul.

Piaget, J., and B. Inhelder. 1967. *Genèse des structures logiques élémentaires.* Paris: Presses Universitaires de France.

Piattelli-Palmarini, M. 1994. *Inevitable illusions: How mistakes of reason rule our minds.* New York: Wiley.

Pinker, S. 1994. *The language instinct.* New York: Harper.

Planck, M. 1968. *Scientific autobiography and other papers.* New York: Citadel Press.

Politzer, G. 2005. Uncertainty and the suppression of inferences. *Thinking & Reasoning,* 11(1): 5–33.

Popper, K. R. 1963. *Conjectures and refutations.* London: Routledge and Kegan Paul.

Prado, J., et al. 2015. Neural interaction between logical reasoning and pragmatic processing in narrative discourse. *Journal of Cognitive Neuroscience,* 27(4): 692–704.

Premack, D., and G. Woodruff. 1978. Does the chimpanzee have a theory of mind? *Behavioral and Brain Sciences,* 1(4): 515–526.

Priestley, J. 1786. *The theological and miscellaneous works of Joseph Priestley.* London: G. Smallfield.

Principe, L. 2004. Reflections on Newton's alchemy in light of the new historiography of alchemy. In *Newton and Newtonianism: New studies.* Dordrecht: Kluwer, pp. 205–219.

Pritchard, D. 2005. *Epistemic luck.* New York: Clarendon.

Pronin, E. 2007. Perception and misperception of bias in human judgment. *Trends in Cognitive Sciences,* 11(1): 37–43.

Pronin, E., T. Gilovich, and L. Ross. 2004. Objectivity in the eye of the beholder: Divergent perceptions of bias in self versus others. *Psychological Review,* 111(3): 781–799.

Proust, J. 2013. *The philosophy of metacognition: Mental agency and self-awareness.* Oxford: Oxford University Press.

Pyke, G. H. 1984. Optimal foraging theory: A critical review. *Annual Review of Ecology and Systematics,* 15: 523–575.

Railton, P. 2004. How to engage reason: The problem of Regress. In R. J. Wallace et al., eds., *Reason and value: Themes from the moral philosophy of Joseph Raz.* New York: Oxford University Press, pp. 176–201.

Rakoczy, H. 2012. Do infants have a theory of mind? *British Journal of Developmental Psychology,* 30(1): 59–74.

Rakoczy, H., F. Warneken, and M. Tomasello. 2008. The sources of normativity: Young children's awareness of the normative structure of games. *Developmental Psychology,* 44(3): 875–881.

Ramachandran, V. S., and E. M. Hubbard. 2001. Synaesthesia–A window into perception, thought and language. *Journal of Consciousness Studies,* 8(12): 3–34.

Raz, J. 2000. *Engaging reason: On the theory of value and action.* New York: Oxford University Press.

———. 2010. Reason, reasons and normativity. In R. Shafer-Landau, ed., *Oxford Studies in Metaethics.* New York: Oxford University Press, p. 524.

Recanati, F. 2012. *Mental files.* Oxford: Oxford University Press.

Reid, T. 2000. *An inquiry into the human mind on the principles of common sense.* Edinburgh: Edinburgh University Press.

Reimer, T., T. Russell, and C. Roland. 2016. Decision making in medical teams. In T. Harrison and E. Williams, eds., *Organizations, communication, and health.* New York: Routledge, pp. 65–81.

Reinach, J. 1903. *Histoire de l'affaire Dreyfus.* Paris: Éditions de la Revue blanche. Kindle edition.

Rescorla, R. A. 1988. Pavlovian conditioning: It's not what you think it is. *American Psychologist,* 43(3): 151.

Resnick, L. B., et al. 1993. Reasoning in conversation. *Cognition and Instruction,* 11(3–4): 347–364.

Ringelmann, M. 1913. Recherches sur les moteurs animés: Travail de l'homme. *Annales de l'Institut National Agronomique 2ème série,* 12: 1–40.

Rips, L. J. 1994. *The psychology of proof: Deductive reasoning in human thinking.* Cambridge, MA: MIT Press.

Rorty, R. 1991. *Objectivity, relativism, and truth: Philosophical papers.* Cambridge: Cambridge University Press.

Rosenthal, D. M. 2005. *Consciousness and mind.* New York: Oxford University Press.

Ross, L., M. R. Lepper, and M. Hubbard. 1975. Perseverance in self-perception and social perception: Biased attributional processes in the debriefing paradigm. *Journal of Personality and Social Psychology,* 32(5): 880–802.

Rowe, G., and G. Wright. 1996. The impact of task characteristics on the performance of structured group forecasting techniques. *International Journal of Forecasting,* 12(1): 73–89.

———. 1999. The Delphi technique as a forecasting tool: Issues and analysis. *International Journal of Forecasting,* 15(4): 353–375.

Royzman, E. B., K. Kim, and R. F. Leeman. 2015. The curious tale of Julie and Mark: Unraveling the moral dumbfounding effect. *Judgment and Decision Making,* 10(4): 296–313.

Rozenblit, L., and F. Keil. 2002. The misunderstood limits of folk science: An illusion of explanatory depth. *Cognitive Science,* 26(5): 521–562.

Rozin, P., L. Millman, and C. Nemeroff. 1986. Operation of the laws of sympathetic magic in disgust and other domain. *Journal of Personality and Social Psychology,* 50(4): 703–712.

Rust, J., and E. Schwitzgebel. 2013. The moral behavior of ethicists and the rationalist delusion. Unpublished manuscript. Available at http://www.faculty.ucr.edu/~eschwitz/SchwitzPapers/RationalistDelusion-130102.pdf.

Sadler, O., and A. Tesser. 1973. Some effects of salience and time upon interpersonal hostility and attraction during social isolation. *Sociometry,* 36(1): 99–112.

Samson, D., et al. 2010. Seeing it their way: Evidence for rapid and involuntary computation of what other people see. *Journal of Experimental Psychology: Human Perception and Performance,* 36(5): 1255–1256.

Sanderson, F. E. 1972. The Liverpool delegates and Sir William Dolben's bill. *Transactions of the Historic Society of Lancashire and Cheshire,* 124: 57–84.

Sandys, M., and C. Dillehay. 1995. First-ballot votes, predeliberation dispositions, and final verdicts in jury trials. *Law and Human Behavior,* 19(2): 175–195.

Sanitioso, R., Z. Kunda, and G. T. Fong. 1990. Motivated recruitment of autobiographical memories. *Journal of Personality and Social Psychology,* 59(2): 229–241.

Schlosser, G., and G. P. Wagner. 2004. *Modularity in development and evolution.* Chicago: University of Chicago Press.

Schmidt, M. F., H. Rakoczy, and M. Tomasello. 2011. Young children attribute normativity to novel actions without pedagogy or normative language. *Developmental Science,* 14(3): 530–539.

Schmidt, M. F., and M. Tomasello. 2012. Young children enforce social norms. *Current Directions in Psychological Science,* 21(4): 232–236.

Schnall, S., et al. 2008. Disgust as embodied moral judgment. *Personality and Social Psychology Bulletin,* 34(8): 1096–1109.

Schulz-Hardt, S., et al. 2006. Group decision making in hidden profile situations: Dissent as a facilitator for decision quality. *Journal of Personality and Social Psychology,* 91(6): 1080–1093.

Schwartz, N. 2015. Metacognition. In M. Mikulincer et al., eds., *APA handbook of personality and social psychology: Attitudes and social cognition.* Washington, DC: American Psychological Association, pp. 203–229.

Scott-Kakures, D. 2001. High anxiety: Barnes on what moves the unwelcome believer. *Philosophical Psychology,* 14(3): 313–326.

Scott-Phillips, T. C. 2014. *Speaking our minds: Why human communication is different, and how language evolved to make it special.* London: Palgrave Macmillan.

Scott-Phillips, T. C., T. E. Dickins, and S. A. West. 2011. Evolutionary theory and the ultimate–proximate distinction in the human behavioral sciences. *Perspectives on Psychological Science,* 6(1): 38–47.

Scribner, S. 1977. Modes of thinking and ways of speaking: Culture and logic reconsidered. In P. N. Johnson-Laird and P. C. Wason, eds., *Thinking: Readings in cognitive science.* New York: Cambridge University Press, pp. 483–500.

Sen, A. 2003. Democracy and its global roots. *New Republic,* October 6, 29.

Serene, E. F. 1979. Robert Grosseteste on induction and demonstrative science. *Synthese,* 40(1): 97–115.

Shapin, S. 1991. "The mind is its own place": Science and solitude in seventeenth-century England. *Science in Context,* 4(1): 191–218.

Shaw, V. F. 1996. The cognitive processes in informal reasoning. *Thinking & Reasoning,* 2(1): 51–80.

Shepard, R. N. 1990. *Mind sights: Original visual illusions, ambiguities, and other anomalies, with a commentary on the play of mind in perception and art.* New York: Freeman.

Silver, N. 2012. *The signal and the noise: Why so many predictions fail—but some don't.* London: Penguin.

Silverman, I. W., and E. Geiringer. 1973. Dyadic interaction and conservation induction: A test of Piaget's equilibration model. *Child Development,* 44(4): 815–820.

Simonson, I. 1989. Choice based on reasons: The case of attraction and compromise effects. *Journal of Consumer Research,* 16(2): 158–174.

———. 1990. The effect of purchase quantity and timing on variety-seeking behavior. *Journal of Marketing Research,* 27(2): 150–162.

Simonson, I., and P. Nye. 1992. The effect of accountability on susceptibility to decision errors. *Organizational Behavior and Human Decision Processes,* 51(3): 416–446.

Skyrms, B. 1996. *Evolution of the social contract.* Cambridge: Cambridge University Press.

———. 2010. *Signals: Evolution, learning, and information.* New York: Oxford University Press.

Slavin, R. E. 1995. *Cooperative learning: Theory, research, and practice.* London: Allyn and Bacon.

———. 1996. Research on cooperative learning and achievement: What we know, what we need to know. *Contemporary Educational Psychology,* 21(1): 43–69.

Sloman, S. A. 1996. The empirical case for two systems of reasoning. *Psychological Bulletin,* 119(1): 3–22.

Smith, A. M. 2001. *Alhacen's theory of visual perception: A critical edition, with English translation and commentary, of the first three books of Alhacen's De Aspectibus, the medieval Latin version of Ibn Al-Haytham's Kitāb Al-Manāẓir.* Philadelphia: American Philosophical Society.

Smith, C. M. 2008. Adding minority status to a source of conflict: An examination of influence processes and product quality in dyads. *European Journal of Social Psychology,* 38(1): 75–83.

Smith, M. 1996. *Ptolemy's theory of visual perception: An English translation of the optics with introduction and commentary.* Philadelphia: American Philosophical Society.

Snyder, M., et al. 1979. Avoidance of the handicapped: An attributional ambiguity analysis. *Journal of Personality and Social Psychology,* 37(12): 2297–2306.

Sober, E. 1993. *Philosophy of biology.* Oxford: Oxford University Press.

Sober, E., and D. S. Wilson. 1998. *Unto others.* Cambridge, MA: Harvard University Press.

Sperber, D. 1994. The modularity of thought and the epidemiology of representations. In L. A. Hirschfeld and S. A. Gelman, eds., *Mapping the mind: Domain specificity in cognition and culture.* Cambridge: Cambridge University Press, pp. 39–67.

———. 1997. Intuitive and reflective beliefs. *Mind & Language,* 12(1): 67–83.

———, ed. 2000. *Metarepresentations: A multidisciplinary perspective.* Oxford: Oxford University Press.

———. 2005. Modularity and relevance: How can a massively modular mind be flexible and context-sensitive? In P. Carruthers, S. Laurence, and S. Stich, eds., *The innate mind: Structure and contents.* New York: Oxford University Press, pp. 53–68.

Sperber, D., and N. Baumard. 2012. Moral reputation: An evolutionary and cognitive perspective. *Mind & Language,* 27(5): 485–518.

Sperber, D., F. Cara, and V. Girotto. 1995. Relevance theory explains the selection task. *Cognition,* 57: 31–95.

Sperber, D., and L. A. Hirschfeld. 2004. The cognitive foundations of cultural stability and diversity. *Trends in Cognitive Sciences,* 8(1): 40–46.

Sperber, D., D. Premack, and A. J. Premack, eds. 1995. *Causal cognition: A multidisciplinary debate.* New York: Oxford University Press.

Sperber, D., and D. Wilson. 1995. *Relevance: Communication and cognition,* 2nd ed. New York: Wiley-Blackwell.

———. 2002. Pragmatics, modularity and mind-reading. *Mind & Language,* 17: 3–23.

Sperber, D., et al. 2010. Epistemic vigilance. *Mind & Language,* 25(4): 359–393.

Stanovich, K. E. 1999. *Who is rational? Studies of individual differences in reasoning.* Mahwah, NJ: Lawrence Erlbaum.

———. 2004. *The robot's rebellion: Finding meaning in the age of Darwin.* Chicago: University of Chicago Press.

———. 2012. On the distinction between rationality and intelligence: Implications for understanding individual differences in reasoning. In K. J. Holyoak and R. Morrison, eds., *The Oxford handbook of thinking and reasoning.* New York: Oxford University Press, pp. 343–365.

Stanovich, K. E., and R. F. West. 2000. Individual differences in reasoning: Implications for the rationality debate. *Behavioral and Brain Sciences,* 23: 645–726.

———. 2008. On the failure of cognitive ability to predict myside and one-sided thinking biases. *Thinking & Reasoning,* 14(2): 129–167.

Stanovich, K. E., R. F. West, and M. E. Toplak. 2013. Myside bias, rational thinking, and intelligence. *Current Directions in Psychological Science,* 22(4): 259–264.

Stanton, L. 1993. "Those who labor for my happiness": Thomas Jefferson and his slaves. In P. S. Onuf, ed., *Jeffersonian legacies.* Charlottesville: University Press of Virginia, pp. 147–180.

Steck, K., B. S. Hansson, and M. Knaden. 2009. Smells like home: Desert ants, *Cataglyphis fortis,* use olfactory landmarks to pinpoint the nest. *Frontiers in Zoology,* 6(1): 5.

Stenning, K., and J. Oberlander. 1995. A cognitive theory of graphical and linguistic reasoning: Logic and implementation. *Cognitive Science,* 19(1): 97–140.

Stenning, K., and M. Van Lambalgen. 2012. *Human reasoning and cognitive science.* Cambridge, MA: MIT Press.

Sterelny, K. 2003. *Thought in a hostile world: The evolution of human cognition.* Oxford: Blackwell.

———. 2012. *The evolved apprentice.* Cambridge MA: MIT Press.

Steven, S., and S. Schaffer. 1985. *Leviathan and the air-pump: Hobbes, Boyle, and the experimental life.* Princeton, NJ: Princeton University Press.

Strickland, B., and F. Keil. 2011. Event completion: Event based inferences distort memory in a matter of seconds. *Cognition,* 121(3): 409–415.

Sunstein, C. R. 2002. The law of group polarization. *Journal of Political Philosophy,* 10(2): 175–195.

Sunstein, C. R., D. Kahneman, and D. Schkade. 1998. Assessing punitive damages (with notes on cognition and valuation in law). *Yale Law Journal,* 107(7): 2071–2153.

Sunstein, C. R., et al. 2007. *Are judges political? An empirical analysis of the federal judiciary.* Washington, DC: Brookings Institution Press.

Surian, L., S. Caldi, and D. Sperber. 2007. Attribution of beliefs by 13-month-old infants. *Psychological Science,* 18(7): 580–586.

Suzuki, T. 2008. Japanese argumentation: Vocabulary and culture. *Argumentation and Advocacy,* 45(1): 49–54.

——. 2012. Why do humans reason sometimes and avoid doing it other times? Kotodama in Japanese culture. *Argumentation and Advocacy,* 48(3): 178–181.

Swaminathan, S. 2016. *Debating the slave trade: Rhetoric of British national identity, 1759–1815.* London: Routledge.

Taber, C. S., and M. Lodge. 2006. Motivated skepticism in the evaluation of political beliefs. *American Journal of Political Science,* 50(3): 755–769.

Tenenbaum, J. B., T. L. Griffiths, and C. Kemp. 2006. Theory-based Bayesian models of inductive learning and reasoning. *Trends in Cognitive Sciences,* 10(7): 309–318.

Tenenbaum, J. B., et al. 2011. How to grow a mind: Statistics, structure, and abstraction. *Science,* 331(6022): 1279–1285.

Tesser, A. 1978. Self-generated attitude change. In L. Berkowitz, ed., *Advances in experimental social psychology.* New York: Academic Press, pp. 289–338.

Tetlock, P. E. 2002. Social functionalist frameworks for judgment and choice: Intuitive politicians, theologians, and prosecutors. *Psychological Review,* 109(3): 451–471.

——. 2005. *Expert political judgment: How good is it? How can we know?* Princeton, NJ: Princeton University Press.

Thaxter, C. B., et al. 2010. Influence of wing loading on the trade-off between pursuit-diving and flight in common guillemots and razorbills. *Journal of Experimental Biology,* 213(7): 1018–1025.

Thomas, K. 1971. *Religion and the decline of magic.* London: Weidenfeld and Nicolson.

Thompson, D. V., R. W. Hamilton, and R. T. Rust. 2005. Feature fatigue: When product capabilities become too much of a good thing. *Journal of Marketing Research,* 42(4): 431–442.

Thompson, D. V., and M. I. Norton. 2011. The social utility of feature creep. *Journal of Marketing Research,* 48(3): 555–565.

Thompson, V. A. 2014. What intuitions are . . . and are not. In B. H. Ross, ed., *The psychology of learning and motivation.* Waltham, MA: Academic Press, pp. 35–75.

Tinbergen, N. 1963. On aims and methods in ethology. *Zeitschrift für Tierpsychologie,* 20: 410–433.

Tomasello, M. 2010. *Origins of human communication.* Cambridge, MA: MIT Press.

Tooby, J., and L. Cosmides. 1989. Evolutionary psychology and the generation of culture, part I: Theoretical considerations. *Ethology and Sociobiology,* 10(1): 29–49.

Toulmin, S. 1958. *The uses of argument.* Cambridge: Cambridge University Press.

Trognon, A. 1993. How does the process of interaction work when two interlocutors try to resolve a logical problem? *Cognition and Instruction,* 11(3&4): 325–345.

Trouche, E., E. Sander, and H. Mercier. 2014. Arguments, more than confidence, explain the good performance of reasoning groups. *Journal of Experimental Psychology: General,* 143(5): 1958–1971.

Trouche, E., J. Shao, and H. Mercier. Submitted. How is argument evaluation biased?

Trouche, E., et al. 2016. The selective laziness of reasoning. *Cognitive Science,* 40(8): 2122–2136.

Tversky, A., and D. Kahneman. 1973. Availability: A heuristic for judging frequency and probability. *Cognitive Psychology,* 5(2): 207–232.

———. 1983. Extensional versus intuitive reasoning: The conjunction fallacy in probability judgment. *Psychological Review,* 90(4): 293–315.

Tversky, A., and E. Shafir. 1992. The disjunction effect in choice under uncertainty. *Psychological Science,* 3(5): 305–309.

Uhlmann, E. L., et al. 2009. The motivated use of moral principles. *Judgment and Decision Making,* 4(6): 476–491.

Urbach, P. 1982. Francis Bacon as a precursor to Popper. *British Journal for the Philosophy of Science,* 33(2): 113–132.

Valdesolo, P., and D. DeSteno. 2008. The duality of virtue: Deconstructing the moral hypocrite. *Journal of Experimental Social Psychology,* 44(5): 1334–1338.

Vouloumanos, A., and J. F. Werker. 2007. Listening to language at birth: Evidence for a bias for speech in neonates. *Developmental Science,* 10(2): 159–164.

Wallace, R. J., et al., eds. 2004. *Reason and value: Themes from the moral philosophy of Joseph Raz.* New York: Oxford University Press.

Wason, P. C. 1968. Reasoning about a rule. *Quarterly Journal of Experimental Psychology,* 20(3): 273–281.

Wason, P. C., and J. St. B. T. Evans. 1975. Dual processes in reasoning? *Cognition,* 3: 141–154.

Watson, J. D. 1997. *The double helix.* London: Hachette. Kindle edition.

Wehner, R. 1997. Prerational intelligence—How insects and birds find their way. In A. B. Scheibel and J. W. Schopf, eds., *The origin and evolution of intelligence.* Boston: Jones and Bartlett, pp. 1–26.

———. 2003. Desert ant navigation: How miniature brains solve complex tasks. *Journal of Comparative Physiology A,* 189(8): 579–588.

Westfall, R. S. 1983. *Never at rest: A biography of Isaac Newton.* Cambridge: Cambridge University Press.

Whitehead, A. N., and B. Russell. 1912. *Principia mathematica.* Cambridge: Cambridge University Press.

Whyte, G. 2008. *The Dreyfus affair: A chronological history.* New York: Palgrave Macmillan.

Wilberforce, W. 1789. *The Speech of William Wilberforce, Esq. Representative for the County of York: On Wednesday the 13th of May, 1789, on the Question of the Abolition of the*

Slave Trade. To which are Added, the Resolutions Then Moved, and a Short Sketch of the Speeches of the Other Members. London: Logographic Press.

Williams, B. 1981. *Moral luck: Philosophical papers 1973–1980.* Cambridge: Cambridge University Press.

Willingham, D. T. 2008. Critical thinking: Why is it so hard to teach? *Arts Education Policy Review,* 109(4): 21–32.

Wilson, D. 2016. Reassessing the conceptual-procedural distinction. *Lingua,* 175: 5–19.

Wilson, D., and D. Sperber. 2002. Truthfulness and relevance. *Mind,* 111(443): 583–632.

Wilson, T. D., et al. 1993. Introspecting about reasons can reduce post-choice satisfaction. *Personality and Social Psychology Bulletin,* 19(3): 331–339.

Wimmer, H., and J. Perner. 1983. Beliefs about beliefs: Representation and constraining function of wrong beliefs in young children's understanding of deception. *Cognition,* 13: 41–68.

Winograd, T. 1975. Frame representation and the declarative procedural controversy. In D. Bobrow and A. Collins, eds., *Representation and understanding.* London: Academic Press, pp. 185–209.

Wittlinger, M., R. Wehner, and H. Wolf. 2006. The ant odometer: Stepping on stilts and stumps. *Science,* 312(5782): 1965–1967.

Wootton, D. 2006. *Bad medicine: Doctors doing harm since Hippocrates.* Oxford: Oxford University Press.

——. 2015. *The invention of science: A new history of the scientific revolution.* London: Harper.

Wray, K. B. 2011. *Kuhn's evolutionary social epistemology.* New York: Cambridge University Press.

Wyneken, J., M. H. Godfrey, and V. Bels, eds. 2007. *Biology of turtles: From structures to strategies of life.* London: Taylor and Francis.

Yaniv, I., and E. Kleinberger. 2000. Advice taking in decision making: Egocentric discounting and reputation formation. *Organizational Behavior and Human Decision Processes,* 83: 260–281.

Zamora Bonilla, J. 2006. Science as a persuasion game: An inferentialist approach. *Episteme,* 2(3): 189–201.

Zaretsky, R., and J. T. Scott. 2009. *The philosophers' quarrel: Rousseau, Hume, and the limits of human understanding.* New Haven, CT: Yale University Press.

Acknowledgments

This book is based on joint work on reasoning that we began in 2005. Work on the book itself started in 2010. Many people and institutions have encouraged and helped us along the way. Both as individuals and together, we have had numerous opportunities to present, develop, and revise our ideas at conferences and talks in departments of Cognitive Science and of Philosophy around the world. In particular, we were invited to present an early version of the main ideas of this book as the Chandaria Lectures at the Institute of Philosophy of the University of London in 2011. Hugo taught classes on reasoning and argumentation at the University of Pennsylvania; in the Cogmaster, an interdisciplinary academic program in Paris; and at the University of Neuchâtel. Dan taught a course based on this work in the departments of Cognitive Science and of Philosophy at the Central European University in Budapest in 2013. On all these occasions and in countless informal conversations, we have benefited from comments, suggestions, and criticisms from more friends, colleagues, and students than we are able to name here. To all of them, we are grateful.

Over all these years, we have enjoyed the personal and intellectual support of many close colleagues and friends, in particular Jean-Baptiste André, Nicolas Baumard, Stéphane Bernard, Cristina Bicchieri, Maurice Bloch, Pascal Boyer, Susan Carey, Coralie Chevallier, Nicolas Claidière, Fabrice Clément, Gergo Csibra, Dan Dennett, Ophelia Deroy, Nadine Fresco, Gyuri Gergely, Abel Gerschenfeld, Christophe Heintz, Larry Hirschfeld, Pierre Jacob, Hélène Landemore, Christine Langlois, Hanna Marno, Olivier Mascaro, Olivier Morin, Tiffany Morisseau, Ira Noveck, Gloria Origgi, Guy Politzer, Jérôme Prado, Louis de Saussure, Thom Scott-Phillips, Barry Smith, Brent Strickland,

Denis Tatone, Radu Umbres, Jean-Baptiste Van der Henst, Deirdre Wilson, and Hiroshi Yama. Thank you all!

We were both inspired by our friendship and collaborations with Vittorio Girotto, a leading figure in the contemporary study of reasoning. He died in April 2016 at the age of fifty-nine, and we miss him.

We are most grateful to colleagues who have read the manuscript or parts of it and provided us with very helpful comments: Clark Barrett, Hannoch Ben-Yami, Pascal Boyer, Peter Carruthers, George Gafalvi, Pierre Jacob, Philip Johnson-Laird, Olivier Morin, Ira Noveck, Josef Perner, Philip Pettit, Thom Scott-Phillips, John Watson, Deirdre Wilson, and one anonymous reviewer for Harvard University Press.

We offer our gratitude to John and Max Brockman, our agents, and to Ian Malcolm, our editor, and his team at Harvard University Press.

Hugo benefited from the financial backing of the Direction Générale de l'Armement (thanks to Didier Bazalgette in particular); the Philosophy, Politics, and Economics program at the University of Pennsylvania (with the generous support of Steven F. Goldstone); the University of Neuchâtel's Cognitive Science group and the Swiss National Science Foundation (Ambizione grant no. PZ00P1_142388); and, finally, the Centre National de la Recherche Scientifique, where he now works, having joined the Institut des Sciences Cognitives Marc Jeannerod. Most of his contributions to the empirical research reported in the book were made in collaboration with his two brilliant doctoral students, Thomas Castelain and Emmanuel Trouche. His deepest thanks go to his families—the family that supported him throughout his wandering years as a young student (and keeps doing so!), and the family that supports him now—Thérèse, Christopher, and Arthur. He should be working less and spending more time with them.

Dan would like to thank the members, staff, and students of the departments of Cognitive Science and of Philosophy at the Central European University in Budapest and of the Institut Jean Nicod in Paris. His research has been supported by a grant from the European Research Council (grant number: ERC-2013-SyG, Constructing Social Minds: Communication, Coordination and Cultural Transmission, ERC, grant agreement no. [609619]). His joie de vivre has been supported throughout these years by his sons, Nathan and Leo.

Illustration Credits

Figure 4. Adelson's checkershadow illusion. © 1995 Edward H. Adelson, Department of Brain and Cognitive Sciences, MIT.

Figure 7. The trajectory of a desert ant. R. Wehner and S. Wehner, "Insect Navigation: Use of Maps or Ariadne's Thread?" *Ethology, Ecology & Evolution* 2, no. 1 (1 May 1990): 27–48. Figure 2. © 1990 Taylor & Francis. Courtesy of Rüdiger Wehner.

Figure 8. Monsters in a tunnel. "Terror Subterra," by Roger N. Shepard. *Mind Sights: Original Visual Illusions, Ambiguities, and Other Anomalies, with a Commentary on the Play of Mind in Perception and Art* (New York: W. H. Freeman and Company, 1990), Figure A1, p. 47. © 1990 Roger N. Shepard.

Figure 9. Three cognitive systems. Daniel Kahneman, "Maps of Bounded Reality: A Perspective on Intuitive Judgment and Choice." Prize Lecture for the Sveriges Riksbank Prize in Economic Sciences in Memory of Alfred Nobel, December 8, 2002, Aula Magna, Stockholm University. Figure 1. © The Nobel Foundation, 2002.

Figure 10. Infants shown the physically impossible event are surprised. Reprinted with permission from Amy Needham and Renee Baillargeon, "Intuitions about Support in 4.5-Month-Old Infants." *Cognition* 47, no. 2 (May 1993): 121–148. Figure 1. © 1993 Elsevier Science Publishers B.V.

Figure 12. Infants see a caterpillar nibble at a piece of cheese. Redrawn from Luca Surian, Stefanie Caldi, and Dan Sperber, "Attribution of Beliefs by 13-Month-Old Infants." *Psychological Science* 18, no. 7 (August 2007): 580–586. Figure 1a. © 2007 Association for Psychological Science.

Index